Springer Theses

Recognizing Outstanding Ph.D. Research

Aims and Scope

The series "Springer Theses" brings together a selection of the very best Ph.D. theses from around the world and across the physical sciences. Nominated and endorsed by two recognized specialists, each published volume has been selected for its scientific excellence and the high impact of its contents for the pertinent field of research. For greater accessibility to non-specialists, the published versions include an extended introduction, as well as a foreword by the student's supervisor explaining the special relevance of the work for the field. As a whole, the series will provide a valuable resource both for newcomers to the research fields described, and for other scientists seeking detailed background information on special questions. Finally, it provides an accredited documentation of the valuable contributions made by today's younger generation of scientists.

Theses are accepted into the series by invited nomination only and must fulfill all of the following criteria

- They must be written in good English.
- The topic should fall within the confines of Chemistry, Physics, Earth Sciences, Engineering and related interdisciplinary fields such as Materials, Nanoscience, Chemical Engineering, Complex Systems and Biophysics.
- The work reported in the thesis must represent a significant scientific advance.
- If the thesis includes previously published material, permission to reproduce this must be gained from the respective copyright holder.
- They must have been examined and passed during the 12 months prior to nomination.
- Each thesis should include a foreword by the supervisor outlining the significance of its content.
- The theses should have a clearly defined structure including an introduction accessible to scientists not expert in that particular field.

More information about this series at http://www.springer.com/series/8790

Maximilian Joost

Synthesis and Original Reactivity of Copper and Gold Complexes

σ-Bond Coordination, Oxidative Addition, Migratory Insertion

Doctoral Thesis accepted by
the Paul Sabatier University, Toulouse, France

Springer

Author
Dr. Maximilian Joost
Laboratoire Hétérochimie Fondamentale et Appliquée
Université Paul Sabatier/CNRS
Toulouse
France

Supervisors
Dr. Abderrahmane Amgoune
Laboratoire Hétérochimie Fondamentale et Appliquée
Université Paul Sabatier/CNRS
Toulouse
France

Dr. Didier Bourissou
Laboratoire Hétérochimie Fondamentale et Appliquée
Université Paul Sabatier/CNRS
Toulouse
France

ISSN 2190-5053 ISSN 2190-5061 (electronic)
Springer Theses
ISBN 978-3-319-18689-4 ISBN 978-3-319-18690-0 (eBook)
DOI 10.1007/978-3-319-18690-0

Library of Congress Control Number: 2015941108

Springer Cham Heidelberg New York Dordrecht London

Printed on acid-free paper

Springer International Publishing AG Switzerland is part of Springer Science+Business Media (www.springer.com)

Parts of this thesis have been published in the following journal articles:

"Oxidative Addition of Carbon–Carbon Bonds to Gold"
M. Joost, L. Estévez, S. Mallet-Ladeira, K. Miqueu, A. Amgoune, D. Bourissou, *Angew. Chem. Int. Ed.* **2015**, *54*, 5236–5240.

"Enhanced π-Backdonation from Gold(I): Isolation of Original Carbonyl and Carbene Complexes"
M. Joost, L. Estévez, S. Mallet-Ladeira, K. Miqueu, A. Amgoune, D. Bourissou, *Angew. Chem. Int. Ed.* **2014**, *53*, 14512–14516.

"Facile Oxidative Addition of Aryl Iodides to Gold(I) by Ligand Design: Bending Turns on Reactivity"
M. Joost, A. Zeineddine, L. Estévez, S. Mallet-Ladeira, K. Miqueu, A. Amgoune, D. Bourissou, *J. Am. Chem. Soc.* **2014**, *136*, 14654–14657.

"Mechanisms of *syn*-Insertion of Alkynes and Allenes into Gold-Silicon Bonds: a Comprehensive Experimental/Theoretical Study"
M. Joost, L. Estévez, S. Mallet-Ladeira, K. Miqueu, A. Amgoune, D. Bourissou, *J. Am. Chem. Soc.* **2014**, *136*, 10373–10382.

"Direct Evidence for Intermolecular Oxidative Addition of σ(Si-Si) Bonds to Gold"
M. Joost, P. Gualco, Y. Coppel, K. Miqueu, C. E. Kefalidis, L. Maron, A. Amgoune, D. Bourissou, *Angew. Chem. Int. Ed.* **2014**, *53*, 747–751.

"Direct syn Insertion of Alkynes and Allenes into Au-Si Bonds"
M. Joost, P. Gualco, S. Mallet-Ladeira, A. Amgoune, D. Bourissou, *Angew. Chem. Int. Ed.* **2013**, *52*, 7160–7163.

"σ-SiH Complexes of Copper: Experimental Evidence and Computational Analysis"
M. Joost, S. Mallet-Ladeira, K. Miqueu, A. Amgoune, D. Bourissou, *Organometallics* **2013**, *32*, 898–902.

For my family

Supervisors' Foreword

Transition metal complexes occupy a forefront position in homogeneous catalysis thanks to their ability to promote a wide range of unique reactions in a selective manner, allowing the synthesis of sophisticated organic molecules and materials. These important achievements have been made possible thanks to intensive and constant fundamental organometallic research seeking to understand how metal complexes work and how to tailor their electronic and geometric properties to a specific need.

In striking contrast, the development of gold complexes in homogeneous catalysis was lagging well behind for a long time. Gold was long considered as chemically inert and thus synthetically useless. This situation changed dramatically a few decades ago and the 2000s have witnessed a real "gold rush" in catalysis. However, all the catalytic applications of gold complexes are essentially based on a unique reactivity, namely the electrophilic activation of CC multiple bonds, meaning that gold behaves as a soft Lewis acid. This noble metal was considered inappropriate to promote the key elementary reactions involved in transition metal catalytic cycles, in particular oxidative addition and migratory insertion processes.

When Dr. Maximilian Joost started his Ph.D. in 2011, very little was known about the reactivity of gold complexes towards these elementary reactions. Maximilian carried out fundamental organometallic studies to gain comprehensive knowledge into the properties of gold complexes, in particular into the parameters governing their reactivity. Thanks to a rational ligand design approach, he showed for the first time that the oxidative addition of aryl halides as well as carbon–carbon bonds is possible with gold, and also evidenced an unprecedented *syn* insertion process with gold.

The experimental work combined with computational investigations has provided valuable information on the bonding, structure and reactivity of new gold complexes. Key parameters controlling the reactivity of gold towards oxidative addition reactions have been precisely identified. This thesis highlights novel reactivity patterns of gold complexes that may guide and inspire the development of new catalytic transformations.

Toulouse, France
March 2015

Dr. Abderrahmane Amgoune
Dr. Didier Bourissou

Acknowledgments

First and foremost, I would like to express my sincere gratitude to my supervisors Dr. Didier Bourissou and Dr. Abderrahmane Amgoune. Their constant encouragement and inspiring guidance in the course of this work contributed invaluably to the successful completion of this thesis.

Furthermore, I would like to thank Dr. Frédéric Leroux and Prof. Dr. Martin Oestreich for judging this work, as well as Prof. Dr. Christopher Russell and Prof. Dr. Michel Etienne for taking part in the thesis committee.

In numerous parts of this thesis, the experimental data is accompanied by theoretical calculations to contribute to the analysis and understanding. I am thankful to Dr. Karinne Miqueu and Dr. Laura Estévez, as well as to Prof. Dr. Laurent Maron and Dr. Christos Kefalidis for these important contributions.

I acknowledge the support of all of the analytical and technical staff of the Institut de Chimie de Toulouse, and I am especially grateful for help and assistance by the members of the NMR and the X-Ray crystallography services.

Furthermore, I am grateful to all the staff at the Laboratoire Hétérochimie Fondamentale et Appliquée for providing a well-honed, efficient research environment. I would like to thank Dr. Ghenwa Bouhadir, Prof. Dr. Blanca Martin-Vaca, Olivier Thillaye de Boullay, Dr. Julien Monot, Dr. Tsuyoshi Kato, Dr. Noel Nebra and Dr. Nicolas Mézailles for valuable scientific discussions, as well as my fellow coworkers in the lab for stimulating and inspiring exchange. Among many, I am especially indebted to Dr. Marc Devillard, Dr. Yannick Escudie, Dr. Johannes Guenther, Feriel Rekhroukh and Dr. Amos Rosenthal. Abdallah Zeineddine is thanked for valuable experimental contributions.

Last but not least, I owe my thanks to my friends, my family, especially my parents and Émilie, who all have been a source of constant support in the course of the last years.

Contents

Abbreviations

Ad	2-Adamantyl group
Alk	Generic alkyl group
Ar	Generic aryl group
$BArF_{20}$	Tetrakis(pentafluorophenyl)borate
$BArF_{24}$	Tetrakis(bis(3,5-trifluoromethyl)phenyl)borate
CMD	Concerted metallation/deprotonation
cod	Cyclooctadiene
CSA	Camphersulfonic acid
Cy	Cyclohexyl group
deg	Degree
DFT	Density functional theory
Dipp	(2,6-diisopropyl)phenyl group
DMAD	Dimethyl acetylendicarboxylate
E	Main group element
Elt. Anal.	Elemental analysis
EPR	Electron paramagnetic resonance
eq.	Equivalent
ESI	Electrospray ionization
EWG	Electron-withdrawing group
FT	Fourier transformed
GIAO	Gauge-including atomic orbital
HOMO	Highest occupied molecular orbital
HRMS	High resolution mass spectrometry
IGLO	Individual gauge for localized orbitals
IMes	*N,N'*-bis(2,4,6-trimethylphenyl)imidazol-2-ylidene
IPr	*N,N'*-bis(2,6-diisopropylphenyl)imidazol-2-ylidene
IR	Infrared
L	Generic neutral, 2-electron donor ligand
LUMO	Lowest unoccupied molecular orbital
M	Transition metal
Mes	Mesityl group

Mp	Melting point
NBO	Natural bond orbital
NHC	*N*-heterocyclic carbene
NLMO	Natural localized molecular orbital
NMR	Nuclear magnetic resonance
NPA	Natural population analysis
PCM	Polarizable continuum model
PES	Potential energy surface
pin	Pinacol
RC	Reaction coordinate
RECP	Relativistic effective core potential
Selectfluor	1-Chloromethyl-4-fluoro-1,4-diazoniabicyclo[2.2.2]octane bis (tetrafluoroborate)
SMD	Universal solvation model based on solute electron density
TEMPO	(2,2,6,6-Tetramethyl-piperidin-1-yl)oxyl
THF	Tetrahydrofurane
THT	Tetrahydrothiophene
Tol	*p*-Tolyl group
TPA	1,3,5-triaza-7-phospha-adamantane
TS	Transition state
Ts	Tosyl (*p*-toluenesulfonyl) group
vdW	van der Waals
X	F, Cl, Br or I
Xanthphos	4,5-Bis(diphenylphosphino)-9,9-dimethylxanthene
XRD	X-ray diffraction

List of Compounds

Chapter 2: σ-SiH Coordination to Cu(I)

1 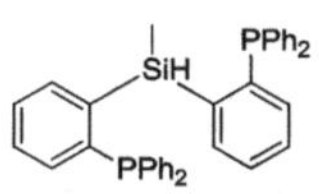

2 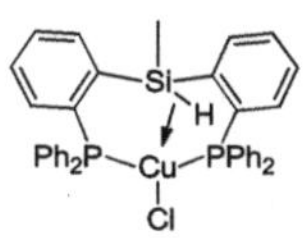

3 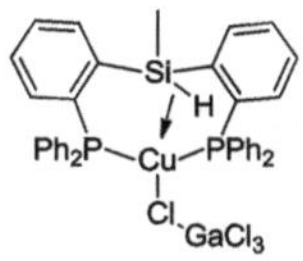

4 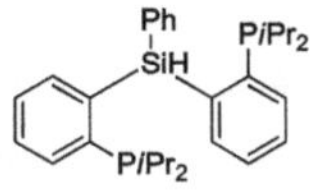

5 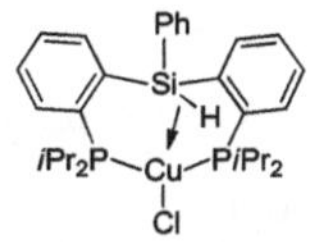

6 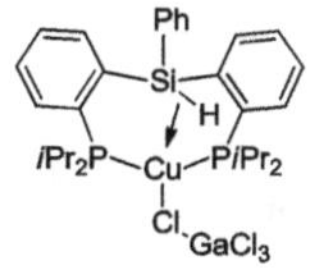

7 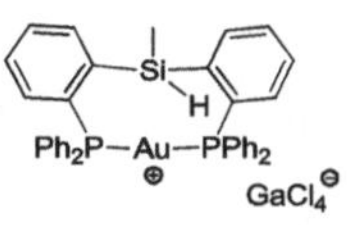

Chapter 4: Migratory Insertion at Au(I)

8 $Ph_3P{-}Au{-}SitBuPh_2$

9

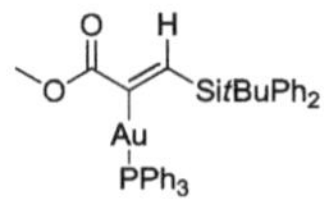

9-Sn

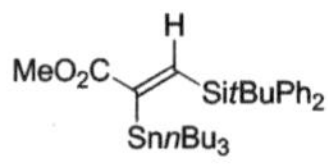

9-Allyl

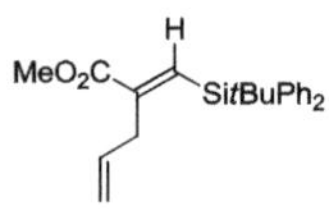

9-Aryl

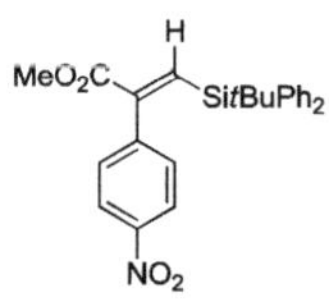

10

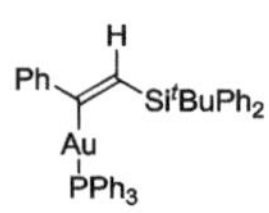

11

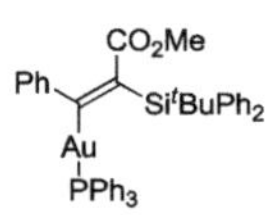

12

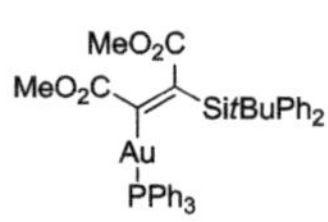

13

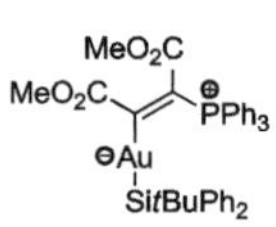

14 $Ph_3P{-}Au{-}SiPh_3$

15

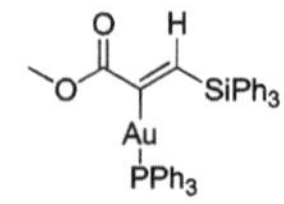

16

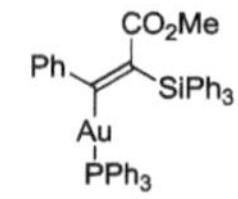

17

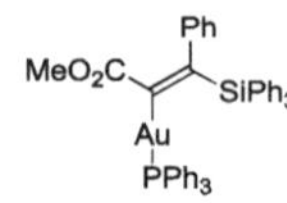

18

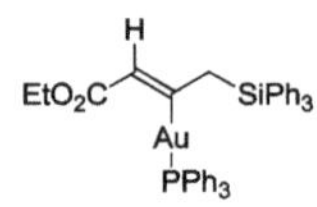

19 $Me_3P{-}Au{-}SitBuPh_2$

20

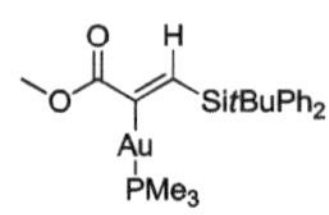

21

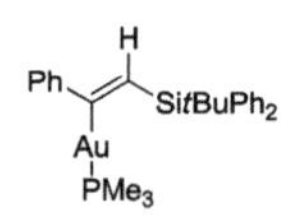

22

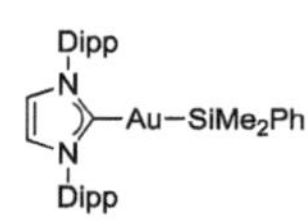

23

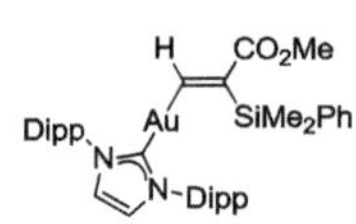

Chapter 5: Oxidative Addition at Gold(I)

24a 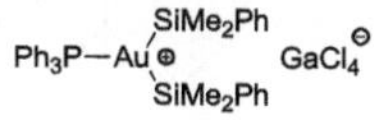

28 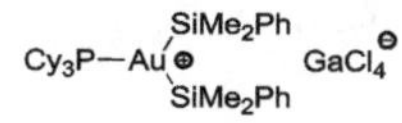

24b 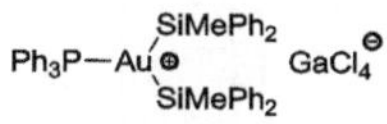

29 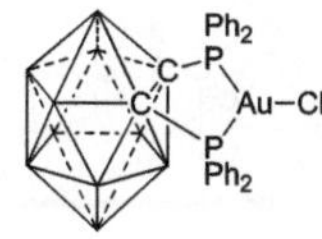

24c 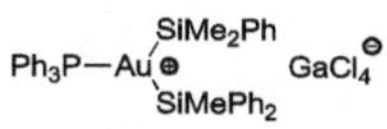

30 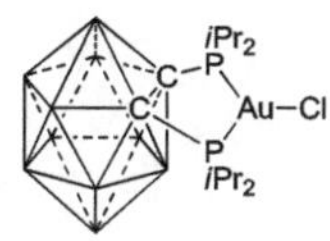

25 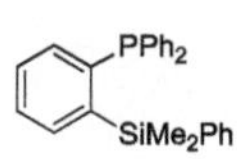

31 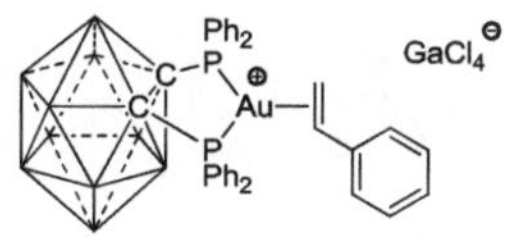

26 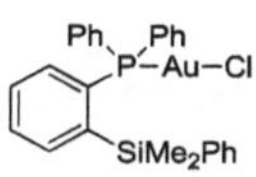

32 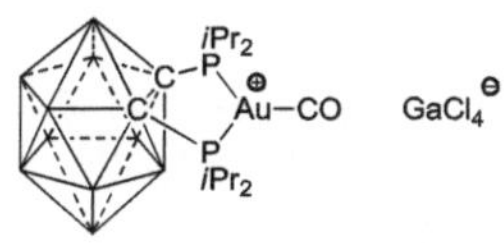

27 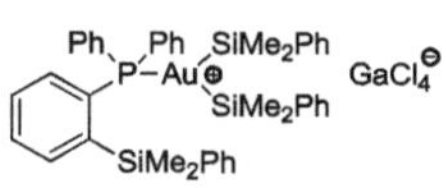

33 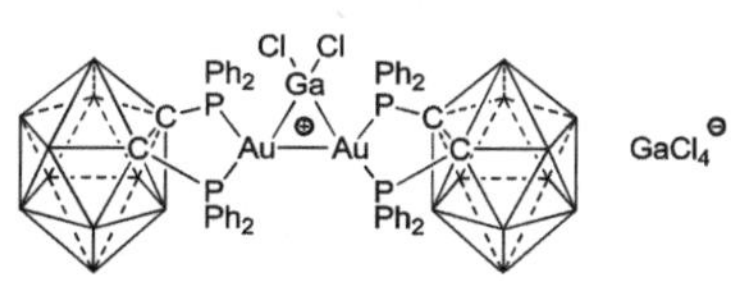

34

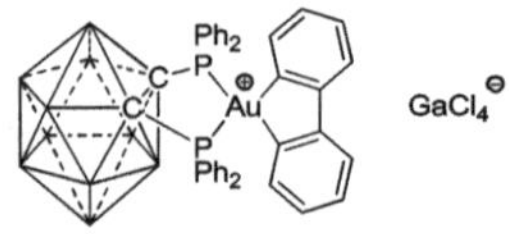

35

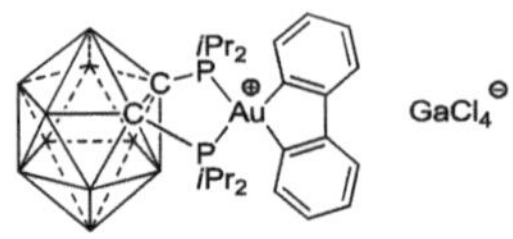

36

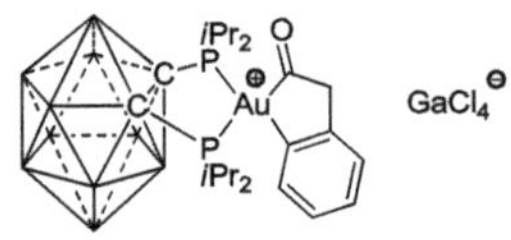

37

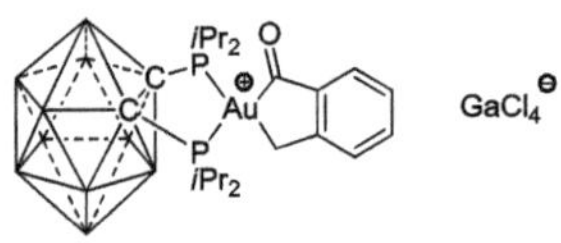

38

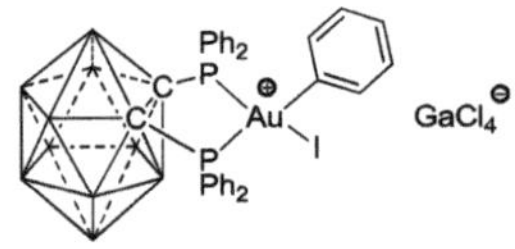

39

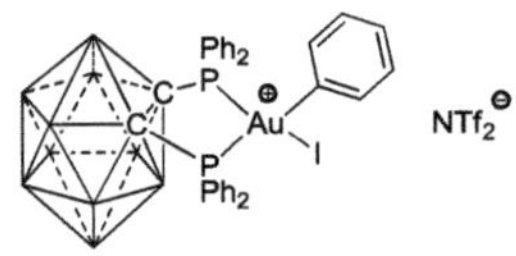

40

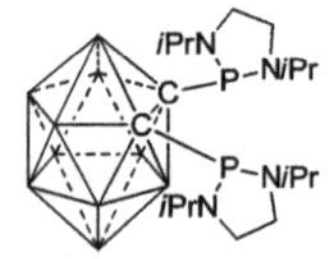

41

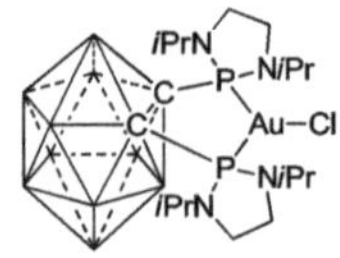

42

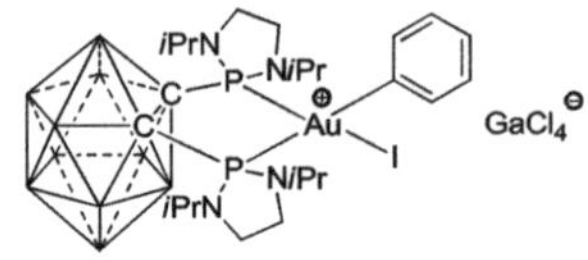

43

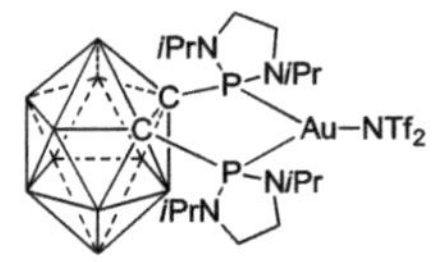

44

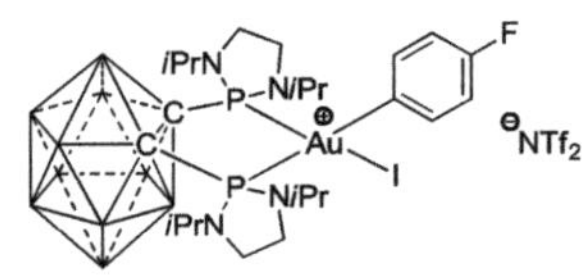

45

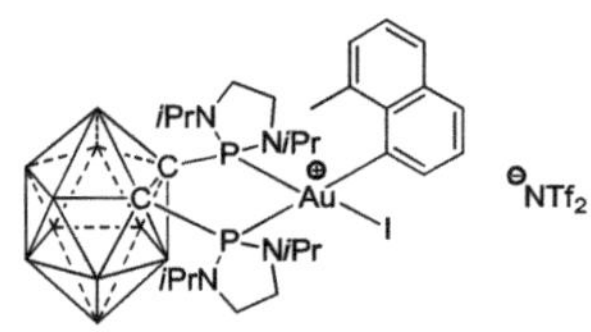

General Remarks

The work presented in this dissertation was carried out in the Laboratoire Hétérochimie Fondamentale et Appliquée at the Université de Toulouse III—Paul Sabatier under the guidance of Dr. Abderrahmane Amgoune and Dr. Didier Bourissou from October 2011 to May 2014. Computational studies accompanying this work were carried out by Dr. Karinne Miqueu and coworkers at the Institut des Sciences Analytiques et de Physicochimie pour l'Environnement et les Matériaux (Université de Pau et des pays de l'Adour) and Prof. Dr. Laurent Maron and coworkers at the Laboratire de Physique et Chimie de Nano-Objets (Université de Toulouse III—Paul Sabatier).

General Procedures

The reactions and manipulations reported in this manuscript were carried out under an atmosphere of dry argon using standard Schlenk techniques or in an argon-filled glovebox, if not otherwise stated. Dichloromethane, diethyl ether, pentane, tetrahydrofuran and toluene were dried by passage through activated molecular sieves (3 Å), using an mBraun solvent purification system. Fluorobenzene and hexamethyldisiloxane were dried by stirring with activated powdered molecular sieves (3 Å) for at least 24 h and subsequent filtration. Solvents were degassed by multiple freeze-pump-thaw cycles. Deuterated solvents were dried by stirring with activated powdered molecular sieves (3 Å) for at least 24 h and subsequent filtration.

o-Lithiated triphenylphosphine, *o*-lithiated phenyldiisopropylphosphine [1], chloro [1,3-bis(2,6-diisopropyl)phenyl)imidazol-2-ylidene)gold(I) [2], 1-bromo-2-diphenylphosphinobenzene [3], 1,1,2-trimethyl-1,2,2-triphenyldisilane [4], 1,2-bis (diphenylphosphino)-1,2-dicarba-*closo*-dodecaborane [5], 1,2-bis(diisopropylphosphino)-1,2-dicarba-*closo*-dodecaborane [6], chloro-[1,2-bis(diphenyl-phosphino)-1,2-dicarba-*closo*-dodecaborane-$\kappa^2 P,P'$]gold(I) [7], biphenylene [8], benzocyclobutenone [9], dilithio-1,2-dicarba-*closo*-dodecaborane [10], 2-chloro-1,3-diisopropyl-1,3,2-diazaphospholidine [11] were prepared according to reported procedures. [AuCl(THT)] was obtained from Umicore (Brussels, Belgium) as a generous gift. 1,2-Dicarba-*closo*-dodecaborane was purchased from KatChem (Prague, Czech Republic). All other starting materials were purchased from Aldrich (Saint-Quentin Fallavier, France),

ABCR (Karlsruhe, Germany) or AlfaAesar (Schiltigheim, France). Chlorosilanes and chlorophosphines were distilled prior to use.

Melting Points

Melting points were determined with a Stuart SMP40 apparatus and are uncorrected.

NMR Spectroscopy

Solution ^{1}H, ^{13}C, ^{11}B, ^{19}F, ^{29}Si, ^{31}P and ^{119}Sn NMR spectra were recorded on Bruker Avance 300, 400 or 500 spectrometers at the indicated frequencies and at 298 K, unless otherwise stated. Chemical shifts (δ) are expressed with a positive sign, in parts per million. ^{1}H and ^{13}C chemical shifts reported are referenced internally to residual protio- (^{1}H) or deutero- (^{13}C) solvent, while ^{11}B, ^{19}F, ^{29}Si, ^{31}P and ^{119}Sn chemical shifts are relative to $BF_3{\cdot}OEt_2$, $CFCl_3$, $SiMe_4$, 85 % H_3PO_4 (aq) and $SnMe_4$ external references, respectively. The following abbreviations and their combinations are used: br, broad; s, singlet; d, doublet; t, triplet; q, quartet, m, multiplet. The ^{1}H and ^{13}C resonance signals were attributed by means of 2D HSQC and HMBC experiments. The *N* values corresponding to ½ [*J*(AX)+*J*(A'X)] are provided when second–order AA'X and AA'XX' systems are observed in the ^{13}C NMR spectra [12, 13].

IR Spectroscopy

IR spectra were recorded on a PerkinElmer Spectrum One FT-IR ATR spectrometer and on a Varian 640-IR FT-IR spectrometer.

Mass Spectrometry

Mass spectra were recorded by the in-house service of the *Institut de Chimie de Toulouse* (118, route de Narbonne, 31062, Toulouse, France) on a Waters UPLC Xevo G2 Q TOF apparatus.

Elemental Analysis

Elemental analyses were performed by the in-house service at the *Laboratoire de Chimie de Coordination* (205, Route de Narbonne, 31077, Toulouse, France) on a PerkinElmer 2400 Series II system and at the *Centre Régional de Mesure Physique de l'Ouest* (Université de Rennes 1, Campus de Beaulieu, Bâtiment 11A, 35042 Rennes, France) on a Thermo Electron FlashEA 1112 CHNS/O elemental analyzer. The Pd trace analysis was performed by the *Service Central d'Analyse* (5, Rue de la Doua, 69100 Villeurbanne, France).

X-ray Crystallography

Crystallographic data were collected at 193(2) K on a Bruker–AXS SMART APEX II or on a Bruker-AXS Kappa APEX II Quazar diffractometer, with Mo K_{α} radiation (λ = 0.71073 Å) using an oil-coated shock-cooled crystal. Temperature-sensitive crystals were selected under the microscope using Stalke's X-Temp 2 device [14]. Phi- and omega-scans were used. Space groups were determined on the basis of systematic absences and intensity statistics. Semi-empirical absorption correction was employed [15, 16]. The structures were solved by direct methods (SHELXS-97) [17], and refined using the least-squares method on F^2. All non-H atoms were refined with anisotropic displacement parameters. Hydrogen atoms were refined isotropically at calculated positions using a riding model with their isotropic displacement parameters constrained to be equal to 1.5 times the equivalent isotropic displacement parameters of their pivot atoms for terminal sp^3 carbon and 1.2 times for all other carbon atoms.

The thermal ellipsoids shown in the presentations of molecular structures determined by single crystal X-ray diffraction in this thesis are at the 50 % probability level. The following color code indicating the respective atom type in the depicted molecular structures is used throughout this manuscript: hydrogen: white, boron: pink, carbon: dark grey, nitrogen: cornflower blue, oxygen: clear purple, fluorine: yellow, silicon: ultramarine blue, phosphorus: red, sulfur: orange, chlorine: clear green, copper: coppery red, gallium: clear blue, iodine: dark purple, gold: golden yellow.

Theoretical Studies

Computational studies were performed by Dr. Karinne Miqueu and Dr. Laura Estevez (Pau) as well as by Prof. Dr. Laurent Maron and Dr. Christos Kefalidis (Toulouse) at the DFT level of theory using the Gaussian 09 program package [18]. Basis sets and functionals are detailed in the respective context. Geometry optimizations were carried out without any symmetry restrictions. The nature of the maxima, minima or saddle points was verified by the absence or presence of only one negative eigenvalue, respectively. The connection between transition states and the corresponding minima was performed by IRC calculations.

Computed compounds have the same number as the analogous experimentally characterized compounds, denoted with an additional asterisk.

References

1. M. Sircoglou, M. Mercy, N. Saffon, Y. Coppel, G. Bouhadir, L. Maron, D. Bourissou, Angew. Chem. Int. Ed. **48**, 3454–3457 (2009)
2. P. de Frémont, N.M. Scott, E.D. Stevens, S.P. Nolan, Organometallics. **24**, 2411–2418 (2005)
3. M.T. Whited, E. Rivard, J.C. Peters, Chem. Commun. (15), 1613–1615 (2006)
4. G. Fritz, B. Grunert, Z. Anorg. Allg. Chem. **473**, 59–79 (1981)
5. H. Nakamura, T. Kamakura, S. Onagi, Org. Lett. **8**, 2095–2098 (2006)
6. F. Teixidor, C. Viñas, M. Mar Abad, R. Nuñez, R. Kivekäs, R. Sillanpäa, J. Organomet. Chem. **503**, 193–203 (1995)
7. O. Crespo, M.C. Gimeno, A. Laguna, P.G. Jones, J. Chem. Soc. Dalton Trans. 1601–1605 (1992)
8. T. Schaub, U. Radius, Tetrahedron Lett. **46**, 8195–8197 (2005)
9. M.S. South, L.S. Liebeskind, J. Org. Chem. **47**, 3815–3821 (1982)
10. B. Wrackmeyer, E.V. Klimkina, W. Milius, Appl. Organomet. Chem. **24**, 25–32 (2010)

11. D. Gau, Synthèse et Réactivité Des Sila-Ylures de Phosphonium, Ph.D. Thesis, Université de Toulouse, Université Toulouse III—Paul Sabatier, 2011.
12. F.A. Bovey, *Nuclear Magnetic Resonance Spectroscopy* (Academic Press, New-York, 1969)
13. R.J. Abraham, H.J. Bernstein, Can. J. Chem. **39**, 216–230 (1961)
14. T. Kottke, D. Stalke, J. Appl. Crystallogr. **26**, 615–619 (1993)
15. Bruker, SADABS, Bruker AXS Inc., Madison, Wisconsin, USA, 2006
16. Bruker, SADABS, Bruker AXS Inc., Madison, Wisconsin, USA, 2008
17. G.M. Sheldrick, Acta Crystallogr. A **64**, 112–122 (2007)
18. M.J. Frisch, G.W. Trucks, H.B. Schlegel, G.E. Scuseria, M.A. Robb, J.R. Cheeseman, G. Scalmani, V. Barone, B. Mennucci, G.A. Petersson et al., Gaussian 09, Revision A.1, Gaussian, Inc., Wallingford, Connecticut, USA, 2009

Chapter 1
General Introduction

1.1 General Introduction

1.1.1 Opening Remarks

Gold chemistry was considered for long time as a rather exotic area of research with relatively few research groups involved. In 1998, the essential stoichiometric and catalytic applications of gold complexes in organic synthesis could be summarized on two pages [1]. This situation changed drastically, when at the end of the 1990s [2–4] the organic community became aware of a quite unique reactivity of gold: the activation of CC multiple bonds by gold salts or complexes towards nucleophilic attack is at the origin of the highly active development of gold catalysis (as illustrated by Fig. 1.1), which has emerged to an immense toolbox allowing for the efficient construction of complex organic structures.

The catalytic application of gold complexes revived the field of gold chemistry and stimulated many important findings. Recent mechanistic investigations helped to gain more insight into the reactivity of gold complexes and to establish new and unexpected activation modes in catalysis [5–7].

Nevertheless, gold differs markedly from other transition metals and its coordination chemistry is still underdeveloped. Fundamental experimental and theoretical studies are required, as some facets of gold chemistry remain uncomprehended. Why does Au(I) shows a reactivity so different compared to its isoelectronic neighbor Pt(0)? Can gold complexes undergo the otherwise ubiquitous insertion reactions in Au-E bonds? Why are Au(I)–Au(III)–Au(I) redox processes so seldom encountered for gold complexes?

To some extent, this situation applies as well to the chemistry of copper: although high levels of activity and selectivity in catalytic reactions have been achieved with this metal, some fundamental aspects of the coordination chemistry of copper are not fully understood.

M. Joost, *Synthesis and Original Reactivity of Copper and Gold Complexes*,
Springer Theses, DOI 10.1007/978-3-319-18690-0_1

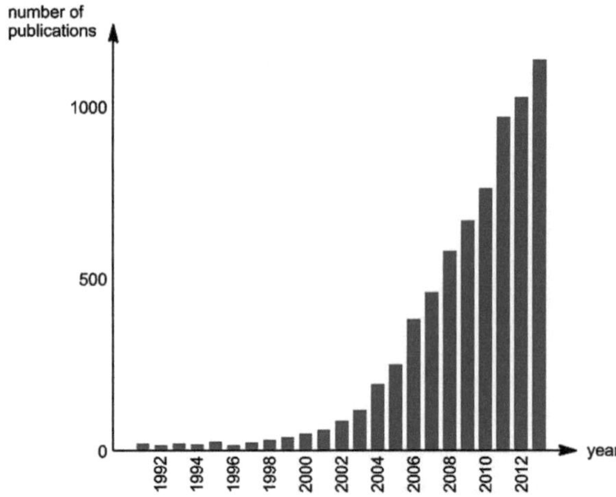

Fig. 1.1 Research publications on gold catalysis (homogeneous and heterogeneous) by year, based on the "web of science" database (August 2014)

1.1.2 Research Objectives and Thesis Outline

As briefly illustrated, the precise understanding of the reactivity of copper and gold complexes lags behind. Therefore, the aim of the research projects presented in this dissertation was to gain further knowledge in regard of fundamental reactions such as migratory insertion or σ-bond activation processes. These reactions are well known for the adjacent group-10 metals and are often encountered in catalytic cycles that are nowadays considered as textbook examples. For Cu and Au, respective information is scarce or even non-existent.

In the first place we strived for the study of a copper-SiH interaction, not observed before, but proposed as intermediate species in various catalytic cycles [8]. Since one should expect a rather weak bonding interaction, we based our investigation on the chelate assistance approach. The use of multifunctional ligands containing a chelating phosphine motif has been already successfully applied in our group [9–11] and elsewhere [12] for the identification of elusive and otherwise difficult characterizable bonding modes. For example, the characterization of the first σ-complex of copper was achieved by employing a diphosphine-disilane ligand [8].

This strategy led to the synthesis of the first σ-SiH complexes of copper that were unambiguously characterized by spectroscopic, structural and computational means.

The major part of this work is dedicated to the chemistry of gold complexes. We aimed at the investigation of two important elementary steps, namely migratory insertion and oxidative addition, often encountered in organometallic chemistry, but poorly understood in the case of gold.

The reactivity of silylgold complexes towards alkynes and allenes was elucidated. It was found previously in our group that methyl propiolate inserts into AuSi bonds [13]. This finding was further probed and shown to be a general process giving rise to a variety of new vinylgold complexes. A combined computational and experimental mechanistic study allowed for the first time the unequivocal demonstration of a *syn* migratory insertion step at a gold(I) complex.

Oxidative additions via the concerted activation of σ-bonds are highly unusual in gold chemistry. With the important finding that σ-SiSi, σ-SnSn and σ-CX bonds undergo spontaneously oxidative addition to a gold(I) center in an intramolecular fashion (supported by the above mentioned chelate assistance) [13, 14], we became further interested in this reactivity. In order to investigate the generality of this process, we envisaged the study of intermolecular oxidative addition to gold(I) complexes.

A combined computational and low temperature NMR spectroscopic study allowed for the characterization of gold(III) bis(silyl) phosphine species stemming from disilanes and gold(I) phosphine complexes. Even though the reaction products proved to be highly unstable, this study demonstrates that oxidative additions at Au (I) can be thermodynamically favorable and kinetically accessible and calls the often assumed redox-neutral behavior of gold complexes into question.

We studied subsequently the oxidative addition of σ-CC and σ-CX bonds to gold (I) complexes. In the course of this investigation we became aware of the importance of the ligand sphere around the metal center for both the redox-activity of gold(I) and stability of the gold(III) complexes stemming from oxidative addition. An analysis of the requirements for these two criteria allowed us the identification of a tailor-made ligand. A diphosphine ligand with an *o*-carborane-based backbone allowed for a small P–Au–P bite angle and proved to be suited for our bond activation studies. Gold(III) complexes stemming from the reactions of cationic gold(I) diphosphine precursors and aryl halides or strained carbocycles have been isolated and fully characterized.

One unifying motivation behind the different research topics presented here is to enhance our knowledge concerning the fundamental coordination chemistry of copper and especially gold. This target is approached by the synthesis and characterization of original complexes and the study of their reactivity. The experimental work was accompanied in close contact by theoretical studies carried out by the groups of Dr. Karinne Miqueu (Pau) and Prof. Dr. Laurent Maron (Toulouse). These computational analyses guided the experimental work and proved invaluable for the understanding of the electronic properties of the prepared compounds. Moreover, the theoretical studies provided a large contribution to the mechanistic investigations of the studied reactions.

References

1. R.V. Parish, Gold Bull. **31**, 14–21 (1998)
2. J.H. Teles, S. Brode, M. Chabanas, Angew. Chem. Int. Ed. **37**, 1415–1418 (1998)
3. A.S.K. Hashmi, L. Schwarz, J.-H. Choi, T.M. Frost, Angew. Chem. Int. Ed. **39**, 2285–2288 (2000)
4. W.A. Nugent, Angew. Chem. Int. Ed. **51**, 8936–8949 (2012)
5. A.S.K. Hashmi, Angew. Chem. Int. Ed. **49**, 5232–5241 (2010)

6. I. Braun, A.M. Asiri, A.S.K. Hashmi, ACS Catal. **3**, 1902–1907 (2013)
7. A.M. Echavarren, C. Obradors, Chem. Commun. **50**, 16–28 (2014)
8. P. Gualco, A. Amgoune, K. Miqueu, S. Ladeira, D. Bourissou, J. Am. Chem. Soc. **133**, 4257–4259 (2011)
9. S. Bontemps, H. Gornitzka, G. Bouhadir, K. Miqueu, D. Bourissou, Angew. Chem. Int. Ed. **45**, 1611–1614 (2006)
10. M. Sircoglou, S. Bontemps, M. Mercy, N. Saffon, M. Takahashi, G. Bouhadir, L. Maron, D. Bourissou, Angew. Chem. Int. Ed. **46**, 8583–8586 (2007)
11. M. Sircoglou, S. Bontemps, G. Bouhadir, N. Saffon, K. Miqueu, W. Gu, M. Mercy, C.-H. Chen, B.M. Foxman, L. Maron et al., J. Am. Chem. Soc. **130**, 16729–16738 (2008)
12. M.-E. Moret, J.C. Peters, Angew. Chem. Int. Ed. **50**, 2063–2067 (2011)
13. P. Gualco, Coordination Originale de Ligands Phosphine-Silanes Sur Le Cuivre, L'argent et L'or, Ph.D. thesis, Université de Toulouse, Université Toulouse III - Paul Sabatier, 2012
14. P. Gualco, S. Ladeira, K. Miqueu, A. Amgoune, D. Bourissou, Angew. Chem. Int. Ed. **50**, 8320–8324 (2011)

Chapter 2
Copper: σ-SiH Coordination to Cu(I)

Abstract This chapter gives firstly an introduction concerning σ-complexes and in particular σ-SiH complexes. Afterwards, the obtained results concerning the coordination of diphosphino-hydrosilanes to Cu(I) will be described.

2.1 Introduction

2.1.1 σ-Bond Coordination to Transition Metals

The activation of σ-bonds by transition metals to undergo oxidative addition is a fundamental reaction in the field of coordination chemistry. Coordination of a σ-bond to a metal center to form a σ-complex is generally involved on the way to complete bond cleavage (Fig. 2.1). Such complexes that may be observed or even isolated, play key roles as intermediates or transition states in oxidative addition processes.

The field of σ-complexes evolved significantly following the pioneering studies of Kubas et al. [1, 2] in 1980s. The finding that an intact dihydrogen molecule coordinates to a transition metal was at first highly debated [2], but the relevance of this discovery for transition metal-catalyzed bond activation processes was soon recognized and stimulated the field of coordination chemistry.

The coordination of a σ-bond is nowadays well understood and can be considered as an early state of bond activation or "arrested oxidative addition" [3]. The precise understanding of the bonding interaction in a σ-complex is therefore of paramount importance for synthetic applications, as knowledge of structure and reactivity of such complexes will lead to more active and selective catalysts. The side-on coordination of σ-bonds has become a textbook example for weak metal-ligand interactions [4] and a number of complexes containing HH [2], HSi [3, 5–8] HB [9], HC [10–12], SiSi [13–16] and even CC [17–19] bonds has been characterized.

A σ-complex can be described as a 3-center 2-electron interaction where a σ-bond acting as a 2-electron donor is coordinated in a slide-on fashion (η^2) to a

M. Joost, *Synthesis and Original Reactivity of Copper and Gold Complexes*,
Springer Theses, DOI 10.1007/978-3-319-18690-0_2

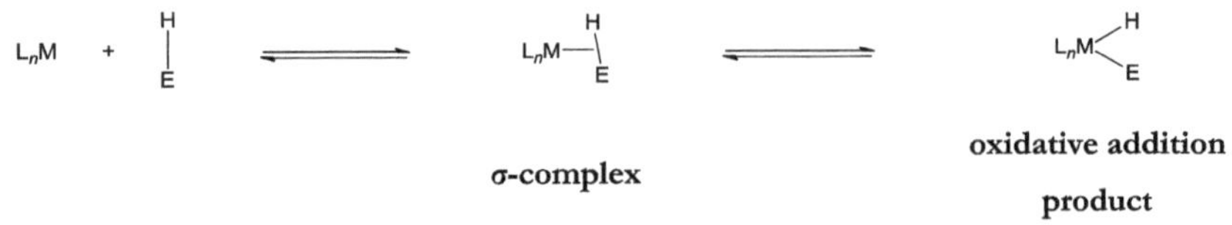

Fig. 2.1 Oxidative addition reaction at a transition metal complex with intermediacy of a σ-complex

metal, in close relationship to the Dewar-Chatt-Duncanson model that rationalizes the bonding interaction of an alkene (or an alkyne) to a metal center to form a π-complex. The σ-orbital of the ligand overlaps with a vacant metal d-orbital which leads to donation of electron density from the ligand to the metal. Furthermore, such a complex may be stabilized by backbonding, i.e. retrodonation of electron density from the metal to the ligand by interaction of a filled metal orbital of suitable symmetry with the σ*-orbital of the ligand (Fig. 2.2).

In analogy to other 3-center 2-electron bonds involving p-block elements like in polyboranes or carbonium ions, the bonding situation of a σ-complex cannot be depicted by a single Lewis structure and is therefore also categorized as a "non-classical interaction".

Equally important and closely related to σ-complexes are agostic interactions. Originally, this weak bonding interaction was defined as the intramolecular 3-center 2-electron coordination of a CH-bond to a transition metal, i.e. the ligand coordinating to a metal center is tethered to a CH-bond that is interacting in a side-on fashion with the same metal [20]. Over time, the term was watered down and used as well for other 3-center 2-electron interactions such as M(σ-SiH), M(σ-BH), M (σ-CC), etc. A recent contribution of Brookhart and Green who originally coined the term "agostic" points out that "*"agostic" is not synonymous with "3-center 2-electron*"" and should be used exclusively for CH bonds [10]. We decided to follow these recommendations and intramolecular σ-interactions other than with CH bonds are not called agostic in this manuscript. Nevertheless, we are well aware of the widespread use of the term agostic and the arguments for it in this terminology issue [2, 3, 5, 17].

Fig. 2.2 General bonding scheme for a σ-complex

2.1.2 σ-SiH Complexes

Although the first reported complex featuring a coordinated σ-bond in 1969 contained actually a hydrosilane [21], the field of σ-complexes and the systematic investigation thereof started to evolve only with the discovery of the first σ-H_2 complex in the early 1980s (vide supra) [1, 3]. With the growing interest in this type of weak bonding interaction, the experimental observation and theoretical understanding of σ-SiH complexes greatly advanced and had direct impact on SiH-bond activation processes such as hydrosilylation reactions.

The η^2-bonding scheme of hydrosilanes is essentially the one described above (see 2.1.1). However, even though a hydrosilane moiety often resembles dihydrogen in terms of reactivity, there are important differences to the more frequently encountered H_2 ligand: the SiH bond is weaker than the one of H_2 (ca. 75–100 kcal/mol vs. 104 kcal/mol), having a bonding SiH orbital higher in energy and an antibonding SiH orbital lower in energy, which makes an SiH moiety a better σ-donor and a better π-acceptor than H_2. Furthermore, the H_2 ligand coordinates in a symmetric fashion, while the SiH-bond coordination is necessarily dissymmetric (Fig. 2.3). The inherent difference in electronegativity of Si and H (1.9 vs. 2.1 on the Pauling scale) as well as the electronic influence of the substitution pattern on silicon modulate the donation and backdonation components.

Furthermore, the approach of a hydrosilane moiety to a metal center is strongly affected by the steric situation determined by the substituents at silicon, in contrast to the small H_2 molecule [3, 5].

2.1.3 Experimental Identification of σ-SiH Interactions

The characterization of σ-SiH complexes is well-established and is based mainly on NMR and IR spectroscopy as well as single crystal X-ray or neutron diffraction analysis.

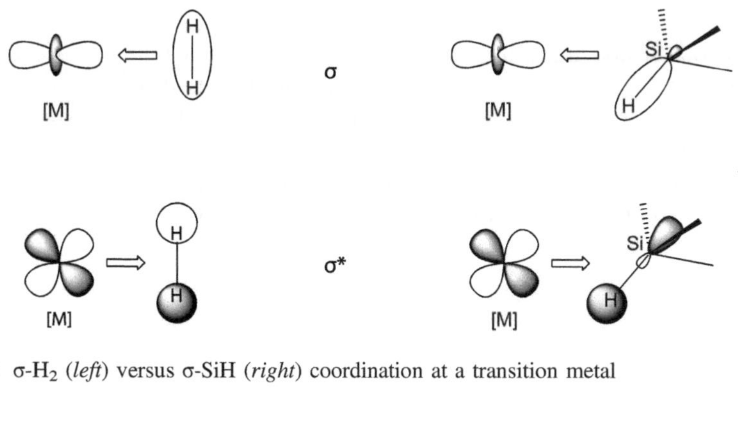

Fig. 2.3 σ-H_2 (*left*) versus σ-SiH (*right*) coordination at a transition metal

A typical NMR signature for the coordination of σ-SiH-bonds to transition metals is the decrease of the silicon-hydrogen coupling constant ($^{1}J_{SiH}$) as compared to the non-coordinated ligand [3, 5]. Known silane complexes are associated with coupling constants down to 40–70 Hz, but σ-SiH complexes featuring $^{1}J_{SiH}$ values at around 100 Hz have been reported as well [22, 23]. In general, it holds that the more activated the σ-SiH-bond is, the lower the $^{1}J_{SiH}$ value.

Another valuable and informative indicator is the Si–H stretching frequency (ν(SiH)): coordination of the σ-SiH bond results in a decrease of its bond strength and therefore ν(SiH) is redshifted upon coordination. Broad, intense bands observed in the region between 1650 and 1800 cm^{-1} are considered to be indicative of η^2-SiH-bond coordination.

Structural information can be obtained by single crystal X-ray diffraction analysis. Close proximity to the metal center of both hydrogen and silicon and an elongation of the SiH bond may hint to coordination. However, the precise localization of hydrogen atoms close to heavy elements is inherently difficult by X-ray crystallography even with high-quality measurements. Single crystal neutron diffraction circumvents this limitation, but crystals of sufficient size and quality are often not easily to obtain. Furthermore, beam time at one of the few neutron sources in Europe is scarce and necessitates not negligible effort to obtain.

Nowadays, high-level DFT computational studies are able to reproduce with high precision experimentally obtained structural and spectroscopic data. The accurate localization of hydrogen atoms in a given structure can be performed by computational means in combination with and complementary to X-ray diffraction analyses [24, 25].

Although the field of σ-SiH complexes is in general well developed and examples exist for virtually all transition metals, there was no example involving a coinage metal to be found in the literature, except for a computational analysis [26, 27]. Therefore we aimed at investigating the coordination of σ-SiH bond to group 11 metal centers.

2.2 σ-SiH Complexes of Cu

Isolation of 1st-row transition metal σ-complexes is generally more challenging than for the heavier metals, due to their less diffuse d-shell that hampers significant overlap with σ-orbitals. In the case of cationic metal centers, the positive charge leads to further contraction of the 3d-shell which makes effective backdonation of electron density from the metal to the σ*-orbital difficult. The isolation of a σ-complex of copper is therefore expected to be non-trivial.

2.2.1 Synthesis and Characterization of Cu(I) Complexes

Ambiphilic chelating phosphine ligands have been used successfully to investigate the coordination of Lewis acids to transition metals [28, 29]. This chelating approach was extrapolated to the coordination and activation of σ-SiSi bonds at Cu (I) and Au(I) [30, 31]. In case of copper, this strategy led to the isolation of the first structurally characterized σ-complex of a coinage metal (Fig. 2.4).

This finding prompted us to investigate the capability of copper to bind other σ-bonds. The chelate assistance approach was extrapolated here to σ-SiH bonds. For that purpose, we turned our interest to a diphosphine-hydrosilane ligand, namely, [$Ph_2P(o\text{-}C_6H_4)SiH(Me)(o\text{-}C_6H_4)PPh_2$] (**1**). This ligand has recently attracted much interest with group 8–10 metals (Ru, Rh, Ir, Ni, Pd, and Pt) [23, 32–39]. The central σ-SiH bond is readily activated by these metals, and the ensuing diphosphine-silyl pincer complexes were found to possess versatile reactivity [33, 34] and interesting catalytic properties [37, 38]. Noteworthily, the σ-SiH complexes en route to the PSiP pincer complexes have recently been intercepted and characterized with Ni, Pd, and Pt [38, 39].

For the purpose of comparison, a palladium complex of 1 (Fig. 2.5), reported by the Iwasa et al., is discussed here in more detail [23]. The key feature of the 1H solution NMR spectrum of this palladium complex is the 1JSiH coupling constant of 110 Hz at the upper range of reported SiH coupling constants for hydrosilane σ-complexes. The molecular structure in the solid state has been determined by single crystal X-ray diffraction analysis: The palladium atom of this compound exhibits a distorted tetrahedral environment, with a triphenylphosphine ligand and the two phosphine donor arms of 1 forming the base of a tetrahedron and the tethered SiH-moiety completing the coordination sphere in the apical position. The SiH bond length is 1.60(3) Å, which is significantly longer than the SiH bond of

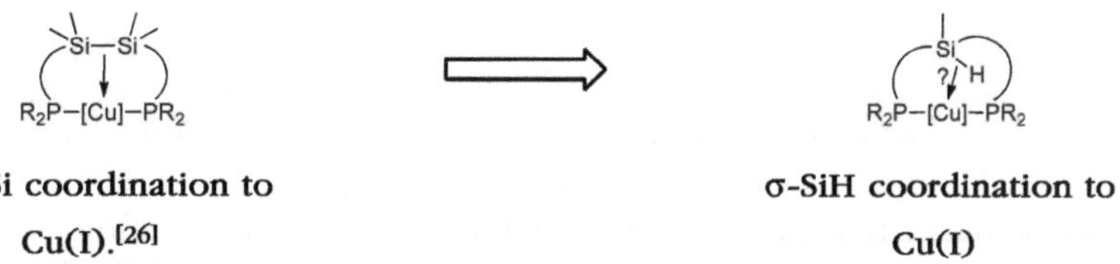

Fig. 2.4 Coordination of a SiSi bond and hypothetical/envisioned coordination of a hydrosilane to Cu(I)

Fig. 2.5 Iwasa's Pd σ-SiH complex and corresponding key experimental data

non-coordinated 1 (1.484 Å, calculated, vide infra), but relatively short in comparison with the majority of reported σ-SiH complexes (ca. 1.6 Å up to 1.9 Å) [3, 5]. Vibrational spectroscopic data were not attributed.

The comparatively small metric and spectroscopic changes in the SiH moiety were attributed to only weak backdonation from Pd to the σ* orbital of the SiH bond.

We prepared the diphosphine-hydrosilane ligand **1** by slight modification of the reported procedure, coupling two equivalents of isolated *ortho*-lithiated triphenylphosphine with dichloromethylsilane (Scheme 2.1) [32]. After purification by column chromatography **1** was obtained in 61 % yield. Coordination to copper was achieved by reacting **1** with 1 eq. of CuCl in dichloromethane (addition at −40 °C and warm-up to room temperature over 1 h). The progress of the reaction is easily followed visually as the initial suspension of CuCl progressively turns to a clear solution. Copper complex **2** was isolated as a white solid (68 % yield) after precipitation with pentane.

Complex **2** displays a single resonance signal at δ = −7.6 ppm in the $^{31}P\{^{1}H\}$ NMR spectrum, in agreement with the symmetric coordination of the two phosphorus atoms. The associated ^{29}Si NMR signal appears as a triplet (J_{SiP} = 23 Hz) at δ = −30 ppm. These data are very similar to those of the free ligand **1** (δ_{Si} = −24 ppm, J_{SiP} = 20 Hz) and indicate the retention of the hydrosilane motif upon coordination. The slight high-field shift of the ^{29}Si NMR signal (by 6 ppm) is hardly informative at this stage. It may result from a weak interaction of the hydrosilane motif with copper or simply from some modification of the geometry around silicon upon coordination. A similar high-field shift was observed upon coordination of a related diphosphine-dimethylsilane ligand to gold, without any interaction of the metal center with the silane motif [40]. The ^{1}H NMR signal for the hydrogen atom bound to silicon is shifted downfield by ~1.1 ppm compared to that of the free ligand (from δ = 5.48 ppm in **1** to δ = 6.63 ppm in **2**). Most indicative is the associated $^{1}J_{SiH}$ coupling constant measured from the ^{29}Si satellites of the ^{1}H NMR signal (Fig. 2.6).

It decreases from 204 Hz in the free ligand **1** to 180 Hz in complex **2**, suggesting some weakening of the SiH bond. However, the relatively large $^{1}J_{SiH}$ coupling constant of **2** compared to reported σ-SiH complexes and in particular compared to Iwasa's related σ-SiH palladium complex (vide supra, $^{1}J_{SiH}$ = 110 Hz) suggests only weak interaction of the σ-SiH-bond with copper. To confirm this bonding picture, we then turned to infrared spectroscopy: The Si–H vibration band is found

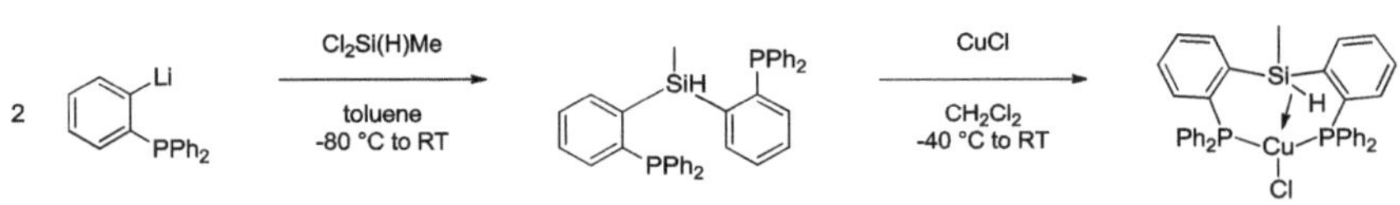

1 (61 %) **2** (68 %)

Scheme 2.1 Synthesis of ligand **1** and Cu(I) complex **2**

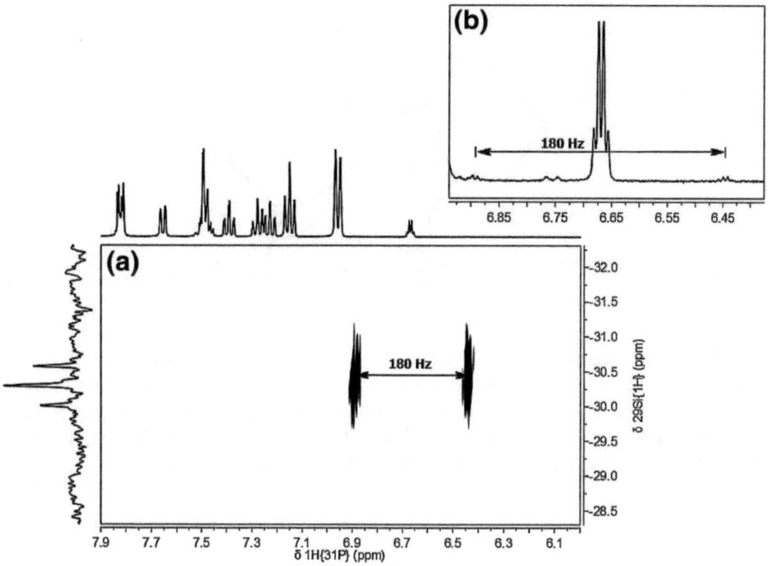

Fig. 2.6 Determination of $^{1}J_{SiH}$ for complex **2**: **a** ^{1}H-^{29}Si-HMBC spectrum; inset **b** magnification of the SiH signal in the ^{1}H{^{31}P} spectrum showing the ^{29}Si satellites

at 1996 cm^{-1} in complex **2**, versus 2142 cm^{-1} in the free ligand. This variation confirms the weakening of the σ-SiH bond upon coordination and again follows the usual trend of η^2-SiH complexes, albeit with a significantly weaker magnitude (bands in the range 1650–1800 cm^{-1} are usually observed) [3, 5]. To gain further insight into the structure of this complex, crystals were grown by slow diffusion of pentane into a dichloromethane solution of **2** at −30 °C, and an X-ray diffraction analysis was performed (Fig. 2.7).

The copper center of **2** is surrounded by the two phosphorus atoms and the chlorine atom, organized in a trigonal-planar arrangement [PCuP = 115.66(2)°, sum of angles around Cu: 359.7°]. Interestingly, the silicon atom is also located close to copper. The Cu–Si distance [2.9977(5) Å] is beyond the sum of the covalent radii (2.43 Å) [41], but well within the sum of the van der Waals radii (4.10 Å) [42]. This is consistent with a possible participation of the σ-SiH bond in the coordination, as suggested by the NMR and IR data. Even though the obtained X-ray data are of acceptable quality, the margin errors for the associated SiH and CuH distances are large and make small changes in the SiH bond distance insignificant. Therefore, discussions of these structural parameters are not meaningful at this point. In order to gain more insight into the structure of copper complex **2** with regard to precise hydrogen atom localization and to obtain subsequently a more detailed picture of the bonding situation in **2** and related complexes, DFT calculations have thus been carried out (see Sect. 2.2.3).

We envisioned increasing the electrophilicity of the metal to explore how this affects the coordination of σ-SiH bonds to copper. For that purpose, **2** was reacted

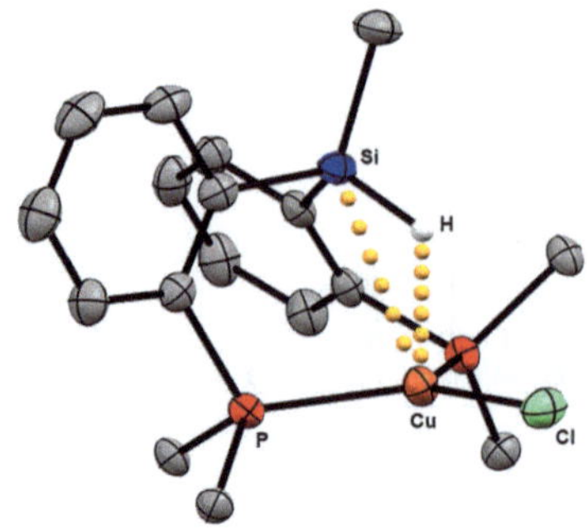

Fig. 2.7 Molecular structure of **2** determined by single crystal X-ray diffraction. Solvent molecules, phenyl substituents (with exception of the *ipso* carbon atoms) at phosphorus and hydrogen atoms, except the one at Si, are omitted for clarity. Selected bond lengths [Å] and angles [°]: CuH 2.00(2), HSi 1.42(2), CuCl 2.2287(5), CuP1 2.2557(6), CuP2 2.2783(5), SiCu 2.9977(5), P1CuP2 115.66(2), P1CuCl 122.10(2), ClCuP2 121.94(2), SiHCu 122(1)

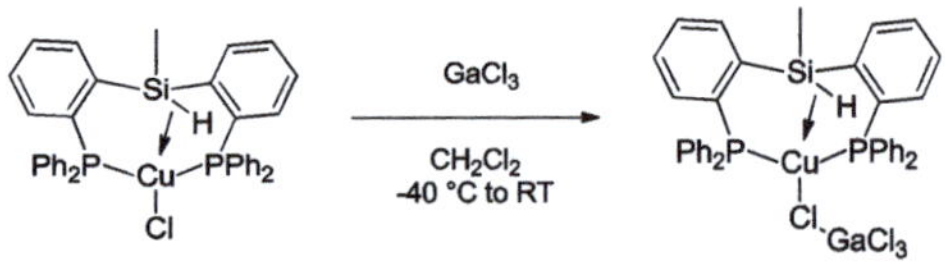

2 **3** (80 %)

Scheme 2.2 Synthesis of Cu(I) complex **3**

with one equivalent of $GaCl_3$ and the new complex **3** was isolated in 80 % yield (Scheme 2.2).

The ^{31}P and ^{29}Si NMR data for **3** (^{31}P: $\delta = -5.0$ ppm and ^{29}Si: $\delta = -30$ ppm, $J_{SiP} = 23$ Hz) very much resemble those of **2**, suggesting that the two complexes adopt similar coordination modes. Notably, the characteristic features suggesting the presence of a weak interaction between the σ-SiH bond and the copper center in **2** are also met in **3**, even to a slightly greater extent: the $^{1}J_{SiH}$ coupling constant is further decreased down to 170 Hz and, concomitantly, the ν(SiH) band is further shifted to lower frequency at 1973 cm^{-1} (Table 2.1).

These data may suggest a slight strengthening of the σ-SiH bond coordination from **2** to **3**. Crystals of **3** were grown from a dichloromethane/pentane solution at −60 °C, and its molecular structure was elucidated by an X-ray diffraction analysis (Fig. 2.8).

Accordingly, the Lewis acid $GaCl_3$ interacts strongly with the chloride at copper, but without completely abstracting it. The CuCl distance is significantly elongated [CuCl = 2.359(1) Å in **3** vs. 2.229(1) Å in **2**], so that complex **3** can be formally considered as a tight ion pair [(**1**)$Cu^{+}\cdots ClGaCl_3^{-}$]. Here also the silicon atom comes close to copper, at an even shorter distance [CuSi = 2.815(1) Å] than in complex **3**. Since the overall geometry around copper in **3** is otherwise similar to

Table 2.1 Selected experimental data for ligand **1** and complexes **2** and **3**

	Free ligand 1	CuCl complex 2	$CuClGaCl_3$ complex 3
$\delta^{29}Si$	−24	−30	−30
J_{PSi}	20	23	23
$\delta^{1}H$	5.48	6.63	6.44
$^{1}J_{SiH}$	204	180	170
ν_{SiH}	2142	1996	1973
MSi	–	2.997(1)	2.815(1)
PMP	–	115.66(2)	120.13(2)

NMR chemical shifts in ppm, coupling constants in Hz, ν(SiH) stretching frequencies in cm^{-1}(IR), MSi distances in Å, and PMP bond angles in deg (X-ray)

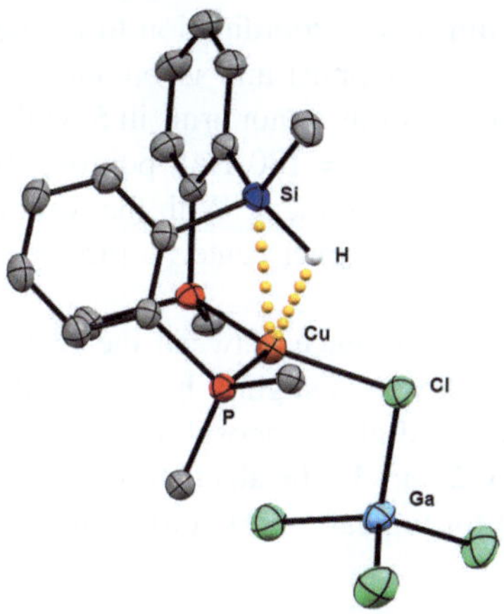

Fig. 2.8 Molecular structure of 3 determined by single crystal X-ray diffraction. Solvent molecules, phenyl substituents (with exception of the ipso carbon atoms) at phosphorus and hydrogen atoms, except the one at Si, are omitted for clarity. Selected bond lengths [Å] and angles [°]: CuCl 2.3587(6), CuH 1.88(2), HSi 1.45(2), CuP1 2.2535(6), CuP2 2.2471(6), CuSi 2.8148(5), ClGa 2.2421(5), P2CuCl 119.18(2), ClCuP1 118.69(2), P2CuP1 120.13(2), SiHCu 115(1), CuClGa 104.10(2)

that of the CuCl complex **2** [trigonal-planar arrangement, quasi-unchanged PCuP bite angle at 120.13(2)°], the shortening of the CuSi distance most likely results from electronic rather than geometric factors. The structural features of **3** are consistent with the spectroscopic data and suggest indeed a slightly stronger coordination of the σ-SiH bond upon decreasing the electron density at copper.

Next, the influence of the substitution pattern at the phosphine donor groups and at the central silicon atom of the ligand on the coordination of the hydrosilane moiety to Cu(I) was investigated. An electron-withdrawing substituent on silicon should diminish the electron density of the SiH bond, while at the same time more basic phosphine sites would electronically enrich the copper center. We expected that both contributions should increase the backdonation from copper to SiH and therefore stabilize the η^2-SiH bond. Ligand **4** ([iPr$_2$P(o-C$_6$H$_4$)SiH(Ph)(o-C$_6$H$_4$) PiPr$_2$] featuring diisopropylphosphine donors and a (phenyl)hydrosilane motif was synthesized in order to fine-tune the strength of the intramolecular σ-bond of the corresponding Cu(I) complexes. **4** was prepared similarly to **1**. The reaction of 2 equivalents of *o*-lithiated diisopropylphenyl phosphine with dichlorophenylsilane gave **4** in 83 % yield after workup by crystallization. As for **2**, the synthesis of the copper(I) complex was performed by reaction of **4** with CuCl at low temperature (Scheme 2.3). Subsequent work-up by precipitation gave CuCl complex **5** in 77 % yield. Generation of a more electrophilic copper species was achieved by reaction of **5** with $GaCl_3$ at −40 °C to give complex **6** in 87 % yield.

The spectroscopic data of **4**, **5** and **6** are somewhat similar to those of **1**, **2** and **3**, respectively (Table 2.2). A notable feature of the ^{1}H NMR spectrum of **4** is the $^1J_{SiH}$ value of 220 Hz. The free ligand **4** exhibits a single resonance at δ = −0.8 ppm in the ^{31}P{^{1}H} NMR spectrum. Upon coordination to copper the ^{31}P NMR signal is shifted to lower field (δ = 16.9 ppm) and is broadened, indicating a symmetric environment for the two phosphorus donor arms in **5** while at the same time the SiH coupling constant decreases ($^1J_{SiH}$ = 180 Hz), pointing at the coordination of the hydrosilane moiety to the CuCl fragment. With the formation of the $GaCl_3$ adduct (**6**) featuring a less electron-rich copper center, a further decrease of $^1J_{SiH}$ down to 170 Hz is observed (Table 2.2).

Even though Δ^1J_{SiH}, the difference between the SiH coupling constant of free ligand and the copper complexes, is slightly bigger for the couple in the isopropyl phosphine series than for the diphenylphosphine complexes (40 and 50 Hz for **5** and **6** vs. 24 and 34 Hz for **2** and **3**), the absolute values of $^1J_{SiH}$ of **5** and **6** are still not in the range expected for typical σ-SiH complexes (vide supra). An analysis of

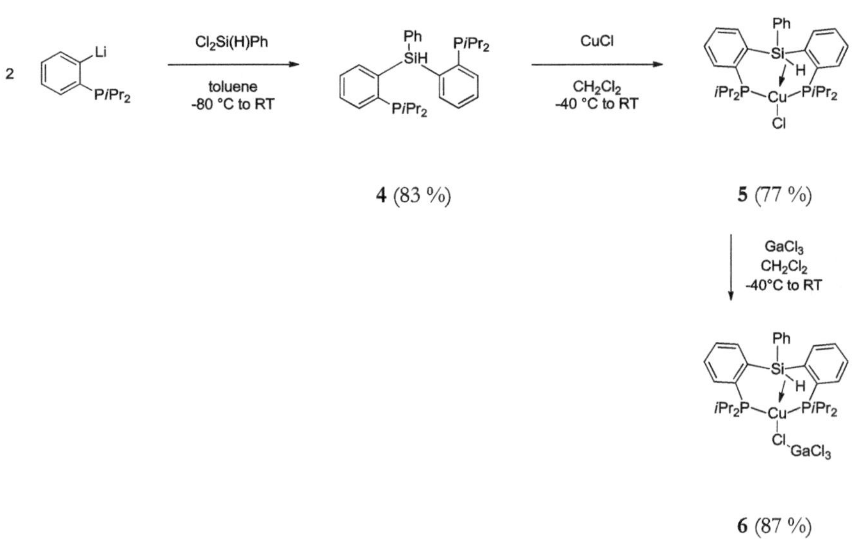

Scheme 2.3 Synthesis of ligand **4** and its corresponding Cu(I) complexes **5** and **6**

Table 2.2 Selected experimental data for ligand **4** and complexes **5** and **6**

	Free ligand 4	CuCl complex 5	$CuClGaCl_3$ complex 6
$\delta^{29}Si$	−27	−34	−35
J_{PSi}	22	19	26
$\delta\ ^1H$	6.42	7.11	7.19
$^1J_{SiH}$	220	180	170
ν_{SiH}	2165	2043	1978
MSi	–	2.8833(6)	2.837(2)
PMP	–	140.19(2)	137.69(8)

NMR chemical shifts in ppm, coupling constants in Hz, ν(SiH) stretching frequencies in cm^{-1} (IR), MSi distances in Å, and PMP bond angles in deg (X-ray)

the IR spectra of **4**, **5** and **6** points into the same direction: ν(SiH) decreases from 2165 cm^{-1} in the free ligand to 2043 cm^{-1} in the CuCl complex and is further redshifted to 1978 cm^{-1} in **6** which is in the same region as for the diphenylphosphino complexes. These spectroscopic characteristics suggest, once again, only weak bonding of the hydrosilane moiety to the copper center and no significant influence of the substitution pattern on silicon and phosphorus on this coordination.

Crystals of **5** and **6** were grown and the molecular structures of these complexes were determined by an X-ray diffraction analysis (Fig. 2.9) in order to analyze as well their structural features.

The coordination environment of Cu in **5** and **6** resembles the complexes of the diphenylphosphine series, both for the copper chloride complex and the $GaCl_3$ adduct, with the copper center being surrounded by the two phosphine arms and the

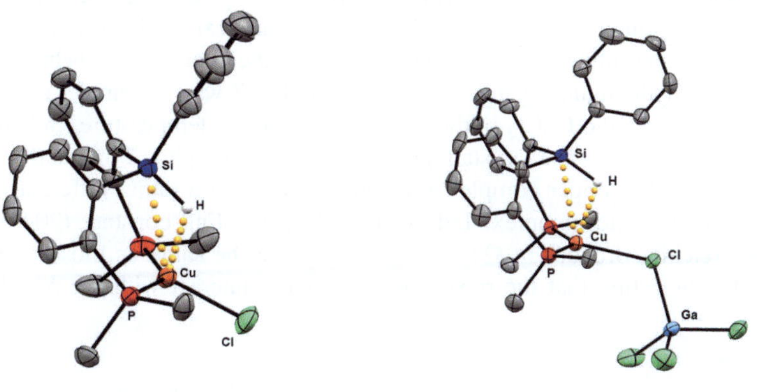

Fig. 2.9 Molecular structures of **5** (*left*) and **6** (*right*) determined by single crystal X-ray diffraction. Isopropyl substituents at phosphorus are simplified. Solvent molecules and hydrogen atoms, except the ones at Si, are omitted for clarity. Selected bond lengths [Å] and angles [°]: **5**: ClCu 2.2681(6), CuH 1.91(2), HSi 1.40(2), CuSi 2.8833(6), CuP1 2.2330(7), CuP2 2.2339(7), ClCuP 107.67(2), P1CuP2 140.19(2), P2CuCl 110.36(2), CuHSi 120(1); **6**: ClCu 2.473(2), CuH 2.08(9), HSi 1.40(6), CuP1 2.230(2), CuP2 2.256(2), ClGa 2.231(2), CuSi 2.837(2), ClCuP1 106.51(7), ClCuP2 115.30(7), P2CuP1 137.69(8), CuHSi 108(5), CuClGa 117.79(9)

chloride atom in a trigonal-planar arrangement. The PCuP angle is significantly opened for **5** [PCuP = 140.19(2)°] and **6** [PCuP = 137.69(8)°], compared to **2** and **3**. The CuCl distances for **5** and **6** are in the same range as observed before: 2.2681 (6) Å for the copper chloride complex and 2.473(2) Å in the $GaCl_3$ adduct, suggesting for **6** as well the formulation as a tight ion pair: $[(\mathbf{4})Cu^{+}\cdots ClGaCl_3^{-}]$.

The most striking feature is the short CuSi distance in **5** (2.8833(6) Å). This distance is further shortened upon formation of the $GaCl_3$ adduct **6** (2.837(2) Å). In accordance with the spectroscopic data, this shortening can be ascribed to a strengthened η^2-SiH coordination to copper.

Overall, the structural parameters confirm the picture of weak bonding of the SiH bond to Cu(I) in complexes **5** and **6**.

Although the modified substitution pattern shows a certain influence on the spectroscopic and structural features of the σ-SiH complexes, its impact is rather small. Even though some characteristics of **5** and **6** may suggest a marginal strengthening of the σ-SiH coordination to Cu, its magnitude is not significant.

The only weak strengthening is maybe due to the negative influence of the electron rich phosphine donor arms on the donation from SiH to Cu(I): Overall, the effects of increased backdonation and decreased donation might cancel out and the bond strength does not change after all.

2.2.2 Comparison to a Corresponding Au(I) Complex

In the group, it was shown previously that the coordination of diphosphine-disilane ligands proceeds very differently with copper and gold: side-on coordination of the σ-SiSi bond was observed with copper [30], while gold undergoes spontaneous oxidative addition and forms bis(silyl) gold(III) complexes [31]. This prompted us to investigate the influence of the metal center on the coordination behavior of the diphosphine-hydrosilane ligand. The gold complex **7** was prepared by reacting **1** with $AuCl(SMe_2)$ and $GaCl_3$ in dichloromethane at low temperature (Scheme 2.4).

The spectroscopic and crystallographic data for complex **7** markedly deviate from those of the copper complexes **2** and **3** and reveal a rather different bonding situation in the gold complex. Indeed, the $^1J_{SiH}$ coupling constant (204 Hz) and ν(SiH) stretching frequency (2141 cm^{-1}) for **7** are the same as those of the free ligand **1**, suggesting that the σ-SiH bond does not interact with the metal in this

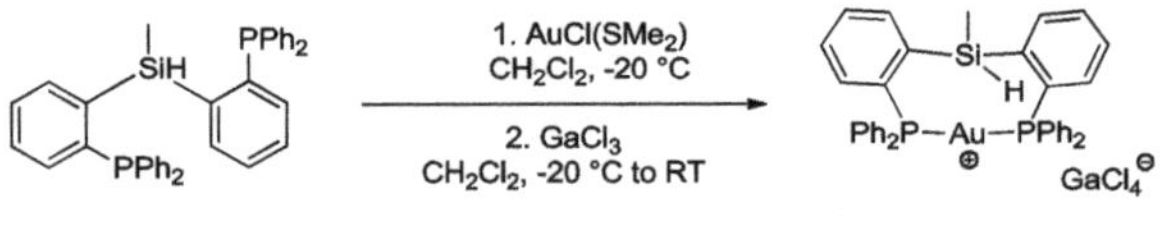

1 **7** (85 %)

Scheme 2.4 Synthesis of Au(I) complex **7**

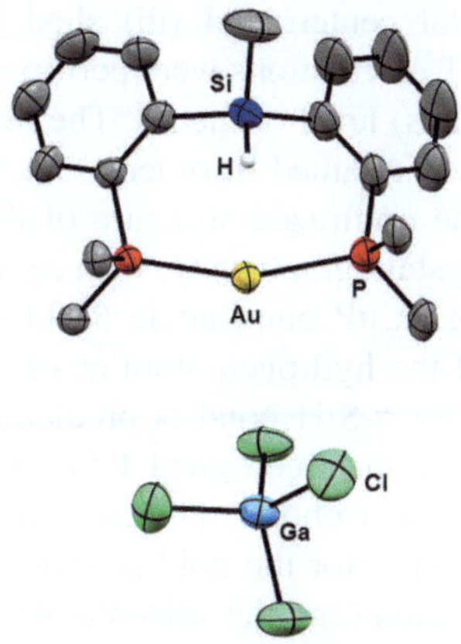

Fig. 2.10 Molecular structure of **7** determined by single crystal X-ray diffraction. Solvent molecules, phenyl substituents (with exception of the *ipso* carbon atoms) at phosphorus and hydrogen atoms, except the one at Si, are omitted for clarity. Selected bond lengths [Å] and angles [°]: AuH 2.63(4), HSi 1.33(4), AuP1 2.294(1), AuP2 2.285(1), AuCl 3.285(1), AuSi 3.166(1), P1AuP2 159.92(4), AuHSi 101(2)

case. This is consistent with the molecular structure of **7** as determined by single-crystal X-ray diffraction analysis (Fig. 2.10).

The gold complex **7** adopts a discrete ion pair structure [(**1**)Au^+,$GaCl_4^-$] [shortest Au···Cl distance at 3.285(1) Å] with a wide PAuP bite angle [159.92(4)°], but the silicon atom remains far away from gold. Indeed, the AuSi distance in **7** [3.166(1) Å] substantially exceeds the CuSi distances in **2** and **3** [2.997(1) and 2.815(1) Å, respectively], although the covalent radii of gold and copper are very similar (1.36 and 1.32 Å, respectively) [41]. The comparison of complexes **2**, **3** and **7** provides further support for the participation of the σ-SiH bond in the coordination to copper, but not to gold. The bonding situation in the gold complex **7** contrasts with the oxidative addition process observed with the σ-SiSi bond to give a disilyl gold(III) complex [31]. In complex **7**, the metal center accommodates a quasi-linear bis(phosphine)Au^+ arrangement. Comparatively, copper is more Lewis acidic [43] and the coordination of the pendant SiH or SiSi bond is favorable. In addition, the fact that the silicon atom remains remote from gold in complex **7** indicates that the coordination of the σ-SiH bond to copper in complexes **2** and **3**, although favored by the two phosphine anchors, is not imposed geometrically.

2.2.3 Computational Analysis

A detailed computational study was carried out in collaboration with the group of Dr. K. Miqueu (Université de Pau) in order to (i) further analyze the structural features of the diphosphine-silane complexes **2** and **3**, (ii) probe the influence of the electron density at the metal center, and (iii) shed light on the nature of the σ-SiH/metal interaction. DFT calculations were performed at the B3PW91/SDD + f (Cu,Au), 6–31G**(other atoms) level of theory. The full substitution pattern of the diphosphine-silane ligand was retained in order to take reliably into account electronic and steric factors. The optimized structure of the CuCl complex **2*** reproduces the one determined crystallographically, with deviations of only 0.03 Å in the CuSi distance and 7° in the PCuP bond angle (Tables 2.1 and 2.3). Most informative is the localization of the hydrogen atom at silicon and the associated Si-H bond length. Accordingly, the σ-SiH bond is predicted to slightly elongate upon coordination, from 1.484 Å in the free ligand **1*** to 1.499 Å in complex **2***. The significance of this variation is corroborated by comparing copper and gold. Indeed, the σ-SiH bond length computed for the gold complex **7*** (1.481 Å) is essentially identical to that of the free ligand (and the silicon atom remains far away from gold at 3.269 Å). This confirms that the side-on coordination of the σ-SiH bond to copper in complex **2** is not a geometric artifact but indeed the result of a bonding interaction.

The next point was to assess how the electrophilicity of copper affects its interaction with the σ-SiH bond. For that purpose, DFT calculations were carried out on complex **3***, featuring the tight $Cu^{+}\cdots ClGaCl_3^{-}$ ion pair structure, and on the cationic complex **3****, featuring a naked copper center (without counteranion contact, Fig. 2.11).

Here, the σ-SiH bond length was found to increase from 1.499 Å in **2*** to 1.513 Å in **3****, and concomitantly, the CuSi distance shortens from 3.021 Å in **2*** to 2.808 Å in **3**** (Table 2.3).

Besides these structural features, spectroscopic data were also computed, in particular the $^1J_{SiH}$ coupling constant. The values measured experimentally by ^1H NMR for complexes **2** and **7** as well as the free ligand **1** were well reproduced

Table 2.3 Selected computational data for ligand **1** and complexes **2***, **3***, **3**** and **7**

	Free ligand 1*	Cu complex 2*	Cu complex 3*	"Naked" Cu complex 3**	Au complex 7*
Si–H	1.484	1.499	1.501	1.513	1.481
M–H	–	2.020	1.987	1.966	2.721
M–Si	–	3.021	2.972	2.808	3.269
PMP	–	122.40	128.41	154.84	155.41
$^1J_{SiH}$	−198.6	−176.4	−164.2	−143	−201.1
ΔE_{NBO}					
$\sigma_{SiH} \rightarrow$ Lp* (M)	–	8.4	12.4	14.3	–
s(M) $\rightarrow \sigma^*_{SiH}$	–	0.9	1.1	1.3	0.25

Bond lengths/distances in Å, bond angles in deg, $^1J_{SiH}$ coupling constants in Hz, and NBO delocalization energies in kcal/mol

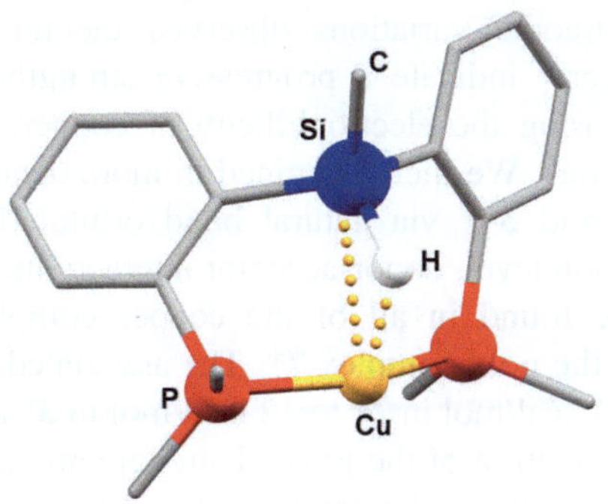

Fig. 2.11 Computed molecular structure of the naked Cu complex **3****, without any counteranion contact. Phenyl substituents (with exception of the *ipso* carbon atoms) at phosphorus and hydrogen atoms, except the one at Si, are omitted for clarity

theoretically using the GIAO method with the IGLOO II basis set (maximum deviation of only 6 Hz; see Tables 2.1 and 2.3).

The geometric variations were compared with the corresponding spectroscopic data. A quasi-linear correlation was found between the σ-SiH bond length and the $^{1}J_{SiH}$ coupling constant (Fig. 2.12). This behavior contrasts with the V-type correlation found between the σ-SiH bond length and the $^{1}J_{SiH}$ coupling constant for cationic silane complexes of ruthenium [44].

The two extreme situations are met with the free ligand **1*** on the one hand (distance SiH = 1.484 Å, $^{1}J_{SiH} = -198.6$ Hz) and the cationic complex **3**** on the other hand (distance SiH = 1.513 Å, $^{1}J_{SiH} = -143.0$ Hz). Although the interaction between the σ-SiH bond and the copper center is weak in all copper complexes, the

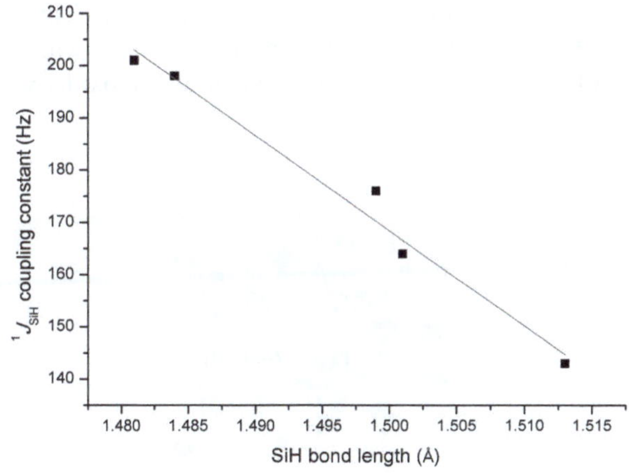

Fig. 2.12 Plot of computed $^{1}J_{SiH}$ coupling constant (using the GIAO method with IGLOO II basis set) versus SiH bond length predicted computationally at the B3PW91/SDD + f(Cu,Au),6-31G** (other atoms) level of theory demonstrating a linear relationship between the SiH bond length and the corresponding $^{1}J_{SiH}$ values for **7***, **1***, **2***, **3*** and **3**** (from *left* to *right*)

structural and spectroscopic variations observed theoretically along the series **2*** → **3*** → **3**** clearly indicate a progressive strengthening of the σ-SiH/Cu interaction upon increasing the electrophilicity of copper, in agreement with the experimental observations. We then examined in more detail the bonding situation in complexes **2***, **3*** and **3**** via natural bond orbital (NBO) analysis. At the second-order perturbation level, donor-acceptor interactions from the σ-SiH bond to the metal center were found in all of the copper complexes **2***, **3***, and **3**** (Fig. 2.13), but not in the gold complex **7***. The associated delocalization energies ΔE_{NBO} increase from 8 kcal/mol in **2*** to 12 kcal/mol in **3*** and 14 kcal/mol in **3****, providing further confirmation of the gradual strengthening of the interaction upon increasing the electrophilicity of the copper center. The relative contributions of Si and H in the involved σ-SiH bond orbital remain essentially the same upon coordination and do not vary significantly with the electron density at copper (36–41 % for Si and 58–61 % for H). Noteworthily, only insignificant backdonation from copper to σ*-SiH was found in **2***, **3*** and **3**** (ΔE_{NBO} < 1.5 kcal/mol), indicating that the coordination of the SiH bond to copper arises essentially, if not exclusively, from σ-SiH → Cu donation.

According to these NBO analyses, coordination of the σ-SiH bond to copper is very similar in nature and magnitude to that of the σ-SiSi bond we reported previously. On the basis of steric and orbital grounds, the formation of σ-complexes is a priori less favorable with disilanes than with hydrosilanes [45], but it is likely that geometric factors also play a role in our systems (the σ-SiSi bond was included within the chelating ligand backbone, while the σ-SiH bond is necessarily external). Finally, as spontaneous oxidative addition of the σ-SiSi bond had been observed at gold [31], a similar process with the σ-SiH bond of the copper complex **3*** and gold complex **7*** was considered theoretically. The structures of the corresponding Cu(III) and Au(III) complexes were optimized, and oxidative addition of the σ-SiH bond was found to be strongly disfavored energetically in both cases (by 34.5 kcal/mol for **3*** and 15.9 kcal/mol for **7***), in agreement with the experimental observations. While σ-SiH coordination is stronger to Cu, oxidative addition of the

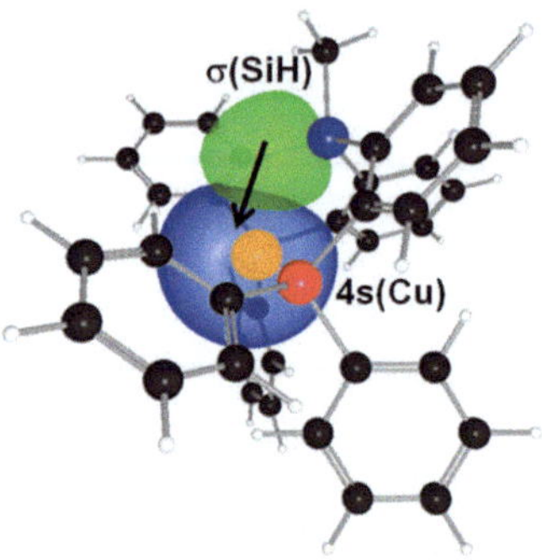

Fig. 2.13 Representation of the NLMOs on Cu and the SiH fragment, determined by an NBO analysis, illustrating the σ-SiH → Cu donation for complex **3****

σ-SiH bond is easier for Au. This contrast illustrates that Lewis acidity of a metal and its tendency to undergo oxidative addition do not necessarily show parallel trends. The reader will encounter these aspects again in Chaps. 3 and 5 where they are discussed in more detail in the context of gold chemistry.

2.2.4 Conclusion

This study provides evidence for the coordination of σ-SiH bonds to copper. The diphosphine-hydrosilanes **1** and **4** are shown experimentally and computationally to engage in weak σ-SiH/Cu interactions. The ensuing copper complexes **2**, **3**, **5** and **6** stand as the first σ-SiH complexes involving a coinage metal. The spectroscopic and geometric features typically associated with the coordination of σ-SiH bonds to transition metals (decrease of the $^1J_{SiH}$ coupling constant and ν(SiH) stretching frequency, elongation of the SiH bond) are also met here with copper, albeit with a substantially weaker magnitude. Variation of the substitution pattern at the phosphine donor arms and at the hydrosilane moiety in **5** and **6** does only have minor influence on the strength of the σ-SiH coordination to copper as evidenced by the similar spectroscopic and structural properties. According to DFT calculations, the coordination of the SiH bond to copper arises from weak σ-SiH → Cu donation, and Cu → σ*-SiH back-donation is negligible. Consistently, the coordination of the σ-SiH bond slightly strengthens when the electrophilicity of copper is increased by (partial) chloride abstraction.

On the other hand, the σ-SiH bond remains pendant in the related gold complex **7**. Neither σ-bond coordination nor oxidative addition of the SiH bond is observed here. More insight into the parameters that govern σ-bond activation processes with gold is given in Chap. 5.

2.3 Experimental Part

2.3.1 Diphosphine-Hydrosilane Ligand (1)

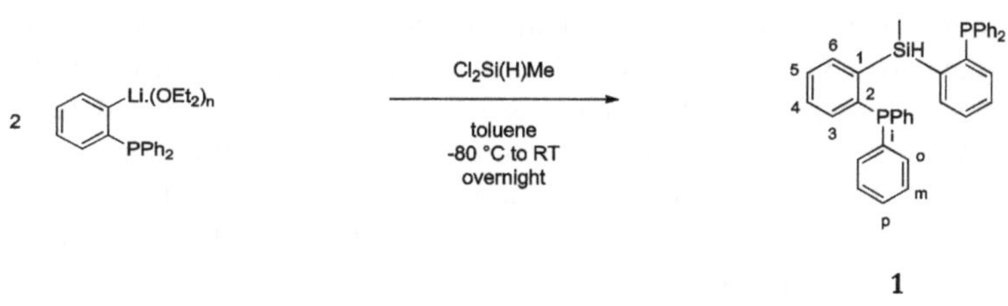

1

A solution of dichloro(methyl)silane (229 μL, 2.2 mmol) in toluene (2 mL) was added dropwise to a solution of *o*-lithiated triphenylphosphine (1.765 g, 4.65 mmol) in toluene (13 mL) at −80 °C. The solution was allowed to warm to room temperature overnight, after which volatiles were removed under vacuum. The resulting orange oil was purified by column chromatography (eluent: pentane/dichloromethane 9:1) to give the desired diphosphine (**1**) as a white solid. Yield: 1.020 g (61 %). **Mp**: 151 °C; **^{1}H NMR** (400 MHz, CD_2Cl_2): δ 7.75–7.57 (m, 2H, H_{ar}), 7.43–7.20 (m, 16H, H_{ar}), 7.19 – 6.96 (m, 10H, H_{ar}), 5.48 (tq, $^4J_{HP}$ = 5.6 Hz, $^3J_{HH}$ = 3.85 Hz, $^1J_{HSi}$ = 204 Hz, 1H, SiH), 0.73 (dt, $^3J_{HH}$ = 3.85 Hz, $^5J_{HP}$ = 0.9 Hz, 3H, $SiCH_3$); **^{31}P{^{1}H} NMR** (162 MHz, CD_2Cl_2): δ −13.2 (s); **^{29}Si{^{1}H} NMR** (60 MHz, C_6D_6): δ –24 (t, $^3J_{SiP}$ = 20 Hz); **^{13}C{^{1}H} NMR** (100 MHz, C_6D_6): δ 145.0 (dd, $^2J_{CP}$ = 46.8 Hz, $^4J_{CP}$ = 3.5 Hz, C_1), 144.4 (d, $^1J_{CP}$ = 11.2 Hz, C_2), 138.4 (d, $^1J_{CP}$ = 12.4 Hz, C_i), 137.3 (dd, $^3J_{CP}$ = 15.4 Hz, $^5J_{CP}$ = 2.4 Hz, C_3), 134.6 (s, C_p), 133.9 (m, C_o, C_m), 129.9 (s, C_5), –2.3 (t, $^4J_{CP}$ = 8.7 Hz, $SiCH_3$), signals of C_4 and C_6 overlap with the solvent signal; **IR**: ν(SiH): 2142 cm^{-1}; **HRMS (ESI+)**: calcd for $[M + H]^+ = C_{37}H_{33}P_2Si^+$: 567.1827. Found: 567.1827.

2.3.2 Coordination of 1 to Copper Chloride (2)

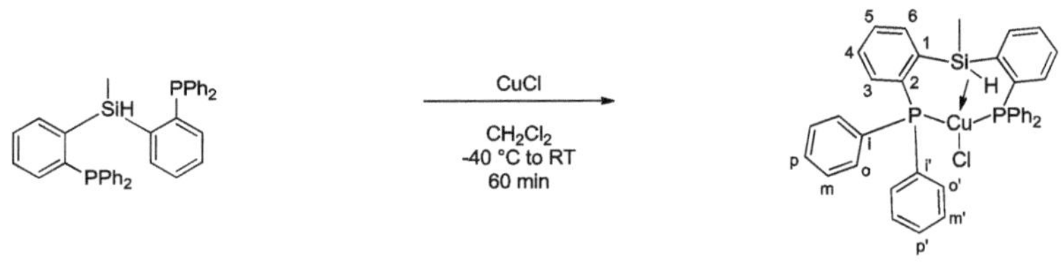

1 2

This atom labeling scheme was as well used for the NMR assignments of complexes **3** and **7**.

A solution of diphosphine-hydrosilane ligand **1** (100 mg, 0.18 mmol) in dichloromethane (3 mL) was added at –40 °C to a dispersion of copper chloride (17.5 mg, 0.18 mmol) in dichloromethane (1 mL). The reaction mixture was stirred for 60 min while warming up to room temperature. The colorless solution was concentrated to a volume of about 1 mL. Upon addition of pentane (5 mL), **2** precipitates as a white solid which was separated by filtration and dried under vacuum. Yield: 80 mg (68 %). Crystals suitable for X-ray diffraction analysis were grown at –30 °C from a saturated solution of **2** in dichloromethane layered with pentane. **Mp**: 215 °C (decomposition); **^{1}H NMR** (400 MHz, CD_2Cl_2): δ 7.85–6.86 (m, 28H, H_{ar}), 6.63 (qt, $^3J_{HH}$ = 3.4 Hz, $^4J_{HP}$ = 6.7 Hz, $^1J_{HSi}$ = 180 Hz, 1H, SiH), 0.46 (d, $^3J_{HH}$ = 3.4 Hz, 3H, $SiCH_3$); **^{31}P{^{1}H} NMR** (162 MHz, CD_2Cl_2): δ –7.6 (s);

^{29}**Si{^{1}H} NMR** (79 MHz, CD_2Cl_2): δ −30 (t, $^3J_{SiP}$ = 23 Hz); ^{13}**C{^{1}H} NMR** (100 MHz, CD_2Cl_2): δ 141.5 (*virtual* t, AA'X, *N* = 19.2 Hz, C_1), 139.6 (*virtual* t, AA'X, *N* = 16.1 Hz, C_2), 135.3 (*virtual* t, AA'X, *N* = 7.5 Hz, C_3), 134.5 (*virtual* t, AA'X, *N* = 7.8 Hz, C_o or C_m), 133.5 (*virtual* t, AA'X, *N* = 18.4 Hz, C_i), 133.0 (*virtual* t, AA'X, *N* = 7.4 Hz, C_o or C_m), 132.7 (s, C_4), 131.7 (*virtual* t, AA'X, *N* = 15.6 Hz, $C_{i'}$), 130.2 (s, C_5), 129.7 (*virtual* t, AA'X, *N* = 2.6 Hz, C_6), 129.4 (s, C_p), 129.3 (s, $C_{p'}$), 128.7 (*virtual* t, AA'X, *N* = 4.9 Hz, C_o or C_m), 128.3 (*virtual* t, AA'X, *N* = 4.6 Hz, C_o or C_m), −4.1 (t, $^4J_{PC}$ = 3.7 Hz, $SiCH_3$); **IR**: ν(SiH): 1996 cm^{-1} (br); **HRMS (ESI+)**: calcd for $[M-(Cl^-)]^+ = C_{37}H_{32}P_2SiCu^+$: 629.1045. Found: 629.1054. **Elt. Anal.**: calcd for $C_{37}H_{32}P_2SiCuCl + CH_2Cl_2$: C, 60.80; H, 4.57. Found: C, 61.28; H, 4.54.

2.3.3 Reaction of 2 with $GaCl_3$ (3)

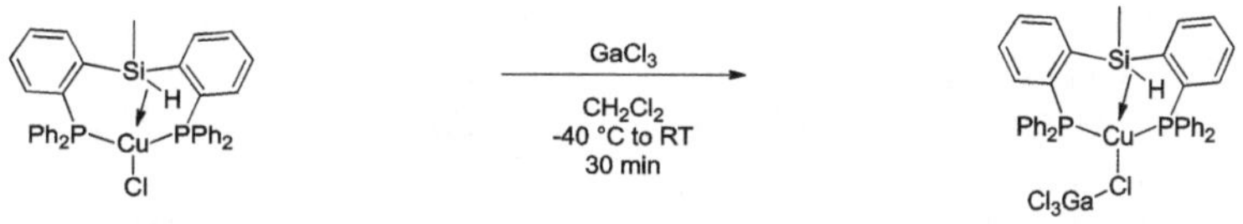

2 **3**

A solution of $GaCl_3$ (13.2 mg, 0.08 mmol) in dichloromethane (1 mL) was added to a solution of **2** (50 mg, 0.08 mmol) in dichloromethane (2 mL) at −40 °C. The mixture was stirred for 30 min while warming up to room temperature, then concentrated to a volume of about 1 mL. Complex **3** precipitated upon addition of pentane (3 mL). The colorless solid was separated from the supernatant by filtration and dried under vacuum. Yield : 51 mg (80 %). Crystals suitable for X-ray diffraction analysis were grown at −60 °C from a saturated solution in dichloromethane layered with pentane. **Mp**: 119 °C; ^{1}**H NMR** (400 MHz, CD_2Cl_2): δ 8.15–6.60 (m, 28H, H_{ar}), 6.44 (mbr, $^1J_{HSi}$ = 170 Hz, 1H, SiH), 0.32 (mbr, 3H, $SiCH_3$); ^{31}**P{^{1}H} NMR** (162 MHz, CD_2Cl_2): δ −5.0 (sbr); ^{29}**Si{^{1}H} NMR** (79 MHz, CD_2Cl_2): δ −30 (t, $^3J_{SiP}$ = 23 Hz); ^{13}**C{^{1}H} NMR** (75 MHz, CD_2Cl_2): δ 140.7 (*virtual* t, AA'X, *N* = 18.7 Hz, C_1), 139.1 (*virtual* t, AA'X, *N* = 16.1 Hz, C_2), 134.2 (*virtual* t, AA'X, *N* = 7.6 Hz, C_3), 133.9 (*virtual* t, AA'X, *N* = 7.9 Hz, C_o or C_m), 132.3 (*virtual* t, AA'X, *N* = 18.1 Hz, C_i), 132.1 (*virtual* t, AA'X, *N* = 7.3 Hz, C_o or C_m), 131.9 (s, C_4), 130.9 (*virtual* t, AA'X, *N* = 15.6 Hz, $C_{i'}$), 129.3 (s, C_5), 128.8 (*virtual* t, AA'X, *N* = 2.6 Hz, C_6), 128.3 (s, C_p), 128.2 (s, $C_{p'}$), 127.9 (*virtual* t, AA'X, *N* = 5.1 Hz, C_o or C_m), 127.3 (*virtual* t, AA'X, *N* = 4.6 Hz, C_o or C_m), −4.8 (t, $^4J_{PC}$ = 3.9 Hz, $SiCH_3$); **IR**: ν(SiH): 1973 cm^{-1} (br); **HRMS (ESI+)**: calcd for $[M-(GaCl_4^-)]^+ = C_{37}H_{32}P_2SiCu^+$: 629.1045. Found: 629.1054. **Elt. Anal.**: calcd for $C_{37}H_{32}P_2SiCuGaCl_4 + C_5H_{12}$: C, 55.20; H, 4.85. Found: C, 54.87; H, 4.52.

2.3.4 Diphosphine-Hydrosilane Ligand (4)

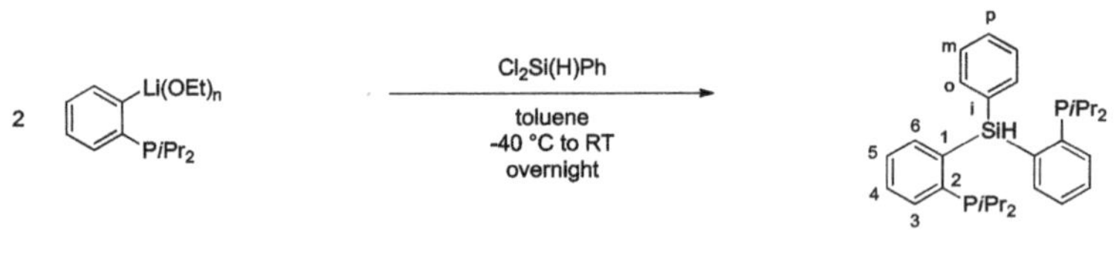

4

A solution of dichloro(phenyl)silane (308 μL, 2.10 mmol) in toluene (2 mL) was added dropwise to a solution of *o*-lithiated diisopropyl(phenyl)phosphine (1.154 g, 4.19 mmol) in toluene (4 mL) at −40 °C. The solution was allowed to warm up to room temperature overnight, after which volatiles were removed under vacuum. The resulting white solid was recrystallized from a saturated dichloromethane solution layered with pentane to give the desired product as colorless crystals. Yield: 940 mg (83 %). **Mp**: 139 °C; **^{1}H NMR** (400 MHz, CD_2Cl_2): δ 7.60–6.95 (m, 13H, H_{ar}), 6.42 (t, $^4J_{HP} = 8$ Hz, $^1J_{HSi} = 220$ Hz, 1H, SiH), 2.10 (septd, $^2J_{HP} = 2$ Hz, $^3J_{HH} = 7$ Hz, 2H, CH_{iPr}), 2.03 (septd, $^2J_{HP} = 2$ Hz, $^3J_{HH} = 7$ Hz, 2H, CH_{iPr}), 1.01 (dd, $^3J_{HH} = 7$ Hz, $^3J_{HP} = 7$ Hz, 6H, CH_{3iPr}), 0.99 (dd, $^3J_{HH} = 7$ Hz, $^3J_{HP} = 7$ Hz, 6H, CH_{3iPr}), 0.77 (dd, $^3J_{HH} = 7$ Hz, $^3J_{HP} = 6$ Hz, 6H, CH_{3iPr}), 0.75 (dd, $^3J_{HH} = 7$ Hz, $^3J_{HP} = 6$ Hz, 6H, CH_{3iPr}); **^{31}P{^{1}H} NMR** (162 MHz, CD_2Cl_2): δ –0.8 (s); **^{29}Si{^{1}H} NMR**: δ –27 (t, $^3J_{SiP} = 22$ Hz) **^{13}C{^{1}H}** NMR (75 MHz, CD_2Cl_2): δ 144.7 (d, $^1J_{CP} = 44.0$ Hz, C_1), 137.4 (d, $^2J_{CP} = 14.3$ Hz, C_3), 136.5 (s, C_5), 132.0 (m, C_2), 129.3 (s, C_4 or C_6), 128.9 (s, C_o or C_m), 128.8 (s, C_o or C_m), 127.9 (s, C_4 or C_6), 127.5 (s, C_p), 25.0 (d, $^1J_{CP} = 20.1$ Hz, CH_{iPr}), 24.9 (d, $^1J_{CP} = 20.1$ Hz, CH_{iPr}), 20.1 (m, CH_{3iPr}); **IR**: ν(SiH): 2165 cm^{-1}; **HRMS (ESI+)**: calcd for $[M + H]^+ = C_{30}H_{43}P_2Si^+$: 493.2609. Found: 493.2604; **Elt. Anal.**: calcd for $C_{30}H_{42}P_2Si$: C, 73.13; H, 8.60. Found: C, 71.33; H, 8.24.

2.3.5 Coordination of 4 to Copper Chloride (5)

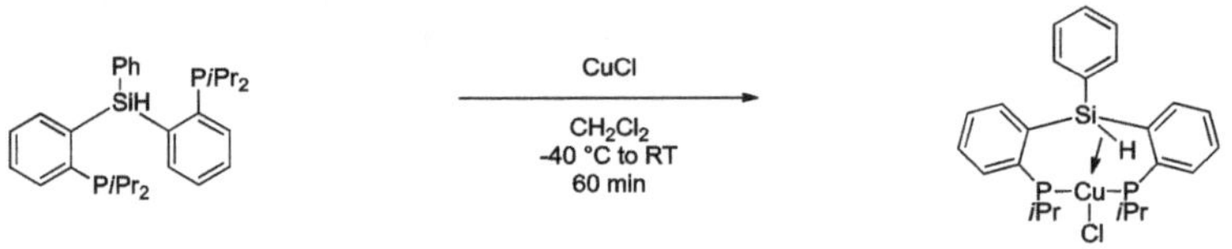

4 **5**

A solution of diphosphine-hydrosilane ligand (**4**) (200 mg, 0.41 mmol) in dichloromethane (3 mL) was added at −40 °C to a dispersion of copper(I) chloride (40.2 mg, 0.41 mmol) in dichloromethane (2 mL). The reaction mixture was stirred for 60 min while warming up to room temperature. The colorless solution was concentrated to a volume of about 1.5 mL. Complex **5** precipitated upon addition of pentane (5 mL). The colorless solid was separated from the supernatant by filtration and dried under vacuum. Yield: 187 mg (77 %). Crystals suitable for X-ray diffraction analysis were grown at −30 °C from a saturated solution of **5** in dichloromethane layered with pentane. **Mp**: 204 °C (decomposition); **^{1}H NMR** (400 MHz, CD_2Cl_2): δ 7.80–7.30 (m, 13H, H_{ar}), 7.11 (t, $^{4}J_{HP}$ = 7 Hz, $^{1}J_{HSi}$ = 180 Hz, 1H, SiH), 2.59 (m, 4H, CH_{iPr}), 1.29–1.23 (m, 12H, CH_{3iPr}), 1.17 (m, 6H, CH_{3iPr}), 0.99 (m, 6H, CH_{3iPr}); **^{31}P{^{1}H} NMR** (162 MHz, CD_2Cl_2): δ 16.9 (s); **^{29}Si{^{1}H} NMR** (79 MHz, CD_2Cl_2): δ –34 (t, $^{3}J_{SiP}$ = 19 Hz); **^{13}C{^{1}H} NMR** (100 MHz, CD_2Cl_2): δ 143.2 (*virtual* t, AA'X, *N* = 18.0 Hz, C_1), 137.9 (*virtual* t, AA'X, *N* = 11.8 Hz, C_2), 136.5 (*virtual* t, AA'X, *N* = 7.3 Hz, C_3), 136.2 (s, C_4 or C_6), 133.8 (*virtual* t, AA'X, *N* = 4.9 Hz, C_i), 132.0 (s, C_5), 129.7 (s, C_p), 129.2 (s, C_o or C_m), 129.1 (s, C_o or C_m), 127.9 (s, C_4 or C_6), 25.2 (*virtual* t, AA'X, *N* = 10.2 Hz, CH_{iPr}), 22.1 (*virtual* t, AA'X, *N* = 8.0 Hz, CH_{iPr}), 19.3 (*virtual* t, AA'X, *N* = 3.1 Hz, CH_{3iPr}), 19.2 (*virtual* t, AA'X, *N* = 6.0 Hz, CH_{3iPr}), 18.9 (*virtual* t, AA'X, *N* = 3.1 Hz, CH_{3iPr}), 16.4 (s, CH_{3iPr}); **IR**: ν(SiH): 2043 cm^{-1} (wbr); **HRMS (ESI+)**: calcd for $[M-(Cl^{-})]^{+}$ = $C_{30}H_{42}P_2SiCu^{+}$: 555.1827. Found: 555.1831; **Elt. Anal.**: calcd for $C_{30}H_{42}P_2SiCuCl$: C, 60.90; H, 7.15. Found: C, 60.17; H, 7.28.

2.3.6 Reaction of 5 with $GaCl_3$ (6)

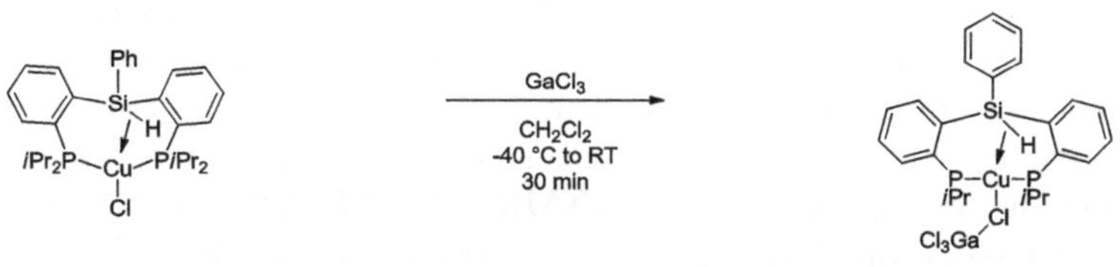

5 **6**

A solution of $GaCl_3$ (29.8 mg, 0.17 mmol) in dichloromethane (2 mL) was added to a solution of **5** (100 mg, 0.17 mmol) in dichloromethane (3 mL) at −40 °C. The mixture was stirred for 30 min while warming up to room temperature, then concentrated to a volume of about 1.5 mL. Complex **6** precipitated upon addition of pentane (4 mL). The colorless solid was separated from the supernatant by filtration and dried under vacuum. Yield: 113 mg (87 %). Crystals suitable for X-ray diffraction analysis were grown at −30 °C from a saturated solution in dichloromethane layered with pentane. **Mp**: 220 °C (decomposition); **^{1}H NMR** (400 MHz, CD_2Cl_2): δ 7.88–7.33 (m, 13H, H_{ar}), 7.19 (t, $^{4}J_{HP}$ = 5 Hz, $^{1}J_{HSi}$ = 170 Hz, 1H, SiH), 2.63 (m, 4H, CH_{iPr}), 1.31 (m, 6H, CH_{3iPr}), 1.16 (m, 12H, CH_{3iPr}), 0.96 (m, 6H, CH_{3iPr}); **^{31}P{^{1}H} NMR** (162 MHz, CD_2Cl_2): δ 19.4 (s); **^{29}Si{^{1}H} NMR** (79 MHz, CD_2Cl_2): δ –35 (t, $^{3}J_{SiP}$ = 26 Hz); **^{13}C{^{1}H} NMR** (75 MHz, CD_2Cl_2): δ 142.0 (*virtual* t, AA'X, *N* = 16.1 Hz, C_1), 137.0 (*virtual* t, AA'X, *N* = 7.0 Hz, C_3), 136.3 (s, C_4 or C_6), 134.8 (*virtual* t, AA'X, *N* = 16.0 Hz, C_2), 132.6 (s, C_5), 131.4 (m, C_i), 130.5 (s, C_p), 130.1 (s, C_o or C_m), 130.0 (s, C_o or C_m), 128.3 (s, C_4 or C_6), 24.5 (*virtual* t, AA'X, *N* = 10.6 Hz, CH_{iPr}), 21.8 (*virtual* t, AA'X, *N* = 10.4 Hz, CH_{iPr}), 19.3 (m, CH_{3iPr}), 16.1 (s, CH_{3iPr}); **IR**: ν(SiH): 1978 cm^{-1} (wbr); **HRMS (ESI+)**: calcd for $[M-(GaCl_4^-)]^+$ = $C_{30}H_{42}P_2SiCu^+$: 555.1827. Found: 555.1832.

2.3.7 Coordination of 1 to Gold(I) (7)

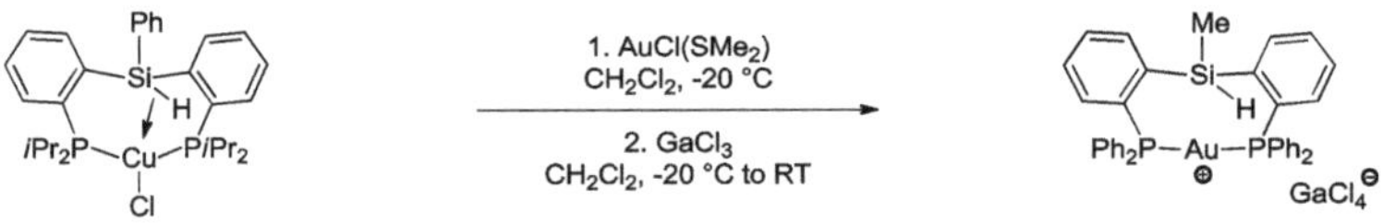

1 **7**

A solution of the diphosphine-hydrosilane ligand **1** (100 mg, 0.18 mmol) in dichloromethane (3 mL) was added at −20 °C to a solution of chloro(dimethylsulfide)gold (52 mg, 0.18 mmol) in dichloromethane (1 mL). The reaction mixture was stirred for 20 min before a solution of $GaCl_3$ (31 mg, 0.18 mmol) in dichloromethane (1 mL) was slowly added at −20 °C. The mixture was allowed to warm up to room temperature under stirring for 30 min. The solution was then concentrated under vacuum to a volume of about 2 mL, filtered and layered with pentane (5 mL) to give **7** as yellowish crystals upon crystallization at −60 °C over 3 days. Yield: 149 mg (85 %). **Mp**: 194 °C; **^{1}H NMR** (400 MHz, CD_2Cl_2): δ 8.15–8.05 (m, 2H, H_{ar}), 7.73–7.38 (m, 24H, H_{ar}), 7.00–7.15 (m, 2H, H_{ar}), 6.88 (q, $^{3}J_{HH}$ = 3.6 Hz, $^{1}J_{HSi}$ = 204 Hz, 1H, SiH), 0.52 (d, $^{3}J_{HH}$ = 3.6 Hz, 3H, $SiCH_3$); **^{31}P{^{1}H} NMR** (162 MHz, CD_2Cl_2): δ 45.4 (s); **^{29}Si{^{1}H} NMR** (79 MHz, CD_2Cl_2): δ –37 (t, $^{3}J_{SiP}$ = 15 Hz); **^{13}C{^{1}H} NMR** (100 MHz, C_6D_6): δ 141.4 (*virtual* t, AA'X, *N* = 14.6 Hz, C_1), 137.1 (*virtual* t, AA'X, *N* = 7.7 Hz, C_3), 134.5 (*virtual* t, AA'X, *N* = 8.2 Hz, C_o or C_m), 133.9 (*virtual* t, AA'X, *N* = 4.4 Hz, C_4 or C_6), 133.2 (*virtual* t, AA'X, *N* = 14.6 Hz, C_o or C_m), 132.7 (s, C_p), 132.4 (s, *N* = 14.6 Hz, $C_{p'}$), 132.3 (*virtual* t, AA'X, *N* = 29.1 Hz, C_2), 131.4 (s, C_5), 130.6 (*virtual* t, AA'X, *N* = 4.4 Hz, C_4 or C_6), 129.7 (*virtual* t, AA'X, *N* = 5.3 Hz, C_o or C_m), 129.6 (*virtual* t, AA'X, *N* = 5.3 Hz, C_o or C_m), 127.1 (*virtual* t, AA'X, *N* = 29.7 Hz, C_i), 126.3

(*virtual* t, AA'X, N = 30.3 Hz, $C_{i'}$), –0.9 (t, $^4J_{CP}$ = 2.3 Hz, $SiCH_3$); **IR**: ν(SiH): 2141 cm^{-1}; **HRMS (ESI+)**: calcd for $[M]^+$ = $C_{37}H_{32}P_2SiAu^+$: 763.1414. Found: 763.1423. **Elt. Anal.**: calcd for $C_{37}H_{32}P_2SiAuGaCl_4$: C, 45.52; H, 3.41. Found: C, 45.38; H, 3.11.

2.3.8 Computational Details

Calculations were carried out with the Gaussian 09 program package [46] on the ligand **1*** and complexes **2***, **3***, **3**** and **7*** at the DFT level of theory using the hybrid functional B3PW91 [47, 48]. B3PW91 is Becke's three parameter functional, with the non-local correlation provided by the Perdew 91 expression. Cu, Au and Ga were treated with the Stuttgart-Dresden set-RECP (relativistic effective core potential) in combination with its adapted basis set [49–52]. The latter has been augmented by a set of polarization functions (f for the metal atoms and d for the Ga atom) [53]. All the other atoms (C, H, Si, P, Cl) have been described with a 6-31G (d, p) double-ζ basis set [54]. Geometry optimizations were carried out without any symmetry restrictions; the nature of the extrema was verified with analytical frequency calculations. The bonding situation in all complexes was studied using Natural Bond Orbital analysis (NBO-5 program) [55–58]. J_{SiH} coupling constants were evaluated by employing the direct implementation of the Gauge Including Atomic Orbitals (GIAO) method at the B3PW91 density functional level of theory [59–61]. The IGLOO II basis set was used for Si and P atoms [62].

References

1. G.J. Kubas, R.R. Ryan, B.I. Swanson, P.J. Vergamini, H.J. Wasserman, J. Am. Chem. Soc. **106**, 451–452 (1984)
2. G.J. Kubas, *Metal Dihydrogen and σ-Bond Complexes* (Kluwer Academic Publishers, New York, 2001)
3. G.I. Nikonov, in *Advanced Organometallic Chemistry*. (Elsevier, Amsterdam, 2005), pp. 217–309
4. J.F. Hartwig, *Organotransition Metal Chemistry: From Bonding to Catalysis* (University Science Books, Sausalito, 2009)
5. J.Y. Corey, Chem. Rev. **111**, 863–1071 (2011)
6. S. Lachaize, S. Sabo-Etienne, Eur. J. Inorg. Chem. **2006**, 2115–2127 (2006)
7. Z. Lin, Chem. Soc. Rev. **31**, 239–245 (2002)
8. U. Schubert, in *Advanced Organometallic Chemistry*. (Elsevier, Amsterdam, 1990), pp. 151–187
9. K. Pandey, Coord. Chem. Rev. **253**, 37–55 (2009)
10. M. Brookhart, M.L.H. Green, G. Parkin, Proc. Natl. Acad. Sci. **104**, 6908–6914 (2007)
11. W.H. Bernskoetter, C.K. Schauer, K.I. Goldberg, M. Brookhart, Science **326**, 553–556 (2009)
12. S.D. Pike, A.L. Thompson, A.G. Algarra, D.C. Apperley, S.A. Macgregor, A.S. Weller, Science **337**, 1648–1651 (2012)
13. G.I. Nikonov, Angew. Chem. Int. Ed. **42**, 1335–1337 (2003)

14. W. Chen, S. Shimada, M. Tanaka, Science **295**, 308–310 (2002)
15. E.C. Sherer, C.R. Kinsinger, B.L. Kormos, J.D. Thompson, C.J. Cramer, Angew. Chem. Int. Ed. **2002**, 41 (1953)
16. G. Aullón, A. Lledós, S. Alvarez, Angew. Chem. Int. Ed. **41**, 1956–1959 (2002)
17. M. Etienne, A.S. Weller, Chem. Soc. Rev. **43**, 242–259 (2013)
18. S.K. Brayshaw, E.L. Sceats, J.C. Green, A.S. Weller, Proc. Natl. Acad. Sci. **104**, 6921–6926 (2007)
19. A.B. Chaplin, J.C. Green, A.S. Weller, J. Am. Chem. Soc. **133**, 13162–13168 (2011)
20. M. Brookhart, M.L.H. Green, J. Organomet. Chem. **250**, 395–408 (1983)
21. J. Hoyano, M. Elder, W.A.G. Graham, J. Am. Chem. Soc. **91**, 4568–4569 (1969)
22. U. Schubert, H. Gilges, Organometallics **15**, 2373–2375 (1996)
23. J. Takaya, N. Iwasawa, Organometallics **28**, 6636–6638 (2009)
24. W. Hieringer, J. Eppinger, R. Anwander, W.A. Herrmann, J. Am. Chem. Soc. **122**, 11983–11994 (2000)
25. M. Lein, J.A. Harrison, A.J. Nielson, Dalton Trans. **40**, 10731 (2011)
26. I. Corral, O. Mó, M. Yáñez, J. Phys. Chem. A **107**, 1370–1376 (2003)
27. I. Corral, O. Mó, M. Yáñez, Int. J. Mass Spectrom. **227**, 401–412 (2003)
28. A. Amgoune, D. Bourissou, Chem. Commun. **47**, 859 (2011)
29. G. Bouhadir, A. Amgoune, D. Bourissou, in *Advance Organometallic Chemistry*, ed. by A.F. Hill and M.J. Fink), (Academic Press, New-York, 2010), pp. 1–107
30. P. Gualco, A. Amgoune, K. Miqueu, S. Ladeira, D. Bourissou, J. Am. Chem. Soc. **133**, 4257–4259 (2011)
31. P. Gualco, S. Ladeira, K. Miqueu, A. Amgoune, D. Bourissou, Angew. Chem. Int. Ed. **50**, 8320–8324 (2011)
32. M.C. MacInnis, D.F. MacLean, R.J. Lundgren, R. McDonald, L. Turculet, Organometallics **26**, 6522–6525 (2007)
33. D.F. MacLean, R. McDonald, M.J. Ferguson, A.J. Caddell, L. Turculet, *Chem. Commun.* 2008, 5146
34. S.J. Mitton, R. McDonald, L. Turculet, Angew. Chem. Int. Ed. **48**, 8568–8571 (2009)
35. S.J. Mitton, R. McDonald, L. Turculet, Organometallics **28**, 5122–5136 (2009)
36. E. Morgan, D.F. MacLean, R. McDonald, L. Turculet, J. Am. Chem. Soc. **131**, 14234–14236 (2009)
37. S.J. Mitton, L. Turculet, *Chem.Eur. J.* ***18***, 15258–15262 (2012)
38. J. Takaya, N. Iwasawa, J. Am. Chem. Soc. **130**, 15254–15255 (2008)
39. J. Takaya, N. Iwasawa, Dalton Trans. **40**, 8814 (2011)
40. P. Gualco, M. Mercy, S. Ladeira, Y. Coppel, L. Maron, A. Amgoune, D. Bourissou, *Chem. Eur. J.* ***16***, 10808–10817 (2010)
41. B. Cordero, V. Gómez, A.E. Platero-Prats, M. Revés, J. Echeverría, E. Cremades, F. Barragán, S. Alvarez, *Dalton Trans.* 2832–2838 (2008)
42. S.S. Batsanov, Inorg. Mater. **37**, 871–885 (2001)
43. M.A. Carvajal, J.J. Novoa, S. Alvarez, J. Am. Chem. Soc. **126**, 1465–1477 (2004)
44. D.V. Gutsulyak, S.F. Vyboishchikov, G.I. Nikonov, J. Am. Chem. Soc. **132**, 5950–5951 (2010)
45. J.E. Bercaw, J.A. Labinger, Proc. Natl. Acad. Sci. **104**, 6899–6900 (2007)
46. M.J. Frisch, G.W. Trucks, H.B. Schlegel, G.E. Scuseria, M.A. Robb, J.R. Cheeseman, G. Scalmani, V. Barone, B. Mennucci, G.A. Petersson, et al., *Gaussian 09, Revision A.1*, Gaussian, Inc. (Wallingford, Connecticut, USA, 2009)
47. A.D. Becke, J. Chem. Phys. **98**, 5648–5652 (1993)
48. K. Burke, J.P. Perdew, and Y. Wang, in *Electron Density Function Theory Recent Program New Directory* (eds.), by J.F. Dobson, G. Vignale, M.P. Das. (Plenum Press, New York, 1998), pp. 81–111
49. B. Metz, H. Stoll, M. Dolg, J. Chem. Phys. **113**, 2563–2569 (2000)
50. M. Dolg, in *Model Methods Algorithms Quantum Chemistry.* (ed.), by J. Grotendorst. (John-von-Neumann-Inst. For Computing, Jülich, 2000), pp. 479–508

51. D. Andrae, U. Häußermann, M. Dolg, H. Stoll, H. Preuß, Theor. Chim. Acta **77**, 123–141 (1990)
52. M. Dolg, U. Wedig, H. Stoll, H. Preuss, J. Chem. Phys. **86**, 866–872 (1987)
53. A.W. Ehlers, M. Böhme, S. Dapprich, A. Gobbi, A. Höllwarth, V. Jonas, K.F. Köhler, R. Stegmann, A. Veldkamp, G. Frenking, Chem. Phys. Lett. **208**, 111–114 (1993)
54. P.C. Hariharan, J.A. Pople, Theor. Chim. Acta **28**, 213–222 (1973)
55. J.K. Glendening, A.E. Badenhoop, J.E. Carpenter, J.A. Bohmann, C.M. Morales, F. Weinhold, *NBO 5.0.* (Theoretical Chemistry Institute, University of Wisconsin, Madison, Wisconsin, USA, 2001)
56. A.E. Reed, L.A. Curtiss, F. Weinhold, Chem. Rev. **88**, 899–926 (1988)
57. J.P. Foster, F. Weinhold, J. Am. Chem. Soc. **102**, 7211–7218 (1980)
58. A.E. Reed, F. Weinhold, J. Chem. Phys. **83**, 1736–1740 (1985)
59. F. London, J. Phys. Radium **8**, 397–409 (1937)
60. R. Ditchfield, Mol. Phys. **27**, 789–807 (1974)
61. K. Wolinski, J.F. Hinton, P. Pulay, J. Am. Chem. Soc. **112**, 8251–8260 (1990)
62. W. Kutzelnigg, U. Fleischer, M. Schindler, in *Deuterium Shift Calculation.* (Springer, Heidelberg, 1991), pp. 165–262

Chapter 3
Fundamental Elementary Steps in Gold Chemistry

3.1 Introduction

The interest in the coordination chemistry of gold was considerably revived following the discovery of the catalytic potential of this metal at the end of the 1990s [1]. The advent and rise of gold catalysis clearly led to a more profound understanding of organogold chemistry. However, the reactivity of gold complexes is essentially dominated by their pronounced π-acidic properties. Other reactivities, as typically known from the chemistry of transition metal compounds, are hardly explored in the case of gold.

Indeed, gold complexes often behave differently compared to other late transition metal complexes. Common elementary steps in transition metal-mediated reactions such as oxidative addition, reductive elimination, transmetallation or migratory insertion are often encountered and well understood for most of the d-block elements. By contrast, in the case of gold, these elementary steps are much less studied, some of them being rather unusual and seldom evidenced.

With regard to these elementary steps, the fundamental organometallic chemistry of gold lags behind. Only in recent years was unambiguous evidence gained that points at a more versatile reactivity of gold complexes than previously anticipated.

This chapter will give a bibliographic overview that summarizes the state-of-the-art concerning fundamental reactivities of gold complexes. The focus is set on elementary steps, in particular oxidative addition/σ-bond activation, reductive elimination, migratory insertion, transmetallation and arylation reactions.

Reports related to typical gold catalysis based on π-bond activation processes, such as for example the mechanistic investigations with regard to nucleophilic attack on π-complexes of gold, accounts on vinylgold complexes generated thereby, studies on geminally diaurated species etc. are excluded. Concerning these topics, the reader is referred to reviews in the recent literature [2–6].

Section 3.2 will briefly recall important characteristics of gold chemistry. Relevant examples of oxidative addition and reductive elimination reactions with

M. Joost, *Synthesis and Original Reactivity of Copper and Gold Complexes*,
Springer Theses, DOI 10.1007/978-3-319-18690-0_3

gold complexes are outlined in Sects. 3.3 and 3.4, respectively. Migratory insertion reactions are described in Sect. 3.5. The related β-hydride elimination follows in Sect. 3.6. Transmetallation reactions will be described in 3.7, the arylation of gold complexes is shortly presented in 3.8. Lastly, Sect. 3.9 sums up catalytic processes involving these elementary steps.

We hope that the assembly and arrangement of the reported literature in this context will reveal certain contributions in a different light which might help to gain a deeper understanding of the reactivity of gold complexes beyond their π-acidity. Even though the origin of the pronounced reluctance for some of the mentioned elementary steps at gold complexes is not yet understood, this overview might help to guide further research in these directions.

3.2 Characteristics of Gold Coordination Chemistry

The coordination chemistry of gold, the noblest of the elements, is best known for gold in its oxidation states I and III, however, examples of gold(0) [7, 8] and gold (V) [9] complexes are known as well. Although formal dinuclear gold(II) complexes featuring Au–Au bonds are abundant, Mößbauer spectroscopy indicates significant gold(III) character [10]. "True" mononuclear, paramagnetic gold(II) complexes are rare [11, 12].

Gold(I) complexes, with a $5d^{10}$ valence shell configuration, are usually colorless in the solid state and in solution. The vast majority of gold(I) complexes are dicoordinated and feature a linear coordination environment [13]. Gold(III) complexes ($5d^8$ valence shell configuration) exhibit often bright colors ranging from orange-red to green and yellow. These complexes are mostly tetracoordinated, square-planar species [14]. Higher coordination numbers have been reported as well, but remain exceptions [15–17].

Gold presents a number of important differences concerning general properties, structure of its compounds and reactivity compared to other transition metals.

These differences result in large part from direct or indirect consequences of relativistic effects. The characteristic properties ascribed to relativistic effects reach in the periodic table of the elements their maximum at element number 79, i.e. gold [18–20]. Origin of these effects is the increase in speed of s-electrons of a given atom that increases roughly with the atomic number Z. The increase in speed results in a relativistic mass increase that is not negligible for the heavy elements; e.g. for gold, the corrected relativistic mass is about 22 % greater than the nonrelativistic rest mass. As the Bohr radius of an electron is inversely proportional to its mass, a mass increase leads to contraction of the s shells and the same is approximately valid as well for p orbitals. The contraction, also called direct relativistic effect, is attended by a second effect: due to the higher electron density around the nucleus, its effective nuclear charge Z_{eff} is reduced, leading to a destabilization, i.e. expansion of all orbitals. This indirect relativistic effect affects in particular the diffuse d- (and f-) orbitals. The d-orbital expansion leads in turn to an indirect

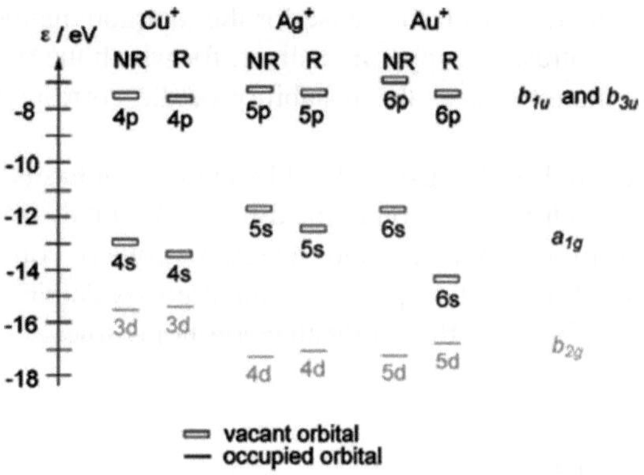

Fig. 3.1 Calculated frontier orbital energies for coinage metal cations. *NR* non-relativistic calculation, *R* relativistic calculation. Reprinted (adapted) with permission from Ref. [22]. Copyright 2004 American Chemical Society

stabilization (relativistic deshielding) of s- and p-orbitals by increasing Z_{eff}. The relativistic deshielding reaches its maximum for a filled d-shell and explains why relativistic effects culminate with the group 10 and 11 metals, and in particular with platinum and gold in the 6th period [21].

As can be seen in Fig. 3.1 where non-relativistic and relativistic calculations on coinage metal cations are compared, the impact of relativistic effects on the frontier orbitals of Au^+ is drastic (a telling example is the relativistic 6s stabilization). While for all of the coinage metals a d-orbital expansion and s-/p-orbital stabilization is observed, these effects are most pronounced for gold.

As a direct result of the notable 6s and 6p orbital contraction, gold is the noblest metal, featuring a high first ionization energy (9.223 eV) [14] and is accordingly difficult to oxidize (E_0: Au(I)/Au(0) = 1.69 V; Au(III)/Au(0) = 1.50 V; Au(III)/Au (I) = 1.41 V) [23].

The bright yellow color of gold metal is as well linked to relativistic effects. Blue visual light is absorbed due to the relatively small energy gap between the 5d and 6s levels. Reflected are the green and red portions of the visual range, making gold appear "golden". This is in contrast to the colorless silver metal, where the 4d–5s gap is larger and the absorption is shifted in the UV region [24].

With regard to coordination compounds of gold, the relativistic orbital contraction results in a small size of the gold atom, and shortened bonds. For example, the covalent radii for copper, silver and gold are 1.32, 1.45 and 1.36 Å, respectively [25].

The contraction of the valence 6s and 6p orbitals of gold is reflected as well in its unusually high electronegativity (2.4 on the Pauling scale) [26]. Closely related to electronegativity is Lewis acidity and cationic gold complexes are indeed strong Lewis acids. Furthermore, the soft character of gold cations, well-suited for the activation of soft electrophiles such as alkynes or alkenes, originates from the expanded, diffuse valence shell, explaining the outstanding catalytic potential of cationic gold complexes [20].

Relativistic effects are as well the cause for the thermodynamic stability of gold (III) complexes, in contrast to copper and silver, for which the oxidation state III is only seldom encountered due to the instability of the corresponding compounds [27, 28].

With regard to the redox chemistry of gold complexes it has been argued that the decreased electron–electron repulsion in the diffuse 5d orbitals leads to an increased effective attraction to the nucleus and therefore gold(I) complexes would be insufficiently nucleophilic metal species to undergo oxidative addition [29]. To date, it is not clear in how far this rationalization holds true.

3.3 Oxidative Addition

It became a common opinion to consider the reactivity of gold as redox-neutral, based on the observation that gold complexes behave indeed not like its close neighbors platinum or palladium that readily cycle between their oxidation states 0 and II. Gold(I) complexes proved to be inert towards the oxidative addition of aryl halides, a crucial stage in many cross coupling reactions, even when an intramolecular carbon–halogen bond is oriented in close proximity to gold in favor for this process (Fig. 3.2) [30].

In this way, gold "earned the reputation" of being a redox-neutral metal. The redox-neutral nature of gold is often considered as an advantage: this "isohypsic reactivity" stands in an orthogonal relationship to the one of platinum and palladium, allowing for reaction sequences initiated by gold-catalyzed nucleophilic activation of a π-system, followed by Pd or Ni catalyzed cross-coupling [30, 31].

However, the origin of the particular averseness of gold to undergo oxidative addition and to cycle between Au(I) and Au(III) oxidation states is not fully

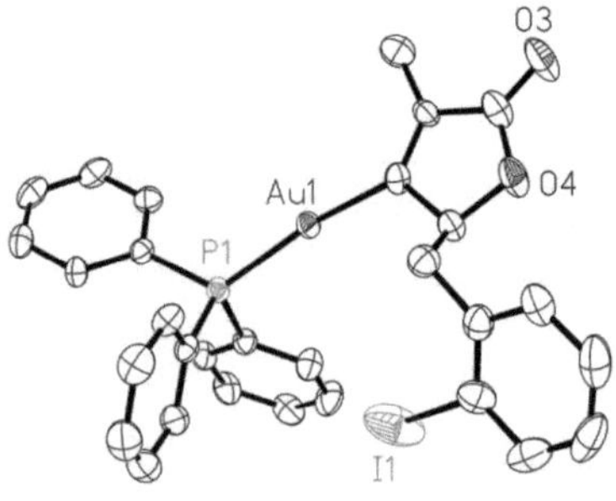

Fig. 3.2 Molecular structure determined by single crystal X-ray diffraction of a stable gold(I) complex featuring an aryl–iodine bond in close proximity to the metal center. Reproduced from Ref. [30] with permission from Wiley-VCH

understood. The following section seeks to summarize the state-of-the-art concerning oxidative addition reactions of molecular gold complexes.

This section reviews the significant contributions concerning stoichiometric oxidative additions to gold(I) complexes and mechanistic studies of these reactions. Examples employing (i) highly polar reagents, such as dihalides and alkyl halides, (ii) nonpolar reagents, and (iii) oxidative addition reactions of aryl halides will be discussed.

3.3.1 Oxidation of Gold(I) with Polar Reagents

From an historic perspective, it becomes evident that the transition from gold(I) to gold(III) is possible under certain circumstances. Already in 1940, the oxidative addition of dihalogen to (phosphine)gold(I) halides has been recognized (Scheme 3.1) [32]. The composition of the formed (phosphine)gold(III)trihalide complexes was confirmed by elemental analyses.

Since then, the reactivity of dihalogens towards gold(I) complexes has been used in numerous examples to access the corresponding gold(III) complexes. The accessibility and stability of the gold(III) complexes depend both on steric and electronic properties of neutral and anionic ligands on gold. The neutral donor ligand is variable and can be a phosphine [32–34], an isonitrile [35], or a carbene [36]. Oxidative addition of chlorine [37, 38], bromine [32, 34] and iodine [33] is known.

These reactions do not follow the σ-bond activation pathway described previously (see Sect. 2.1.1). Although mechanistic details concerning the oxidation of gold(I) with dihalogens are scarce, an S_N2-type pathway is generally accepted to account for these transformations. In the case of the oxidation of an (NHC)methylgold(I) complex with iodine, Labinger, Bercaw and coworkers confirmed this picture based on a detailed kinetic study (Scheme 3.2) [38].

Cyclic dinuclear ylide complexes, first reported by Schmidbaur [39, 40], are known as well to oxidatively add dihalogen (Scheme 3.3). Reaction with an equimolar amount of Cl_2, Br_2 or I_2, gave rise to formal gold(II) complexes, that

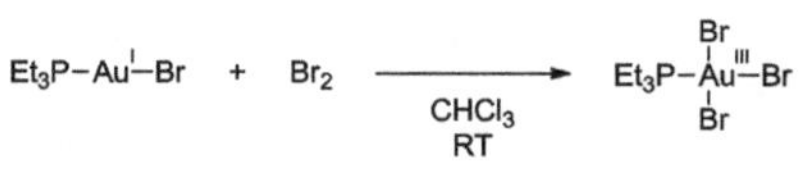

Scheme 3.1 An early example of a gold(I)–gold(III) redox process reported by Mann and Purdie

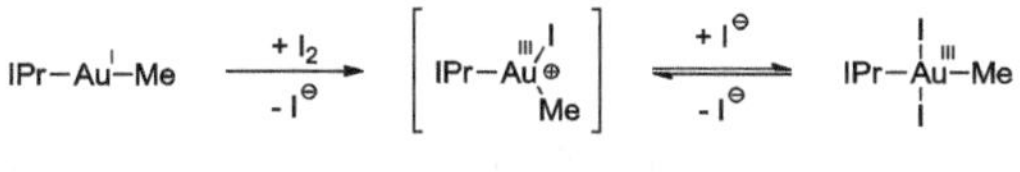

Scheme 3.2 Synthesis of trans-[AuI_2(Me)(IPr)] via stepwise addition of iodine

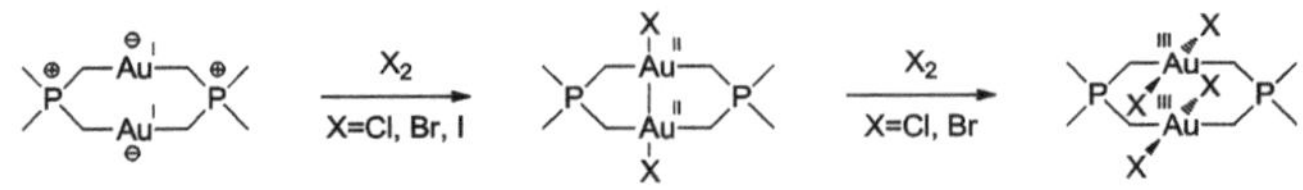

Scheme 3.3 Oxidative addition of dihalogen at dinuclear ylide complexes yielding Au(II)–Au(II) and Au(III)–Au(III) complexes

could be further oxidized by a second equivalent of the dihalogen to the corresponding dinuclear gold(III)–gold(III) complexes.

More recently, the potential of other oxidants has been evaluated in the context of gold-catalyzed redox-processes (see Sect. 3.9). 1-Chloromethyl-4-fluoro-1,4-diazoniabicyclo-[2.2.2]-octane bis(tetrafluoroborate) ("Selectfluor"), containing a highly electrophilic fluorine atom, xenon difluoride and hypervalent iodine reagents, such as iodobenzene dichloride are capable of promoting the gold(I)–gold(III) transition. Toste and coworkers isolated an NHC gold(III) fluoride complex employing XeF_2 (Scheme 3.4).

The above mentioned reactions with polar reagents demonstrate that the transition from oxidation state I to III for organogold compounds is feasible. Although formally to be considered as oxidative additions, they do not follow σ-bond activation pathways.

3.3.2 *Oxidative Addition of Alkyl Halides*

Kochi and Schmidbaur independently reported in 1972 on the reaction of methyl iodide with (phosphine)methylgold(I) complexes yielding the (phosphine)gold(I) iodide and ethane (Scheme 3.5) [41, 42].

A detailed study by the Kochi group [41, 43] and others [44, 45] showed that a catalytic process is involved in this homocoupling reaction (Scheme 3.6). The (phosphine)methylgold(I) complex undergoes oxidative addition of methyl iodide to form a dimethyl gold(III) species. The methylgold complex serves as well as a transmetallating agent: a methyl transfer takes place from the gold(I) to the gold(III) species, with loss of one equivalent of (phosphine)gold iodide to form a

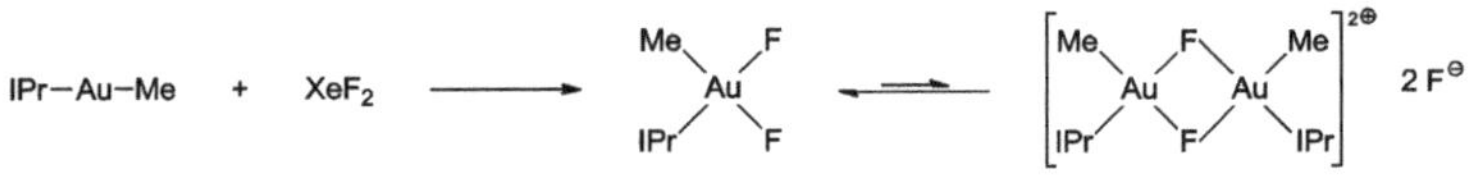

Scheme 3.4 Synthesis of a gold(III) fluoride complex by oxidation of a methylgold(I) complex with XeF_2

Scheme 3.5 Ethane formation by reaction of $[Au(CH_3)(PPh_3)]$ with methyl iodide as reported by Schmidbaur and Kochi

Scheme 3.6 Gold-catalyzed synthesis of ethane from methyl iodide. L = PPh_3

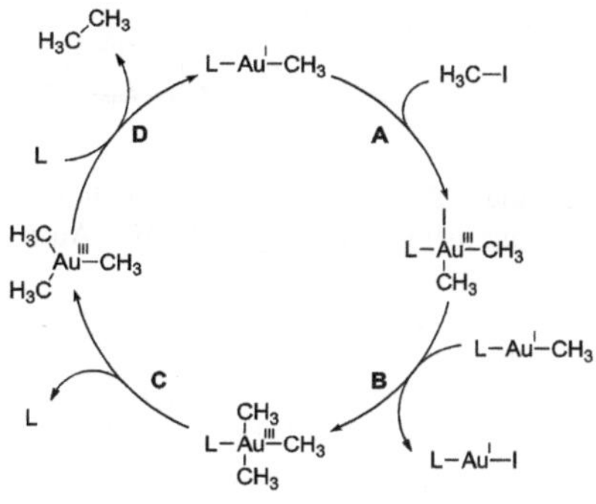

(phosphine)trimethylgold(III) complex. Subsequent decoordination of the phosphine ligand from this complex generates a short-lived tricoordinated species that undergoes rapidly reductive elimination with ethane liberation. Re-coordination of triphenylphosphine regenerates the initial catalytic species.

The precise mechanism for the oxidative addition step was not established, as the reaction order with respect to methyl iodide could not be determined, [45] but an S_N2-type mechanisms for the oxidative addition of alkyl halides to transition metals is generally accepted [46]. The stability of the (phosphine)trimethylgold(III) complex is dependent on the substitution pattern of the phosphine donor ligand [43–45].

As for dihalogen (vide supra), oxidative addition of methyl halides at dinuclear gold(I) ylide complexes is feasible and yields the corresponding Au(II)–Au(II) and Au (III)–Au(III) complexes, depending on the employed stoichiometry (Scheme 3.7). This reactivity has been extensively studied by Schmidbaur, Fackler and others [47, 48]. Kinetic data point at an S_N2-type pathway [49].

Methylene-bridged gold(III)–gold(III) complexes have been obtained by double oxidative addition of methylene dihalides [50] (Scheme 3.8).

Recently, Toste and coworkers [51] reported on the oxidative addition of trifluoroiodomethane to gold(I). Near-ultraviolet light was shown to promote the reaction of CF_3I with phosphine arylgold(I) complexes to give the corresponding gold(III) complexes (Scheme 3.9). These compounds were shown to be remarkably stable; their molecular structures in the solid-state were determined by X-ray crystallography.

Mechanistic investigations concerning the formation of these gold(III) complexes pointed at a photoinitiated radical chain mechanism (Scheme 3.10).

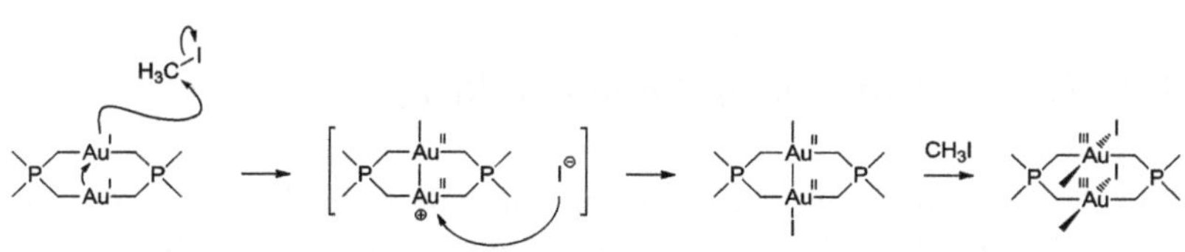

Scheme 3.7 Likely mechanism for the oxidative addition of methyl iodide at dinuclear ylide complexes yielding Au(II)–Au(II) and Au(III)–Au(III) complexes

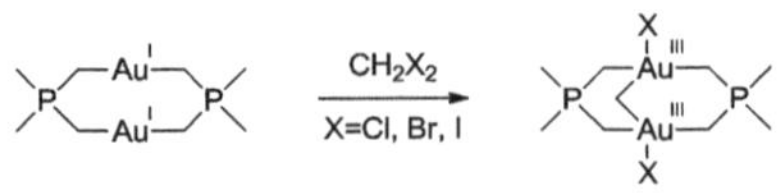

Scheme 3.8 Double oxidative addition of methylene dihalides at dinuclear ylide complexes yielding methylene bridged Au(III)–Au(III) complexes

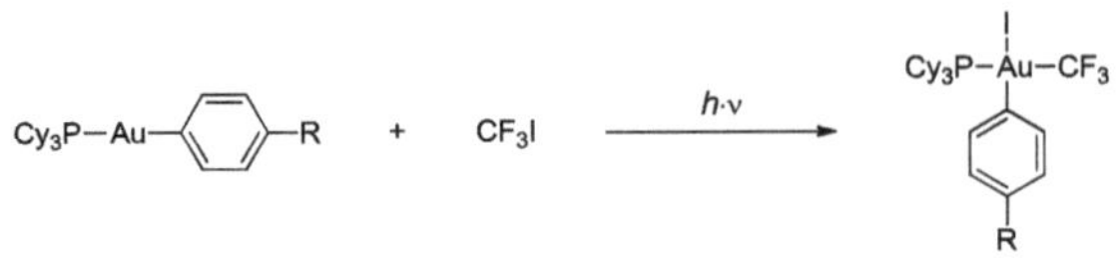

Scheme 3.9 Oxidative addition of CF_3I to arylgold(I) complexes. R = H, F, Me, OMe

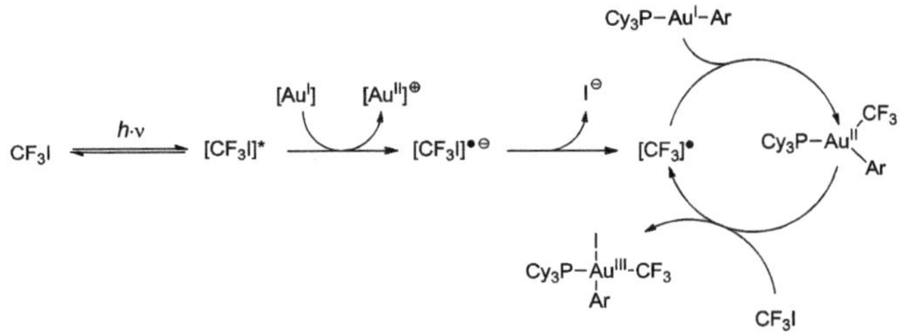

Scheme 3.10 Proposed radical chain mechanism for the oxidative addition of CF_3I to arylgold(I) complexes

Photoexcited CF_3I oxidizes the arylgold(I) complex to give an intermediate gold(II) complex and the radical-anion $[CF_3I]^{\cdot -}$ which subsequently undergoes homolytic bond cleavage generating iodide and $[CF_3I]^{\cdot}$. The latter reacts with $[Au(Ar)(PCy_3)]$ forming a tricoordinate gold(II) complex which is then further oxidized by CF_3I yielding the trifluoromethyl-arylgold(III) complex and a $[CF_3]^{\cdot}$-radical that closes the propagating cycle.

When a phosphine methylgold(I) complex was reacted with CF_3I under otherwise identical conditions, the corresponding gold(III) compound was not detected. Instead, methyl iodide and a trifluoromethylgold(I) complex was formed, most likely via CI bond-forming reductive elimination.

3.3.3 Oxidative Addition of Nonpolar Reagents

As all of the above-mentioned reactions present only formally oxidative additions, the ability of molecular gold(I) complexes to undergo oxidative addition following a σ-bond activation pathway (see Sect. 2.1.1) was questionable for long time. Oxidative addition reactions at gold(I) complexes employing non-polar reagents (for which S_N2-like mechanisms are excluded) were almost unknown until the beginning of this decade.

While the bond activation of less or even non-polarized bonds by other transition metals, such as palladium or platinum was intensively investigated, the analogous gold chemistry has been neglected over a long period. With the rising interest in gold catalysis and in the quest for new or overlooked reactivities, σ-bond activation by gold(I) complexes is being revisited and experiencing increasing interest. Mentioned below are contributions dealing with the activation of nonpolar σ-bonds at molecular, mononuclear gold species.

3.3.3.1 Oxidative Addition of Disulfides to Gold(I)

Limited knowledge with regard to oxidative addition of disulfides to yield thiolate gold(III) complexes has been gained:

The reaction of bis(trifluoromethyl)-1,2-dithiethene with (triphenylphosphine) gold(I) chloride to yield a chlorotriphenylphosphonium tetrathiolatoaurate(III) salt was first reported by Davison [52]. Shortly after, Iberson elucidated the molecular structure of the product by means of an X-ray crystallographic study [53] (Scheme 3.11).

Bachman reported in 2008 that highly fluorinated disulfides undergo facile and reversible oxidative addition to dithiolate gold(I) complexes [54] (Scheme 3.12).

Both the gold(I) and the gold(III) thiolate complexes were structurally characterized. Information concerning mechanistic details of the disulfide bond activation process is not available.

3.3.3.2 Oxidative Addition of σ-SiSi and σ-SnSn Bonds to Gold(I)

In our group, the spontaneous intramolecular oxidative addition of σ-SiSi [55] and σ-SnSn [56] bonds tethered to a diphosphine ligand by coordination to gold(I) was

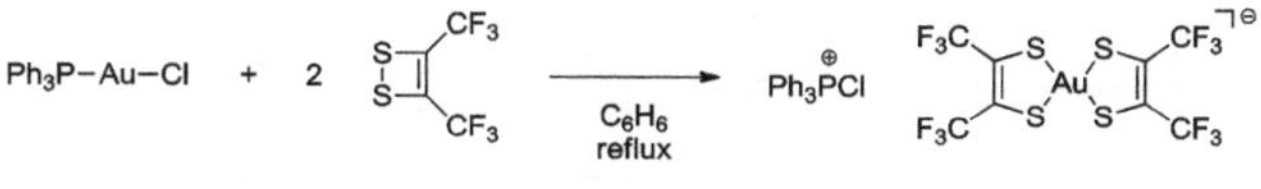

Scheme 3.11 Oxidative addition of a diethene at gold(I)

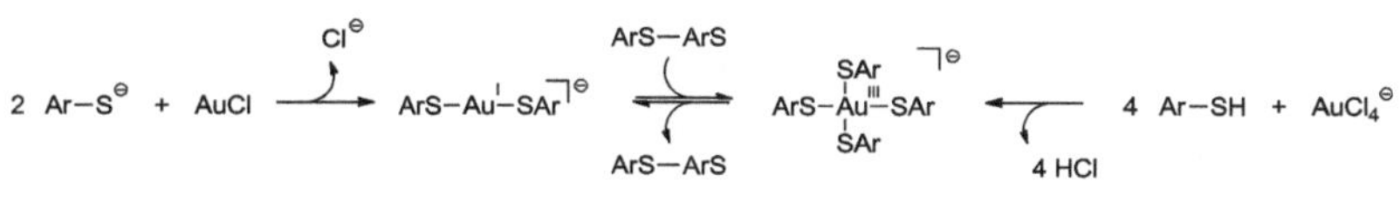

Scheme 3.12 Reversible oxidative addition/reductive elimination of disulfides at gold. Ar = C_6F_5, 4-HC_6F_4

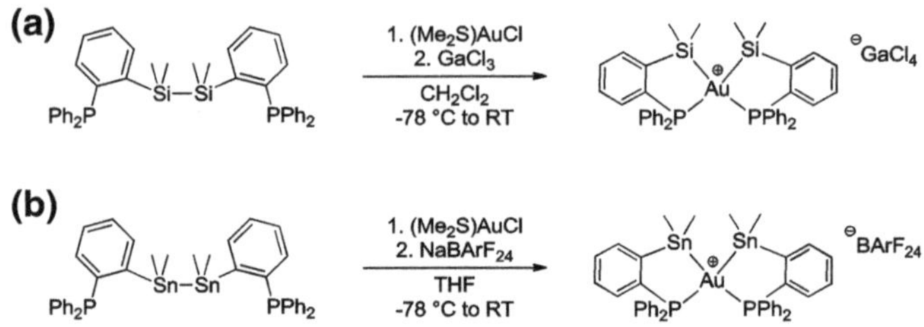

Scheme 3.13 Synthesis of a bis(silyl)gold(III) complex (**a**) and a bis(stannyl) gold(III) complex (**b**) by oxidative addition of SiSi /SnSn bonds at gold(I)

observed, yielding bis(silyl) and bis(stannyl)gold(III) complexes, respectively (Scheme 3.13). These gold(III) complexes were characterized by multinuclear NMR spectroscopy and X-ray diffraction crystallography.

Both reactions were analyzed computationally and were found to proceed via concerted σ-bond activation pathways with symmetric geometries in the transition state, featuring low activation barriers of 10.4 kcal/mol for the SiSi and only 4.1 kcal/mol for the SnSn bond, in accord with the weaker bond strength of the latter.

Reaction of water or oxygen with a related, in situ prepared bis(silyl)gold(III) complex featuring a chloride anion resulted in the reduction of the metal center to give a diphosphine gold(I) complex featuring a net insertion of an oxygen atom in the SiSi bond (Scheme 3.14) [57]. This sequence presents a rare example of a gold (I)–gold(III)–gold(I) redox process.

It was found that oxidative addition still proceeds with only one phosphine donor arm orientating the gold(I) center and the SiSi bond (Scheme 3.15) [55]. The product is less stable than its analogue stemming from the diphosphine-disilane ligand and decomposes at temperatures above −30 °C, but was characterized unambiguously by multinuclear NMR spectroscopy.

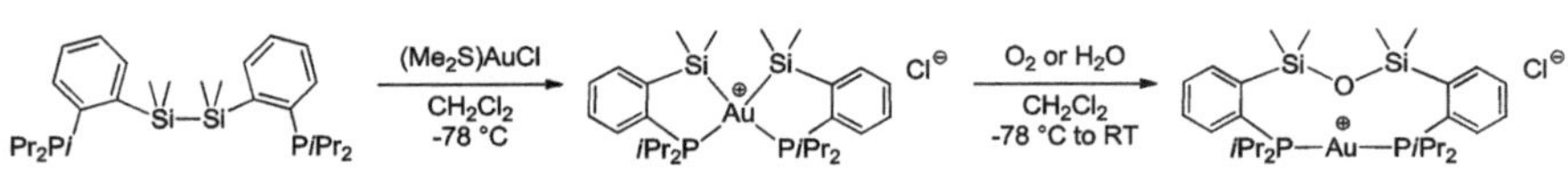

Scheme 3.14 Chelate-assisted, gold-mediated insertion of oxygen into the SiSi bond of a disilane framework via a gold(I)–gold(III)–gold(I) redox process

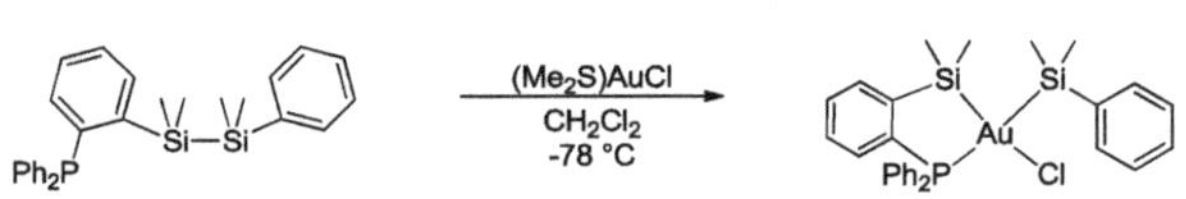

Scheme 3.15 Synthesis of a silyl(phosphinosilyl)gold(III) chloride complex

3.3.4 Oxidative Addition of Aryl Halides to Gold(I)

As mentioned in the introduction to this section, gold(I) complexes are not considered to be able to undergo oxidative addition of aryl halides. However, in 2005, a contribution by Corma and coworkers [58] reported on the unexpected cross-coupling of aryl iodides with arylboronic acids catalyzed by gold(I) Schiff-base complexes. Curiously, gold(III) complexes featuring identical imine ligands were not capable of performing the Suzuki-type cross-coupling, but only the previously observed homocoupling of arylboronic acids [59, 60]. This methodology was extended to the Sonogashira reaction by Corma and others (Scheme 3.16) [61, 62].

These reports provoked a lively debate in the organometallic community. Echavarren et al. [63] failed to reproduce the results obtained by the Corma group and found palladium impurities at low ppm levels to be involved in the catalytic activity.

Hereupon, Corma [64] analyzed his own results more thoroughly and conducted as well a computational study, pointing at the in situ formation of small gold clusters by decomposition of the Schiff-base catalysts. The catalytic activity of these subnanometric gold species was reviewed [65].

In the course of this controversy concerning the ability of gold(I) complexes to undergo oxidative addition of aryl halides, O'Hair and coworkers [66] investigated the reaction of iodobenzene with gold(I) complexes and small gold clusters by means of mass spectrometry and DFT-computations. While cationic bis(phosphine) gold(I) complexes proved to be inert towards iodobenzene, the monoligated cationic phosphine complexes were shown to undergo C–I activation reactions: iodobenzene and trimethylphosphine or triphenylphosphine gold(I) complexes formed adducts, as indicated by detection of the corresponding cations in the gas phase. These species were shown to fragment subsequently to the respective phenyl phosphonium ions and gold iodide with phosphorus–carbon bond formation

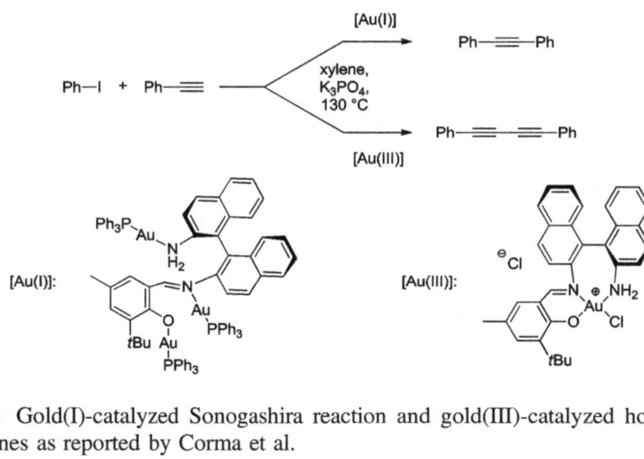

Scheme 3.16 Gold(I)-catalyzed Sonogashira reaction and gold(III)-catalyzed homocoupling of terminal alkynes as reported by Corma et al.

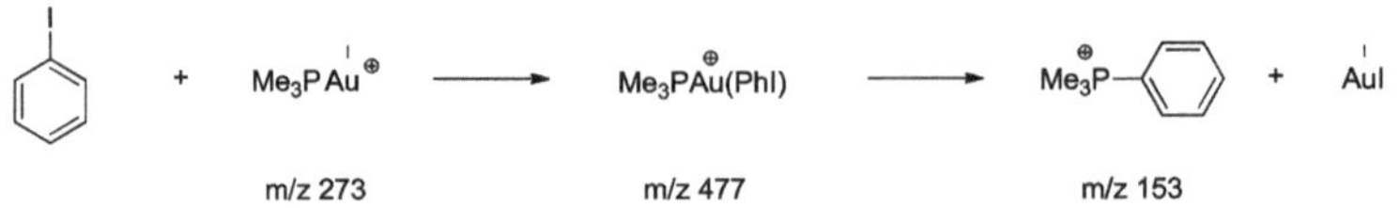

Scheme 3.17 Reaction of a cationic, monoligated phosphine gold(I) complex with phenyl iodide in the gas-phase and detected products

(Scheme 3.17). The overall-process may be explained by an oxidative addition/reductive elimination sequence via a gold(III) species.

The accompanying calculations were consistent with the experimental results: oxidative addition of iodobenzene at bis(trimethylphosphine) gold(I) was predicted to be an endothermic process by 5.9 kcal/mol. By contrast, the analogous reaction with the monoligated cationic complex $[Au(PMe_3)]^+$ features an accessible activation barrier (24.7 kcal/mol), and both the formation of the gold(III) species and subsequent reductive elimination reactions are shown to be exothermic, in line with the experimental observations in the gas-phase.

Further evidence for the feasibility of oxidative addition of aryl halides to gold(I) was provided by our group [67]. A suitable model system based on a naphthyl-phosphine ligand allowed for the intramolecular oxidative addition of aryl-halide bonds to a gold(I) center (Scheme 3.18).

While the $C(sp^2)$–I bond oxidatively adds to the metal center at room temperature, activation of the corresponding $C(sp^2)$–Br bond necessitates higher temperatures. Oxidative addition of the $C(sp^2)$–Cl bond was not observed even at elevated temperatures.

A similar chelate-assisted approach revealed a practical methodology for the synthesis of original gold(III) "PCP"pincer complexes. A bis(phosphine) aryl bromide reacts cleanly to a cyclometallated bis(phosphine) gold(III) bromide complex after coordination to $[AuBr(Me_2S)]$ (Scheme 3.19). Although the chelate

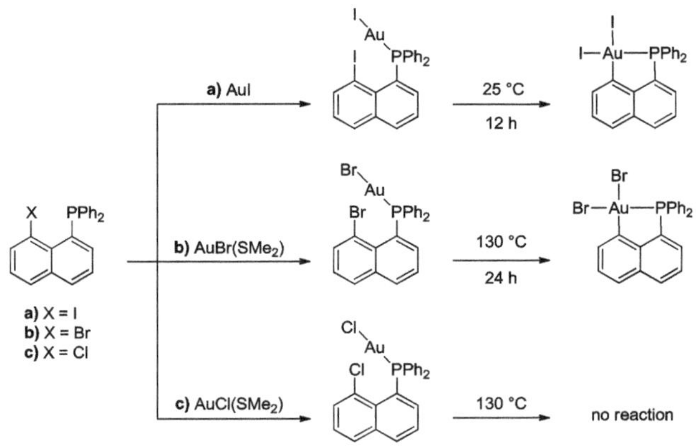

Scheme 3.18 Intramolecular oxidative addition of Ar–I and Ar–Br bonds at Au(I)

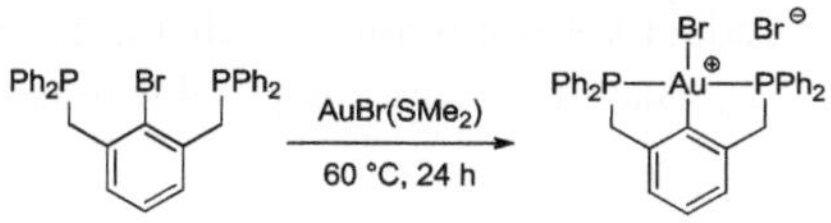

Scheme 3.19 Synthesis of a cyclometallated bis(phosphine)gold(III) bromide complex via oxidative addition of a C(*sp*^{2})–Br bond at gold(I)

system facilitates the oxidative addition reaction (the reaction proceeds already at 60 °C), C(*sp*^{2})–Cl bond activation could not be achieved for the respective ligand.

The mechanism of this transformation was explored in detail by DFT calculations. Interestingly, coordination of both phosphine donor arms lowers the activation barrier for the CBr bond cleavage significantly, in contrast to the usual observed reluctance of (bis)phosphine gold(I) complexes to undergo oxidative addition (vide supra) [66].

Echavarren et al. [68] showed that the second phosphine arm is important for the intramolecular oxidative addition of the CBr bond at (2-bromobenzyl)phosphine gold(I) complexes. Even after prolonged heating, monodentate tris(2-bromobenzyl) phosphine gold(I) chloride does not undergo C(*sp*^{2})-Br bond activation (Scheme 3.20). Calculations indicate that a high-lying transition state precludes the formation of a thermodynamically favorable five-membered metallacycle. It was suggested that the origin of the elevated barrier is closely linked to the high deformation energy of the linear gold(I) complex necessary to achieve the square-planar geometry of the gold(III) product.

As shown above, gold(I) complexes do not only undergo oxidative addition with polar bonds. To date, solid evidence has been gained, confirming that—at least intramolecularly—oxidative addition of nonpolar bonds or aryl halides to gold(I) via concerted σ-bond activation is possible. However, the averseness of gold(I) complexes to this transformation remains opaque.

3.4 Reductive Elimination

The following section will sum up the literature concerning reductive elimination at gold(III) complexes which is closely linked to the oxidative addition as its microscopic reverse. As above, only stoichiometric transformations are included. The focus is set on transformations that give rise to carbon-heteroatom or carbon-carbon

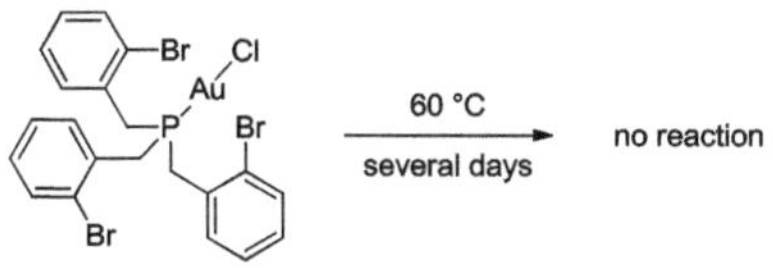

Scheme 3.20 Attempted intramolecular oxidative addition of the aryl bromide bond at gold(I)

coupling products. The reductive elimination of disulfides from tetrathiolate gold (III) complexes reported by Bachman was already discussed (vide supra) [54].

3.4.1 CX Bond-Forming Reactions (X = Halogen, NR_2, …)

3.4.1.1 C(sp^3)-X

One of the earliest publications in the field of organometallic chemistry concerning a gold(III)–gold(I) redox process dates back as far as 1934: Burawoy and Gibson studied the reactivity of [$(AuBr_2Et)_2$] which undergoes upon heating reductive elimination to yield ethyl bromide and gold(I) bromide, as proven by elemental analyses, showing that carbon–bromine bond formation at gold is feasible [69, 70].

Bercaw and Labinger studied the synthesis of (IPr)AuI_2Me by reaction of (IPr) AuMe with I_2, and its decomposition pathway to methyl iodide and (IPr)AuI. A detailed kinetic study indicated that the formation of MeI proceeds via reductive elimination at a three-coordinate intermediate after reversible decoordination of iodide (Scheme 3.21) [38].

Toste investigated the reactivity of (IPr)alkylgold(III) difluoride complexes [71]. Complexes featuring bulky alkyl chains readily underwent reductive elimination with formation of fluoroalkanes (Scheme 3.22).

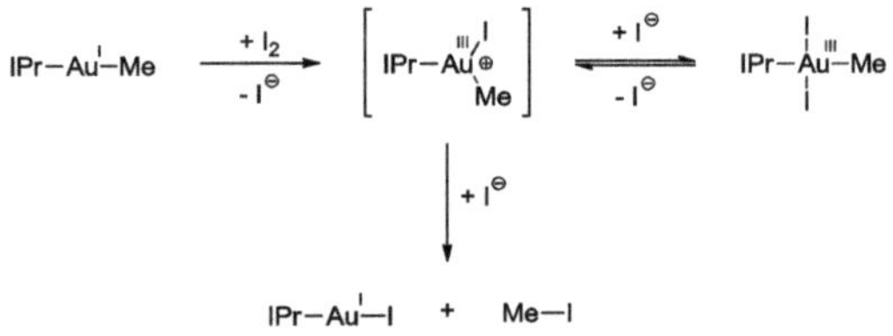

Scheme 3.21 Mechanistic pathway for the formation of methyl iodide from (IPr)AuI_2Me

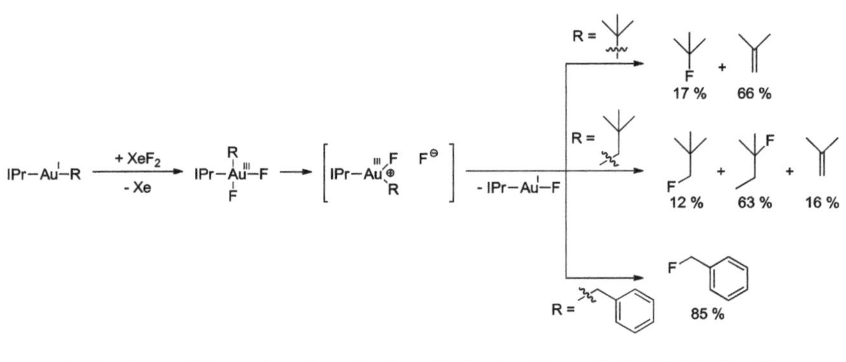

Scheme 3.22 CF bond formation via reductive elimination from alkylgold(III) fluorides

The experimental data obtained in the course of the study are in accord with decoordination of fluoride from the tetracoordinated gold(III) complex to give a tricoordinated intermediate which subsequently undergoes C(sp^3)–F bond-forming reductive elimination. Depending on the alkyl group bound to gold, this reductive elimination step competes with side-reactions (β-hydride eliminations, Wagner-Meerwein rearrangements). However, *cis*-[AuF_2(Me)(IPr)] did not undergo reductive elimination, even at elevated temperatures, in contrast to the reactivity of the corresponding iodide complex which undergoes facile reductive elimination (see above).

3.4.1.2 C(sp^2)-X Bond-Forming Reactions

Evidence for gold-mediated C(sp^2)–Br bond-formation was reported early on by Aresta and Vasapollo. Spontaneous reductive elimination from a phosphine (*o*-($H_2C{=}CH$)C_6H_4)gold(III) complex which was isolated at low temperature and characterized, yielded (phosphine)gold(I) bromide and (*o*-bromo)styrene (Scheme 3.23) [72].

Similar aryl halide formations were observed as well upon reaction of several perhalophenyl phosphine or arsine gold(I) complexes with dihalogens (Cl_2, Br_2, I_2), although the supposed intermediate gold(III) complexes were not characterized [73, 74].

Rosenthal et al. reported recently on aryl chloride bond forming reactions via reductive elimination from gold(III) complexes. (IPr)arylgold(I) complexes were oxidized with phenyliodoso dichloride to the corresponding (IPr)arylgold(III) dichlorides. Depending on the electronic properties of the aryl group, either thermal (Ar = Ph) or photochemical (Ar = C_6F_5) aryl chloride formation was observed (Scheme 3.24) [75, 76]. The intermediacy of a tricoordinated species formed by decoordination of chloride was proposed. This tricoordinate intermediate was suggested to undergo subsequently CCl bond-forming reductive elimination.

The difference in reactivity was attributed to the strength of the AuCl bond in the arylgold(III) chloride complexes: If Ar = phenyl, decoordination of chloride to give the corresponding tricoordinate intermediate would be facile and reductive elimination occurs immediately even at low temperatures. However, if the aryl ligand is a pentafluorophenyl group, its highly electron-withdrawing character would perturb the electronic situation at the metal center, leading to a strong gold–chloride bond that is only cleaved photochemically.

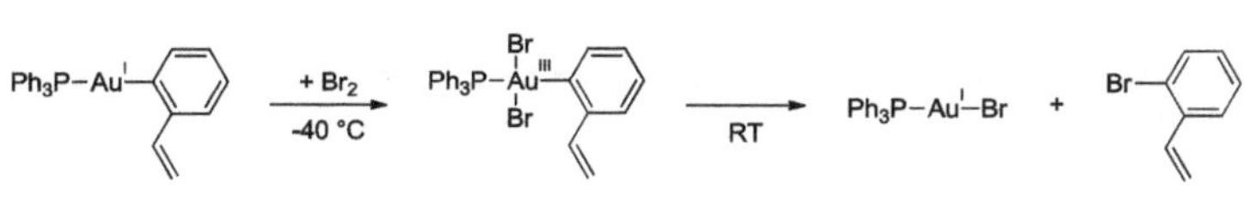

Scheme 3.23 CBr bond-formation by reductive elimination at a dibromo(aryl)gold(III) complex

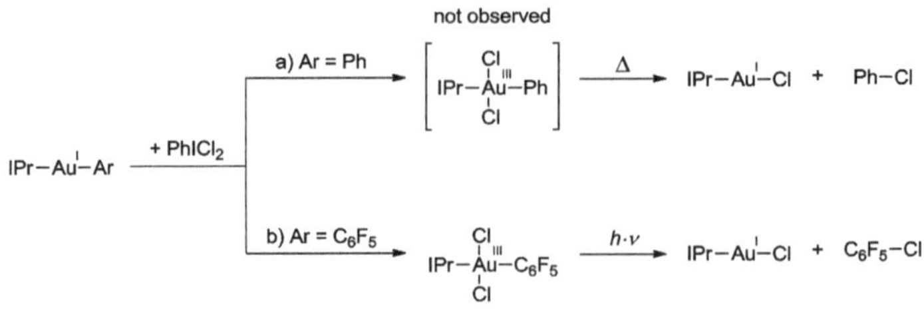

Scheme 3.24 CCl bond-forming reactions via thermal (pathway *a*) or photochemical (pathway *b*) reductive elimination from dichloroarylgold(III) complexes

Limbach and coworkers [77] investigated stoichiometric CH bond activation by gold(III) complexes (see Sect. 3.8) followed by reductive C(sp^2)-N bond formation. A dichloro(morpholine)phenylgold(III) complex was isolated and characterized. This complex reacted upon addition of a base (NaOAc) to form *N*-phenylmorpholine. The proposed mechanism involves deprotonation of the morpholine ligand and subsequent loss of a chloride ligand from the dichloro(morpholide)phenylgold (III) complex. The formed three-coordinate gold(III) intermediate undergoes subsequently reductive elimination yielding the aniline as the cross-coupled product and a gold(I) complex (Scheme 3.25). Although the proposed pathway is reasonable, none of the key intermediates was detected, nor could the final gold(I) product be observed.

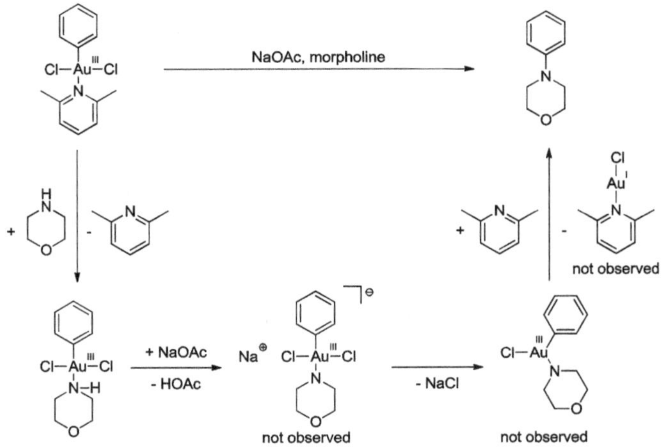

Scheme 3.25 Mechanistic rationale for a CN bond-forming reaction via reductive elimination from a gold(III) complex, as established by Limbach and coworkers

3.4.2 CC Bond-Forming Reactions

3.4.2.1 Dinuclear Complexes Involved in CC Bond-Forming Reactions

Schmidbaur reported on CC-bond forming reductive eliminations at digold ylide complexes; these reactions were limited to ethane or propane formation [40, 78]. Bennett et al. [79–81] reported on biphenyl formation stemming from reductive elimination at a dinuclear cyclometallated gold(II)–gold(II) complex (Scheme 3.26). The reaction is supposed to proceed via a migration step to position both aryl ligands on the same gold atom, followed by CC bond coupling.

3.4.2.2 Alkane Synthesis by Reductive Elimination from Dimethylgold Species

Dialkylgold(III) complexes undergo C(sp^3)–C(sp^3) bond forming reductive elimination to give alkanes. This reactivity was observed early on for $Me_4Au_2I_2$ which loses ethane upon heating (Scheme 3.27) [82].

The mechanism for ethane formation by reductive elimination at [$AuMe_2X(PR_3)$] complexes was extensively investigated by Komiya and Kochi [83] and others [44, 45]. It was found that the reductive elimination for complexes of this type occurs preferably at a tricoordinated species which forms by decoordination of the phosphine ligand (see Sect. 3.3.2). Excess phosphine considerably retards the rate of reductive elimination, in favor of a dissociative mechanism. A theoretical analysis further supported this mechanism involving a tricoordinate intermediate [84].

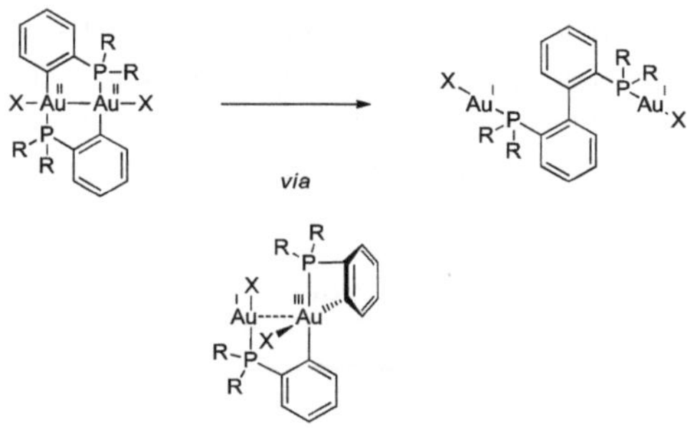

Scheme 3.26 CC bond formation by reductive elimination at (organo)dimethylgold(III) complexes

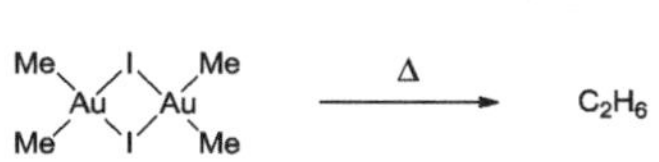

Scheme 3.27 Ethane formation at a iododimethylgold(III) complex

Tobias and Kuch [85] studied ethane formation from cationic diphosphine dimethylgold(III) complexes, and furnished as well evidence for a dissociative mechanism.

3.4.2.3 CC(sp^2) Coupling

Komiya and coworkers [86, 87] studied the selectivity for the reductive elimination from different dimethyl gold(III) complexes. (Phosphine)(R)dimethyl gold(III) complexes were shown to react upon heating to different coupling products and the corresponding gold(I) complexes (Scheme 3.28). In case of R = alkyl, alkenyl or alkynyl, formation of ethane competed with cross-coupling. The ratio of both products was found to depend on the electronic properties of R. If R is an aryl group, the reaction yielded selectively the methyl–aryl coupling product. Kinetic data were in line with a dissociative mechanism. Bulky phosphine ligands or electron withdrawing substituents on the aryl ligand on the gold center favored the formation of the reactive tricoordinate intermediate species which underwent subsequently reductive elimination to form a methyl–C(sp^2) bond.

In the course of their investigations concerning the reactivity of gold fluoride complexes, Toste [88] reported on the coupling of arylboronic acids with a methylgold(III) fluoride complex. Based on competition experiments which revealed an electronic insensitivity of the reaction with regard to the nature of the employed arylboronic acid, it was proposed that a concerted bimolecular outer-sphere mechanism is operating, where CC and BF bonds form simultaneously, instead of the conventional transmetallation/reductive elimination pathway (Scheme 3.29).

Toste et al. described recently the reactivity of trifluoromethyl–aryl gold(III) complexes prepared by oxidative addition of CF_3I to gold(I) (see Sect. 3.3.2). [Au(4-F-C_6H_4)(CF_3)I(PPh_3)] was found to undergo C(sp^2)–I bond-forming reductive elimination at elevated temperatures yielding aryl iodides and the corresponding phosphine trifluoromethylgold(I) complex (Scheme 3.30, upper path). Reductive elimination occurs likely at a tricoordinated intermediate complex, formed by

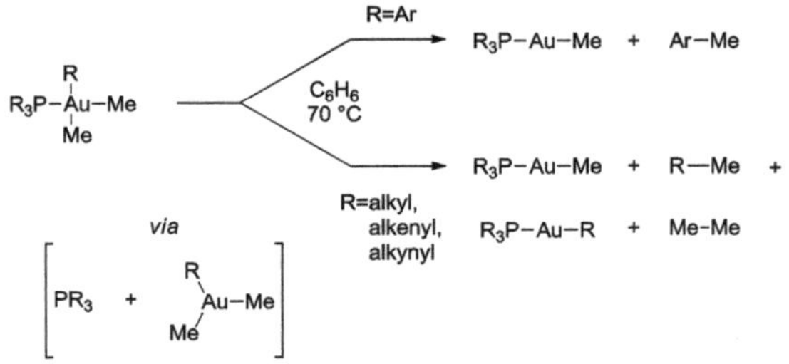

Scheme 3.28 CC bond formation by reductive elimination at (organo)dimethylgold(III) complexes. Ar = phenyl, 2-furyl, 4-FC_6H_4

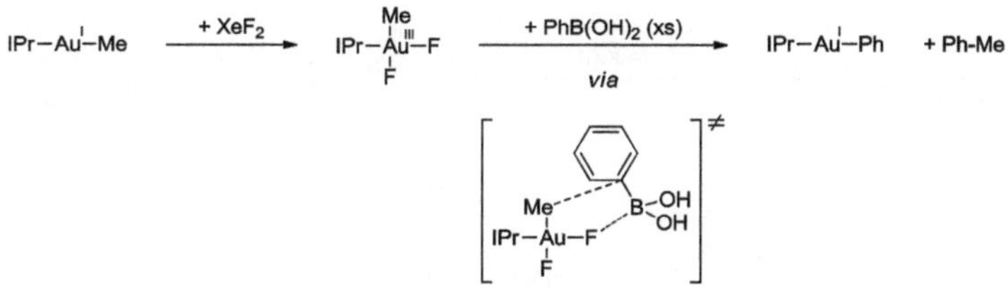

Scheme 3.29 Bimolecular reductive elimination of toluene from a gold(III) fluoride and phenylboronic acid via a concerted outer-sphere mechanism with simultaneous formation of CC and BF bonds

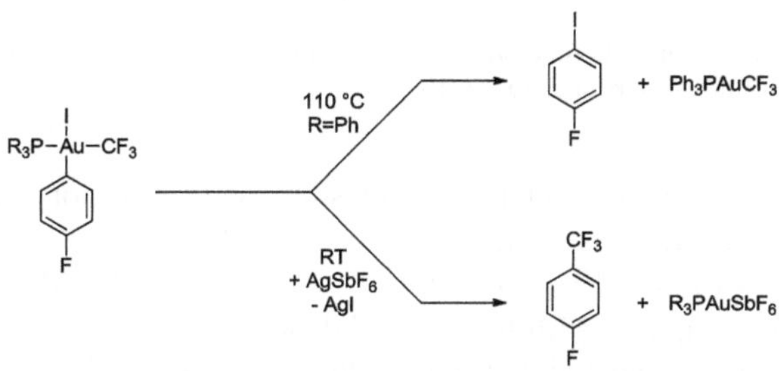

Scheme 3.30 Reductive elimination at a trifluoromethyl-arylgold(III) complex. R = Ph, Cy

dissociation of the phosphine ligand, in agreement with the results of Komiya and coworkers (vide supra). Accordingly, [Au(*p*-F-C_6H_4)(CF_3)I(PCy_3)] did not undergo thermal reductive elimination, likely due to the better σ-donor properties of PCy_3, preventing the dissociation from gold(III). Free triphenylphosphine ligand in the reaction media led to complete inhibition of the reductive elimination at [Au (4-F-C_6H_4)(CF_3)I(PPh_3)].

As no C(sp^2)–CF_3 bond-forming was observed for the thermally initiated reductive elimination, C(sp^2)–I reductive elimination had to be significantly faster. However, reductive elimination also proceeded at a cationic tricoordinated complex upon abstraction of the iodide ligand of [Au(aryl)(CF_3)I(PR_3)] (R = Cy, Ph) by a silver salt, leading exclusively to trifluoromethylated arenes (Scheme 3.30, lower path). This reaction occurred remarkably fast (even at −10 °C within seconds in case of tricyclohexylphosphine gold(III) complexes).

The capability of gold(III) salts to promote the coupling of aromatic heterocycles was first discovered by Constable et al. [89, 90]: when studying the coordination chemistry of (thienyl)pyridines towards $NaAuCl_4$, CC-homocoupling products were obtained (Scheme 3.31). Similar observations were made by the group of Fuchita [91]. Lippert and coworkers [92] studied the coupling of uracil bases using gold(III) salts. Kim and coworkers [93] employed $AuCl_3$ in stoichiometric quantities in the synthesis of diporphyrins. Product formation is proposed to occur via double arylation of the gold chloride and subsequent reductive elimination of two *cis* orientated

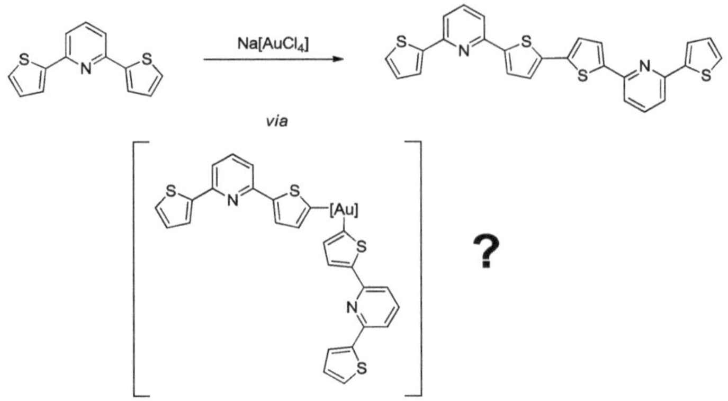

Scheme 3.31 Isolation of a homocoupled product in the course of a study concerning the coordination of bis(thienyl)pyridines to Au(III)

aryl moieties at the gold(III) center, although solid evidence is not available. Radical pathways were excluded based on the absence of EPR signals [92].

Wendt and coworkers [94] investigated the reaction of (IMes)AuPh (phenyl (1,3-bis(2,4,6-trimethylphenyl)imidazol-2-ylidene)gold(I) complex) with methyl iodide, methyl triflate and iodobenzene (Scheme 3.32). While methyl iodide and methyl triflate gave rise to mixtures of CC coupling products (i.e. toluene, biphenyl and ethane), reaction with iodobenzene yielded exclusively biphenyl and (IMes) AuI, after 50 h at 110 °C. For none of the performed experiments a gold(III) intermediate complex was detected.

Catalysis by trace amounts of colloidal palladium or radical pathways were likely to be excluded, as neither was the reaction inhibited by mercury in the reaction mixture nor by added TEMPO as a radical scavenger. Catalysis by in situ formed gold nanoparticles was ruled out as well, because no induction period was observed. The reaction involving methyl triflate was investigated computationally and an oxidative addition/reductive elimination pathway was identified. However, both the barrier for the formation of the gold(III) complex and the subsequent reductive

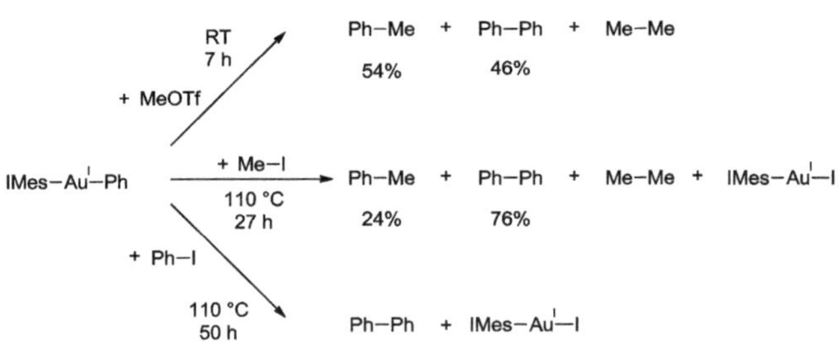

Scheme 3.32 Reaction of (IMes)AuPh with MeOTf, MeI and PhI, giving rise to CC coupling products

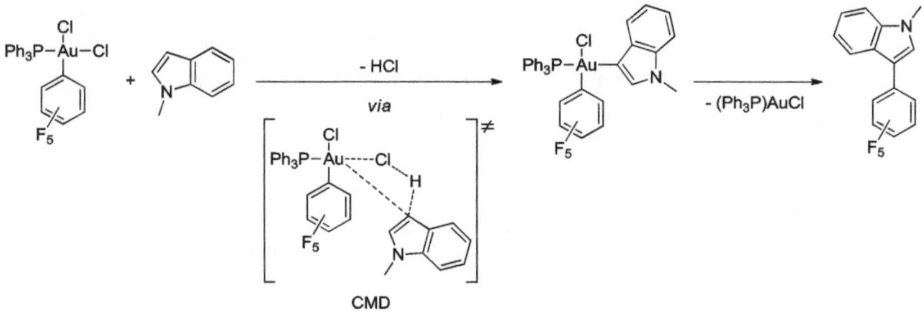

Scheme 3.33 A C(sp^2)–C(sp^2) bond-forming reaction via an auration/reductive elimination sequence at gold(III)

elimination at this species were calculated to be prohibitively high in energy with $\Delta G^{\neq}$ values of 40.4 kcal/mol and 51.6 kcal/mol, respectively. The mechanism of this transformation remains unclear.

Nevado [95] studied the reactivity of a (phosphine)(pentafluorophenyl)gold(III) dichloride complex with 1-methylindole and observed C(sp^2)–C(sp^2) cross-coupling (Scheme 3.33). Reductive elimination from a bis(aryl)gold(III) complex was proposed to account for the formation of 1-methyl-3-(pentafluorophenyl)-indole.

The chloride–aryl exchange could be performed as well by employing (fluoroaryl)boronic acids under harsh conditions [96], The ensuing *cis*-bis(aryl)gold(III) complexes were shown to undergo reductive elimination with diarene formation (Scheme 3.34).

Vicente and coworkers reported on the synthesis of bis(aryl)gold(III) or aryl((α-keto)alkyl)gold(III) complexes featuring a chelating *o*-metallated phenylazo ligand. These complexes underwent reductive elimination at room temperature upon addition of triphenylphosphine to yield the CC-coupled products and the corresponding gold(I) complexes (Scheme 3.35) [97, 98].

Some examples of the intermediate bis(aryl)gold(III) or aryl(alkyl)gold(III) complexes were isolated and structurally characterized [99]. The proposed mechanism for the observed cross-coupling involves displacement of the chloride and the azo-group of the chelating aryl ligand by triphenylphosphine, followed by reductive elimination. This mechanistic picture in which reductive elimination proceeds directly at a four-coordinated gold(III) species stands in stark contrast to reductive

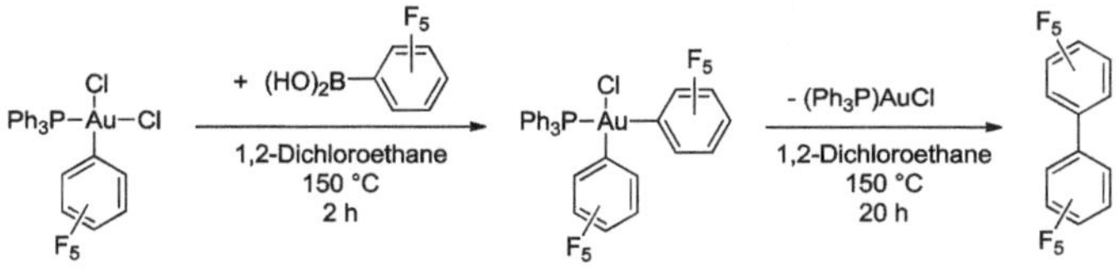

Scheme 3.34 A transmetallation/reductive elimination sequence to give fluorinated diarenes under C(sp^2)–C(sp^2) bond-formation at gold(III)

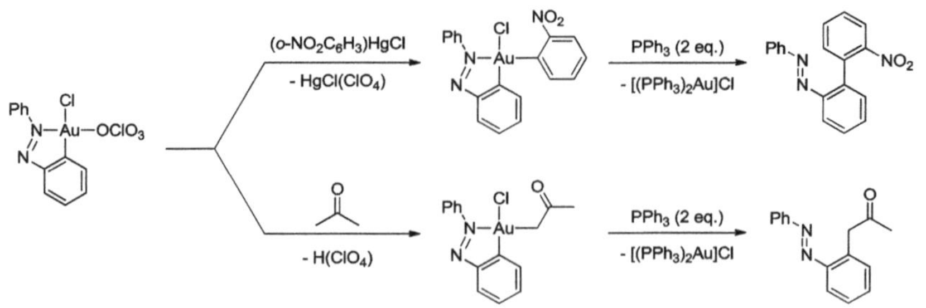

Scheme 3.35 Gold-mediated C(sp^2)–C(sp^2) and C(sp^2)–C(sp^3) coupling by reductive elimination at gold(III) complexes

eliminations at alkylgold(III) complexes which likely proceed via three-coordinate intermediates and where traces of added phosphine effectively inhibit this process (vide supra).

Toste et al. [100] communicated a similar reactivity. In contrast to Nevado's (triphenylphosphine)bis(C_6F_5)gold(III) chloride (Scheme 3.34), an analogous bis (monofluorophenyl) complex undergoes facile reductive elimination even at temperatures as low as −50 °C. The bis(aryl)gold(III) species was prepared in situ by oxidation of [Au(*p*-F-C_6H_4)(PPh_3)] with phenyliodoso dichloride, followed by transmetallation employing a second equivalent of the arylgold(I) complex at −78 °C. *cis*-[AuCl(*p*-F-C_6H_4)$_2$Cl(PPh_3)] was identified by ^{31}P and ^{19}F NMR spectroscopy and shown to decompose to [AuCl(Ph_3P)] and difluorobiphenyl (Scheme 3.36).

Particularly noteworthy with regard to this reaction is the finding that excess phosphine led to an increase in the rate of product formation. This behavior is rationalized by fast chloride–phosphine ligand exchange, followed by even more facile reductive elimination at the cationic (diphosphine)bis(aryl)gold(III) species. These results confirm that reductive elimination of aryl groups at gold(III) to form a biaryl can occur at a four-coordinate gold(III) species.

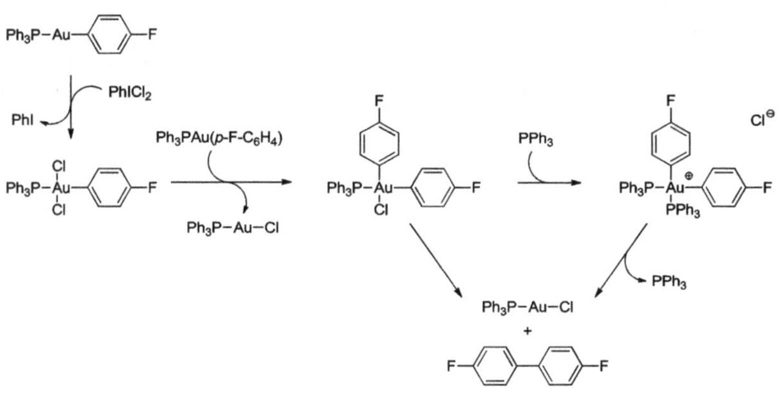

Scheme 3.36 Facile bis(aryl) reductive elimination at gold(III) phosphine complexes

As a whole, it was shown that two mechanisms for reductive elimination at gold (III) complexes can be operative: alkyl- and (alkyl)(aryl)gold(III) complexes undergo reductive elimination most likely from three-coordinate intermediates after ligand dissociation, while four-coordinate (bis)arylgold(III) complexes may react directly. Pentafluorophenyl ligands lead to high stability of gold(III) compounds. In order to initiate reductive eliminations at such complexes, photochemical or harsh thermal conditions are required.

3.5 Migratory Insertion

Insertion reactions of CC multiple bonds in metal–element bonds are ubiquitous in transition metal chemistry and are encountered as fundamental elementary steps in many catalytic cycles. However, the migratory *syn* insertion reaction (Scheme 3.37) remains elusive for gold complexes.

Instead, gold complexes usually react with alkynes and alkenes to form π-complexes which are activated for subsequent nucleophilic attack to form vinylgold complexes. In general, the outcome of nucleophilic additions at C≡C bonds catalyzed by gold complexes are *trans* configurated products, suggesting that nucleophilic attack proceeds in an *anti* fashion (Scheme 3.38).

The majority of mechanisms for homogeneously gold-catalyzed reactions are initiated by this general reactivity scheme [101].

3.5.1 syn *Insertion in Gold(I) Catalysis?*

Although a *syn* addition/isomerization scheme was proposed several times as an alternative pathway to the *anti* addition in order to rationalize the observed selectivities of nucleophilic addition reactions catalyzed by gold [102–108], no direct evidence for this pathway is available [2].

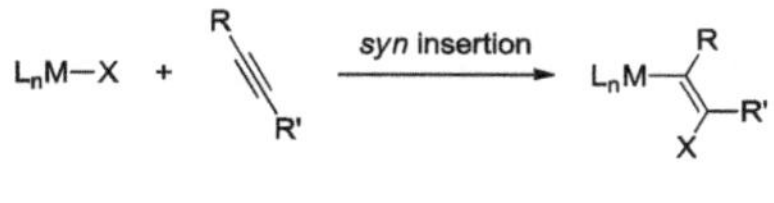

Scheme 3.37 General scheme for a migratory syn insertion reaction of an alkyne at a transition metal. Reported examples for M = Au(I) are scarce

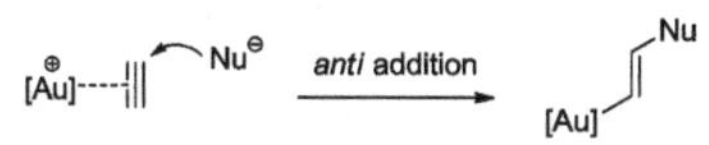

Scheme 3.38 General reactivity scheme of cationic gold(I) π-complexes with nucleophiles

A telling example on the gold-catalyzed alkoxylation of alkynes was reported by the Teles group at BASF, proposing an associative mechanism based on ab initio gas-phase calculations (Scheme 3.39) [109]. Reaction of propyne with $[Au(PMe_3)]^+$ gives rise to the corresponding π-complex and is followed by coordination of methanol to the gold center via the oxygen atom to result in a tricoordinated cationic intermediate. Next, addition of the coordinated methanol molecule to the Markownikow carbon of the alkyne in a *syn* fashion was predicted. After proton transfer assisted by gold and isomerization around the CC double bond, the catalytic cycle closes with product release.

The *syn* addition of methanol to propyne can be as well considered as a *syn* insertion of an alkyne into the Au–O bond (Step B in Scheme 3.39).

However, the mechanistic picture established by Teles et al. does not account for the experimentally observed high *anti* selectivity for the addition of methanol to internal alkynes. Hashmi et al. [110] revised the mechanism using a more sophisticated computational model including solvent effects and found that the attack of methanol to the coordinated alkyne proceeds indeed in an external *anti* fashion with proton transfer assisted by the solvent (water or methanol) (Fig. 3.3).

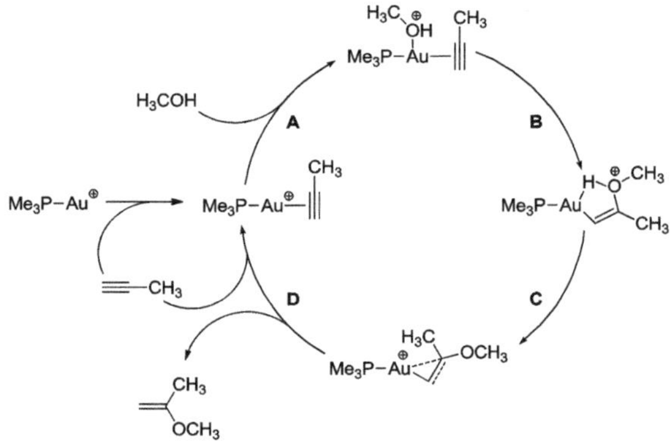

Scheme 3.39 Catalytic cycle for the methoxylation of propyne as proposed by Teles et al. Step *B* presents a syn insertion reaction

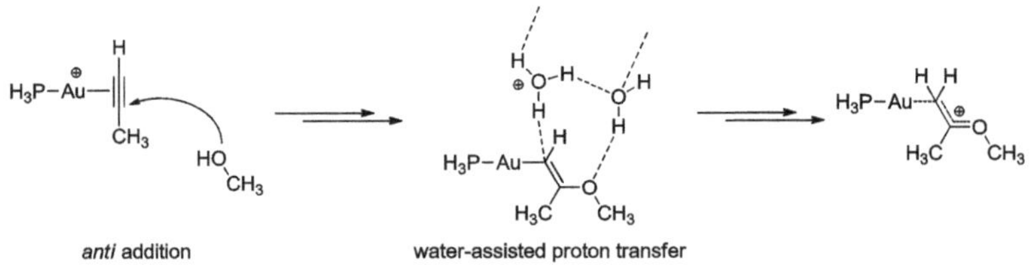

Fig. 3.3 Key intermediates in the calculated mechanism for the gold-catalyzed addition of methanol to propyne in water accounting for the experimentally observed anti selectivity. Not all of the water molecules that were included in the calculations are drawn here

3.5.2 Insertion Reactions in Gold Chemistry

Among the relevant literature dealing with stoichiometric insertion reactions at gold the following contributions to this field are particularly noteworthy:

Puddephatt et al. [111, 112] reported early on the reaction of Me_3PAuMe with hexafluorobutyne to give rapidly a mixed-oxidation state dinuclear Au(I)/Au(III) complex that was structurally characterized (Scheme 3.40). The Au(III) fragment of this species undergoes slowly reductive elimination and provides, depending on the solvent, either the formal product of a *syn* insertion reaction of the alkyne into the Au–Me bond (pathway a)) or ethane and a corresponding Au(I)/Au(I) dinuclear complex (pathway b)).

Although the mixed oxidation-state intermediate has been isolated and gives valuable insight, further mechanistic details are unclear. Furthermore, this reaction is limited to hexafluorobutyne in combination with methylgold complexes.

The reaction of bromine with (diphenyl(*o*-vinylphenyl)phosphine)gold(I) bromide (Scheme 3.41) or (diphenyl(*o*-allylphenyl)phosphine)gold(I) bromide leads to 5- or 6-membered metallacycles, respectively [113]. The authors suggest a mechanism involving first addition of bromine to the gold(I) center with subsequent migration of a bromide to the terminal carbon atom of the side chain and AuC bond formation to account for the observed products [114]. Unfortunately no information concerning the stereochemical outcome of this reaction is available.

More recently, Sadighi et al. [115] reported the reversible addition of an (NHC) gold fluoride across several alkynes. The vinylgold complex isolated from the reaction of (NHC)gold fluoride with phenylacetylene displays a *trans* arrangement

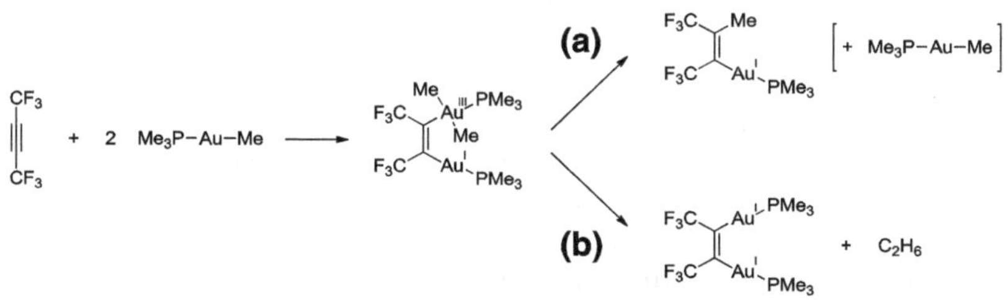

Scheme 3.40 Reaction of hexafluorobutyne with Me_3PAuMe; *a* in ether, *b* in acetone

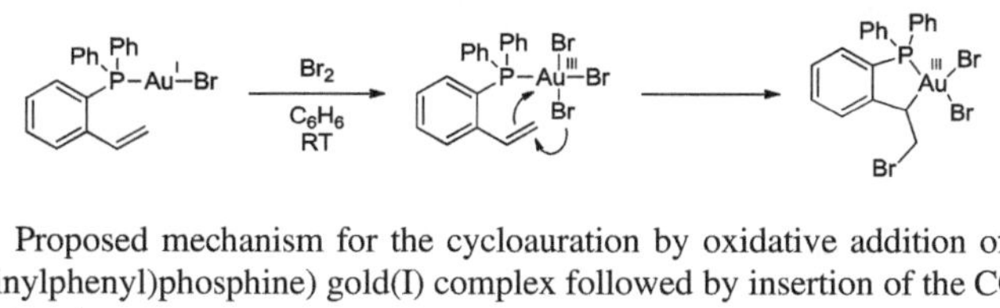

Scheme 3.41 Proposed mechanism for the cycloauration by oxidative addition of bromine to a (diphenyl(o-vinylphenyl)phosphine) gold(I) complex followed by insertion of the CC double bond into the AuBr bond

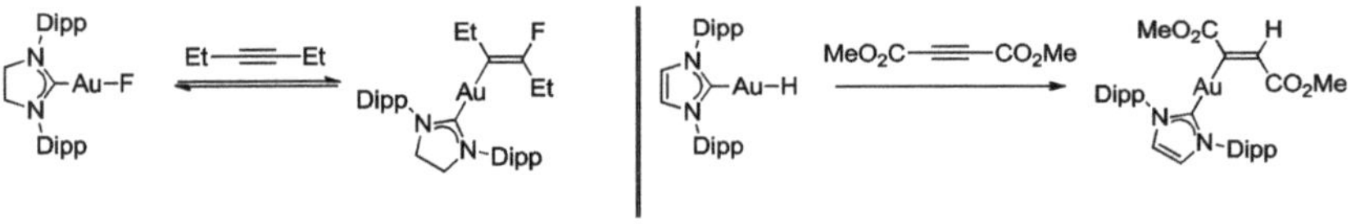

Scheme 3.42 *Left* Reaction of 3-hexyne with Sadighi's (NHC)gold fluoride yielding a vinylgold complex with Au and F atoms in trans arrangement. *Right* Reaction of dimethyl acetylenedicarboxylate with an (NHC)gold hydride yielding a vinylgold complex with Au and H atoms in trans arrangement. Dipp = (2,6-diisopropyl)phenyl

of the gold and fluoride atoms (Scheme 3.42, left). The reaction was thus proposed to proceed via displacement of the fluoride, followed by nucleophilic *anti* addition to the π-coordinated alkyne.

The same stereochemical configuration was observed in the product of the reaction of (NHC)gold(I) hydride with dimethyl acetylenedicarboxylate reported as well by the Sadighi group (Scheme 3.42, right): a *trans* configurated vinylgold complex was formed which was tentatively rationalized either by a radical mechanism or by a *syn* insertion of the alkyne in the AuH bond with subsequent isomerization [116]. However, experimental data to either confirm or refute these mechanistic schemes is not available.

The groups of Hashmi and Köppel investigated in a combined experimental and computational study the reactivity of Sadighi's Au(I) hydride towards ethylene and strain-activated norbornene and concluded that the insertion of olefins into AuH bonds at gold(I) complexes is not a favorable process (Scheme 3.43) [117].

In the context of gold-catalyzed addition of amines to CC π-bonds, the insertion of alkynes [102–105] and allenes [104, 106–108] into the AuN bond of an intermediate gold–amide or gold–amine species leading to hydroamination products, was proposed. Toste and Goddard investigated related gold-catalyzed hydroamination reactions of alkenes and allenes by experimental and computational methods, suggesting that inner-sphere *syn*-insertion mechanisms are not operative for these transformations [118, 119]. Toste, Bergman et al. reported on the synthesis of (NHC)gold(I) amides and studied the reactivity of these complexes [120]. Although highly polar, electrophilic reagents such as carbon dioxide and acrylonitrile were shown to undergo facile insertion into the AuN bond, no reactivity towards alkynes was observed (Scheme 3.44). This finding shows that the involvement of gold(I) amides in catalytic hydroamination reactions is very unlikely.

A gold-catalyzed hydroamination of alkynes with ammonia was developed by Bertrand and coworkers (Scheme 3.45) [103].

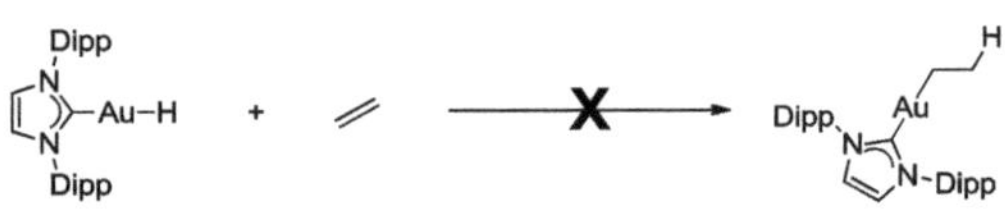

Scheme 3.43 Attempted insertion of ethylene into the AuH bond of an (NHC)gold hydride. Dipp = (2,6-diisopropyl)phenyl

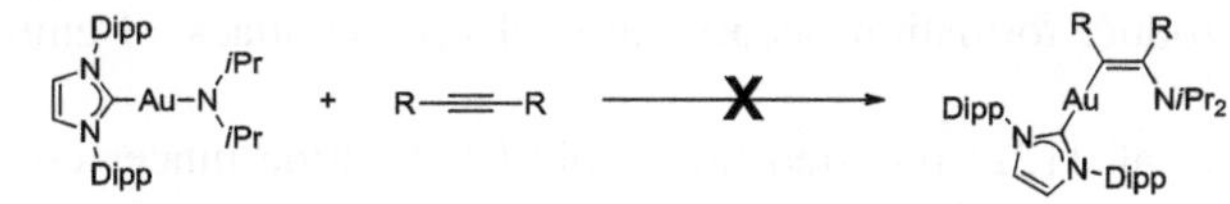

Scheme 3.44 Attempted insertion of 3-hexyne (R = Et) or diphenylacetylene (R = Ph). Dipp = (2,6-diisopropyl)phenyl

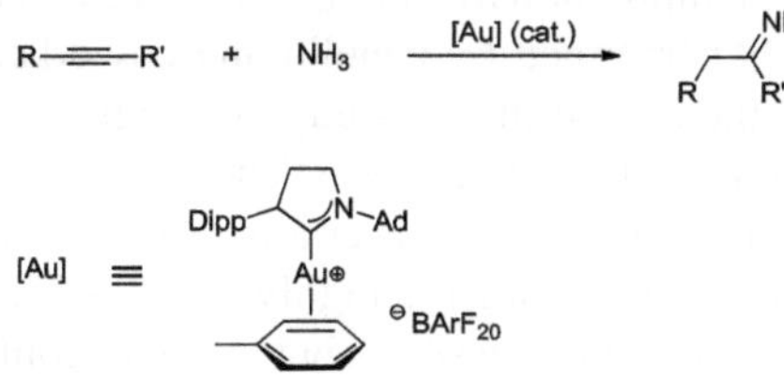

Scheme 3.45 Gold-catalyzed hydroamination of alkynes. R, R′ = alkyl, aryl

Lledós and Ujaque studied the mechanism of this reaction in detail by means of DFT calculations [121]. The results of this work are notable in the present context concerning the CN and AuC bond-forming step. In agreement with the experiment, [103] a Werner-type ammonia gold complex was found to be thermodynamically more stable (by 10.8 kcal/mol) compared to a 2-butyne π-complex (Fig. 3.4). However, direct *syn* insertion of the alkyne into the AuN bond of the ammonia complex (i.e. inner-sphere nucleophilic addition) is unlikely, as this pathway features a high barrier of 28.9 kcal/mol, lying 9.9 kcal/mol above the transition state for an outer-sphere *anti* addition. These results indicate that the ammonia gold complex is only a resting state of the catalytic cycle for the hydroamination

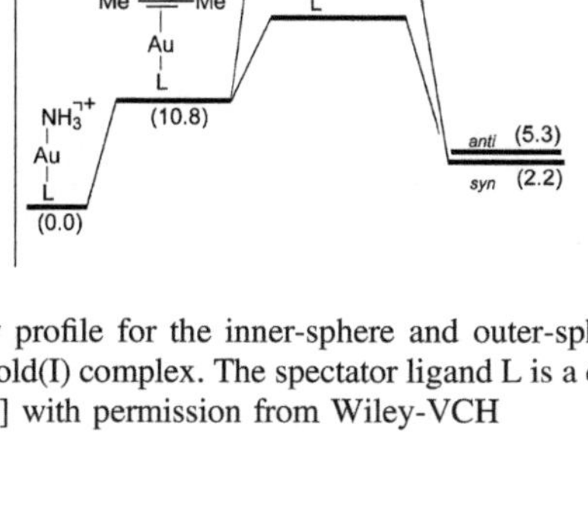

Fig. 3.4 Calculated energy profile for the inner-sphere and outer-sphere nucleophilic attack of ammonia on 2-butyne at a gold(I) complex. The spectator ligand L is a cyclic alkyl-amino carbene. Reproduced from Ref. [121] with permission from Wiley-VCH

reaction, but product formation occurs via nucleophilic attack of ammonia on the π-complex.

Bochmann et al. [122] reported on a gold(III) hydride pincer complex that is unreactive towards several alkynes and alkenes, but undergoes insertion reactions with allenes, yielding regioselectively Au(III)vinyl complexes (Scheme 3.46).

The above-mentioned examples highlight that gold complexes exhibit a most particular reactivity with regard to insertion reactions. However, a *syn* insertion reaction has not been unambiguously identified to date. An investigation concerning gold-oxo species performed by Cinellu and coworkers may be considered as a borderline case: In the course of this study, an original auraoxetane stemming from the *syn* oxyauration of norbornene (Scheme 3.47) was isolated [123]. Although this example is indeed a *syn* insertion reaction, an alternative pathway leading to the *trans*-configurated isomer is highly unfavorable. Not only is the *endo* face of norbornene sterically inaccessible, but a *trans*-configurated auraoxetane would be also too strained to exist.

A definite proof for a general migratory *syn* insertion at gold is still missing and therefore the ability of gold complexes to undergo this reaction can be legitimately questioned.

3.5.3 Alternative Routes to Vinylgold Complexes

Apart from the π-coordination/*anti* addition scheme which allowed in several cases for the isolation of stable vinylgold(I) complexes [2], these compounds are to date in general prepared by classic organometallic procedures starting from gold(I)

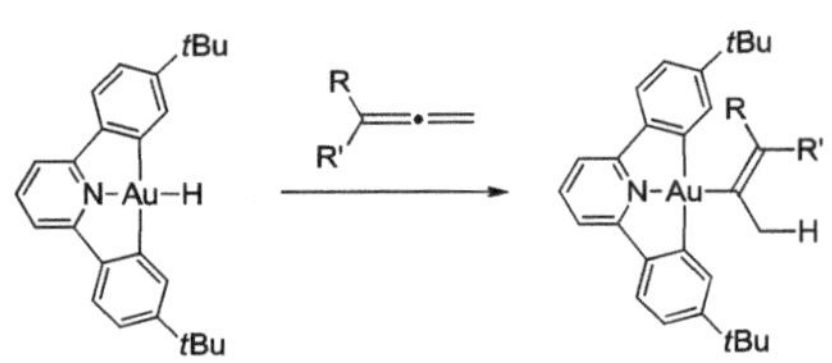

Scheme 3.46 Insertion of allenes into the AuH bond of a gold(III) hydride. R,R′ = (Me, Me) or (Cy, H)

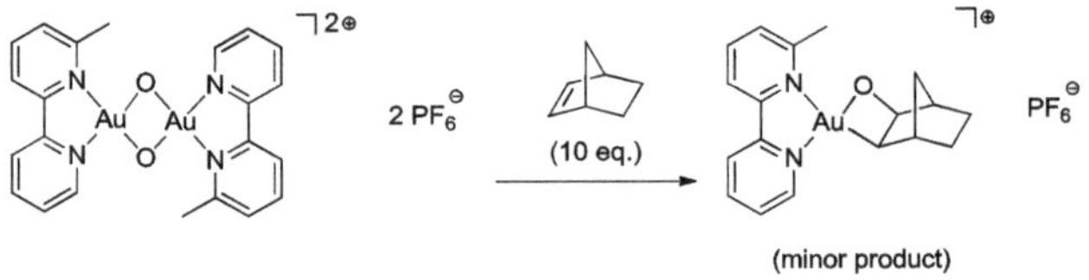

Scheme 3.47 Insertion of norbornene in the AuO bond

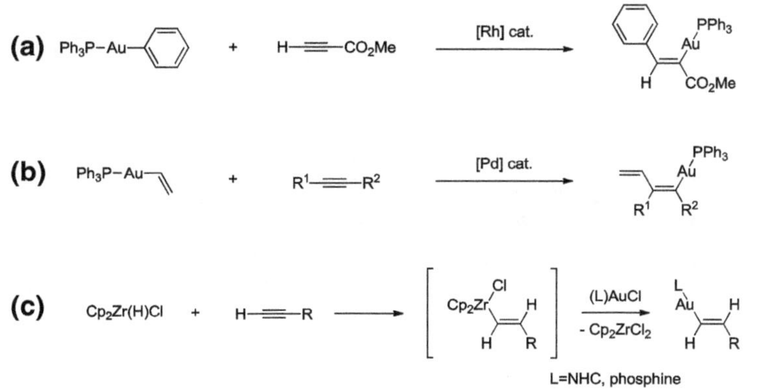

Scheme 3.48 Procedures for the synthesis of vinylgold complexes developed the Blum group: **a** Rh catalyzed carboauration, **b** Pd catalyzed carboauration, **c** Stoichiometric hydrozirconation followed by transmetallation

chloride complexes either via salt metathesis with Grignard/organolithium reagents [124, 125], or via transmetallation with boronic acids (see Sect. 3.7) [126]. The Blum group developed rhodium, palladium- or zirconium-mediated protocols, furnishing diastereo- and regioselectively vinylgold complexes (Scheme 3.48) [127–129].

A protocol for the synthesis of α-boryl vinylgold(I) complexes which is not based on a second transition metal was developed by Li, Wang and coworkers [130]. Reacting a phenylethynylborohydride salt with triphenylphosphine gold(I) chloride yielded selectively the corresponding *cis*-alkene (Scheme 3.49). The mechanism of this transformation was studied by computational means and was proposed to proceed via coordination of $[Au(PPh_3)]^+$ to the alkyne followed by a concerted 1,2-hydride shift from boron to carbon with concomitant AuC bond formation.

However, a general and direct procedure for the synthesis of vinylgold complexes is not available. The scarce information on migratory insertion reactions with gold that would present a synthetic entry for well-defined vinylgold complexes, stands in contrast to their potential for synthetic applications, as underlined by the

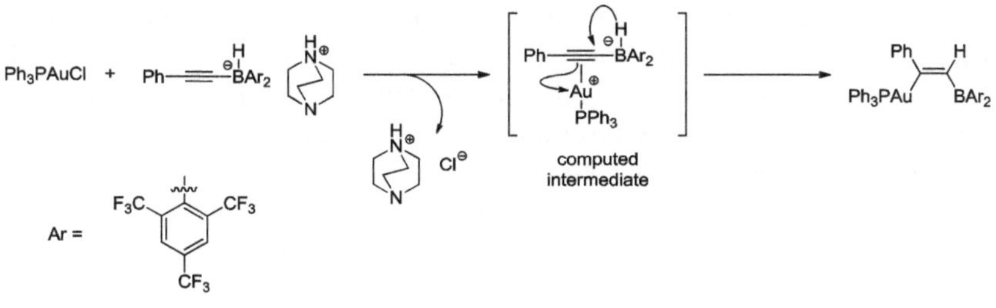

Scheme 3.49 Synthesis of an α-boryl vinylgold(I) complex

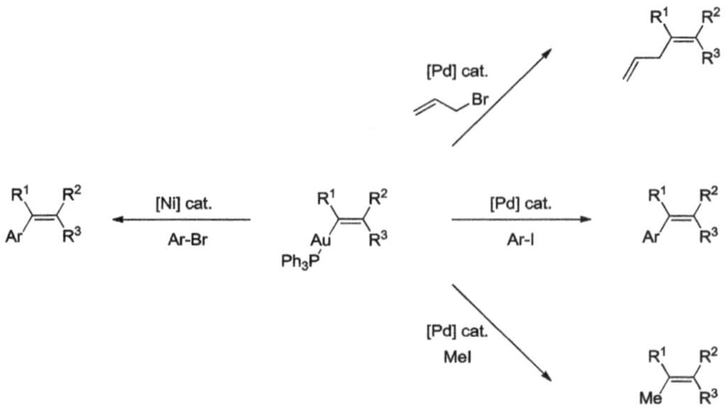

Scheme 3.50 Synthetic application of vinylgold complexes for Pd-catalyzed cross-coupling reactions

use of vinylgold species in palladium-catalyzed cross-coupling reactions. The vinyl moiety on gold can be readily transmetallated to a palladium or nickel species without the need for additives [31, 125, 128, 131]. Subsequent reaction with various organic halides gives cross-coupled products with high stereoselectivity (Scheme 3.50).

These examples illustrate that direct access to vinylgold complexes via migratory insertion of alkynes would be advantageous. The particular reluctance of organogold complexes towards this reaction remains to be explored.

3.6 β-Hydride Elimination

Alkylgold complexes are reluctant towards β-hydride elimination, an elementary step typically observed for other transition metals, for example with complexes of the valence-isoelectronic palladium [132]. Consequently, reports on β-hydride elimination at gold complexes are very rare.

Toste observed the formation of alkenes derived from a formal β-hydride elimination process at an NHC alkylgold(III) fluoride complex (see Sect. 3.4.1), but the corresponding NHC gold hydride complex was not detected [71].

The only systematic investigation of this process is a contribution by Hashmi and Köppel who studied experimentally and computationally the β-hydride elimination of ethylene from an (NHC)ethylgold(I) complex (i.e. the reverse reaction of the above discussed migratory insertion) [117]. DFT calculations predicted a high barrier (49.7 kcal/mol) for the elimination of ethylene from (IPr)Au(Et). This complex was shown experimentally to decompose at around 180 °C, but not via the envisioned β-hydride elimination pathway (Scheme 3.51).

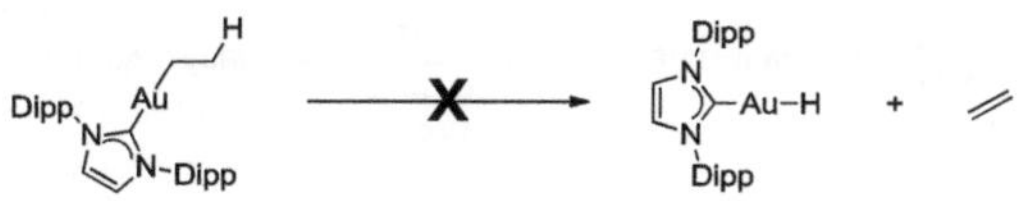

Scheme 3.51 Attempted β-hydride elimination from an (NHC)ethylgold(I) complex. Dipp = (2,6-diisopropyl)phenyl

The lack of reactivity was attributed to the filled 5d-shell of gold(I) complexes, preventing the necessary hydrogen-metal interaction required for the elimination to take place.

This explanation may be generally valid and account for the reluctance of gold complexes to undergo β-hydride eliminations. According to a detailed analysis by Schmidbaur, Raubenheimer and Dobrzańska who analyzed gold-hydrogen interactions reported in the literature, "*there is currently no evidence for agostic interactions in organogold compounds* […]" [133]. Nevertheless, as suggested by Hashmi et al., these interactions may exist in case of gold(III) compounds for which empty 5d-orbitals are available [117].

3.7 Transmetallation

The following section summarizes the relevant literature with regard to transmetallation reactions in which an organic fragment is transferred from a p-block element to gold. Simple salt metathesis reactions with gold halides and organolithium or Grignard reagents are excluded, as well as transmetallations involving transition metals. A gold-to-gold transmetallation where an alkyl group is transferred from a gold(I) to a gold(III) complex was already shortly described above (see Sect. 3.3.2).

3.7.1 B-to-Au

Early evidence for transmetallation of an aryl group from boron to gold was given by Schmidbaur and Fackler [134, 135]. Sodium tetraphenylborate was shown to transfer a phenyl group to gold chloride and oxide complexes (Scheme 3.52).

A closely related reaction was reported by Straub and coworkers. Attempting to isolate a monoligated cationic NHC gold(I) complex with the weakly coordinating tetrakis(bis(3,5-trifluoromethyl)phenyl)borate anion ($BArF_{24}^-$) that was detected by NMR spectroscopy, a (trifluoromethyl)phenylgold(I) complex was obtained instead (Scheme 3.53) [136].

Gray and coworkers developed a protocol for the efficient synthesis of arylgold complexes. Arylboronic acids react with phosphine or NHC gold(I) chlorides or bromides in the presence of a mild base in alcoholic media to yield the corresponding arylgold(I) complexes (Scheme 3.54) [137, 138]. Further optimization of

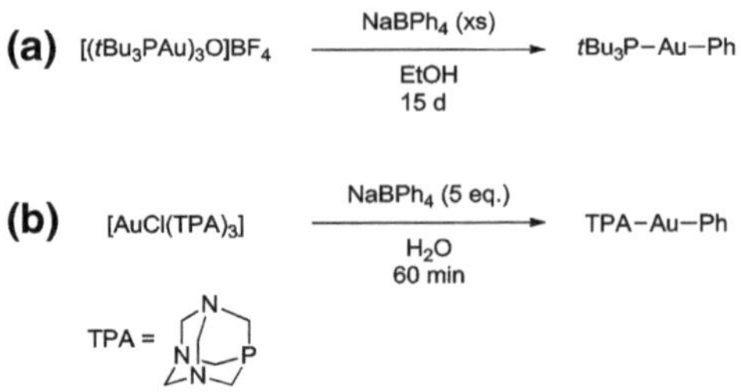

Scheme 3.52 Seminal reports on boron-to-gold transmetallations: **a** Schmidbaur, Ref. [134]; **b** Fackler, Ref. [135]

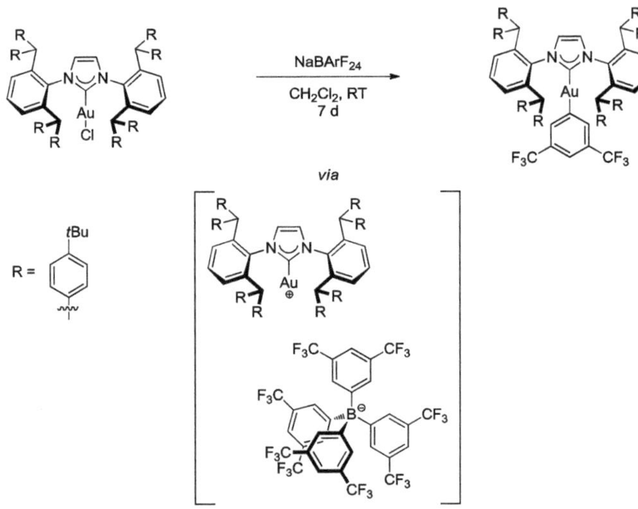

Scheme 3.53 Transmetallation of a bis(trifluoromethyl)phenyl group from boron to gold(I)

the reaction conditions employing microwave irradiation led to shorter reaction times [139]. A large variety of aryl and vinyl groups was shown to undergo the transmetallation with phosphine gold(I) halide complexes [126, 138, 140–143].

Nolan studied the reaction of arylboronic acids and [AuCl(IPr)] in more detail [144]. The synthetic protocol was improved by using potassium hydroxide as the base, allowing for the transmetallation to be performed at room temperature. The mechanism for this transformation was investigated by experimental and computational means, and found to involve most likely a gold(I) hydroxo species (Scheme 3.55), although an alternative route with participation of a borate was not excluded by the authors.

Fürstner reported on the synthesis of *gem*-digold vinyl complexes via B-to-Au transmetallation. By reacting two equivalents of a phosphine gold(I) triflimidate complex with a vinylboronic acid *gem*-digold complexes were obtained [145]. Gray and coworkers [146] prepared similar *gem*-digold aryl species (Scheme 3.56).

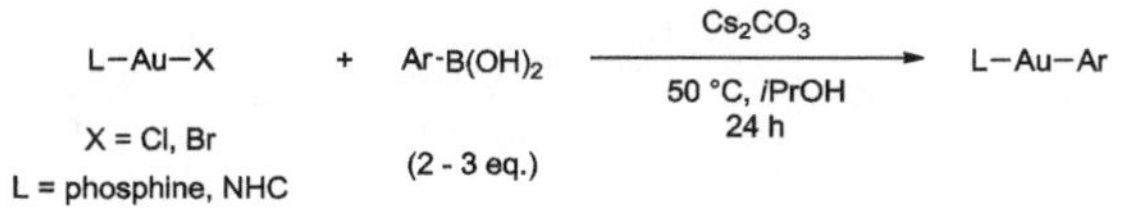

Scheme 3.54 Synthesis of arylgold(I) complexes via B-to-Au transmetallation

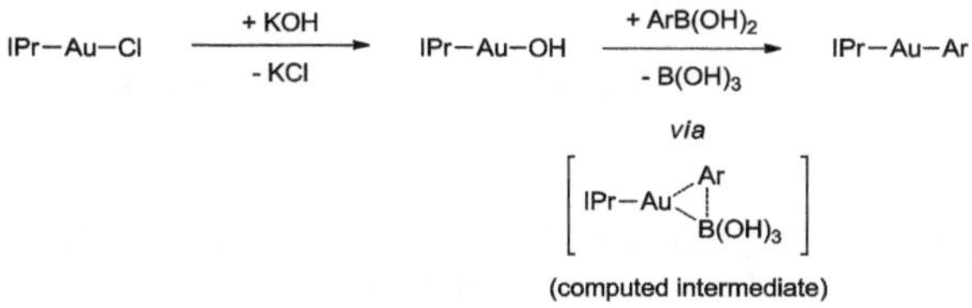

Scheme 3.55 Proposed mechanism for the reaction of gold(I) chloride complexes with arylboronic acids in the presence of a base as reported by Nolan and coworkers

As shown by Nevado, electron-poor arylboronic acids react in the absence of a base in harsh conditions with arylgold(III) chloride complexes to yield the corresponding diarylgold(III) complexes (Scheme 3.57). Arylboronic acids lacking electron-withdrawing substituents on the aromatic ring do not undergo the transmetallation step under these conditions.

3.7.2 *Si-to-Au*

Transmetallation reactions of organic fragments from silicon to gold are comparatively rare. Gold-catalyzed reactions involving arylsilanes as coupling partners will

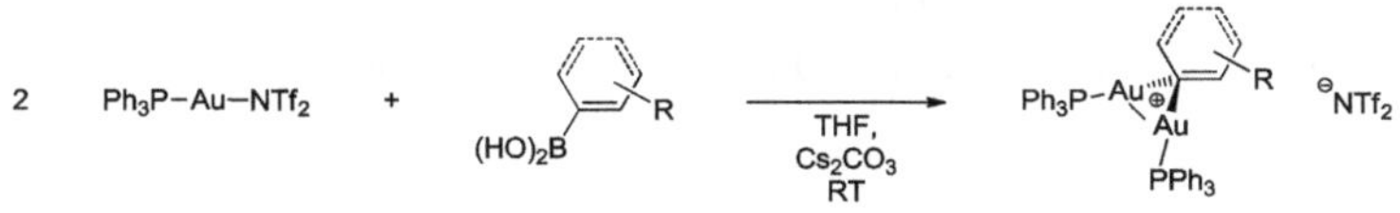

Scheme 3.56 Synthesis of gem-digold vinyl- and aryl complexes

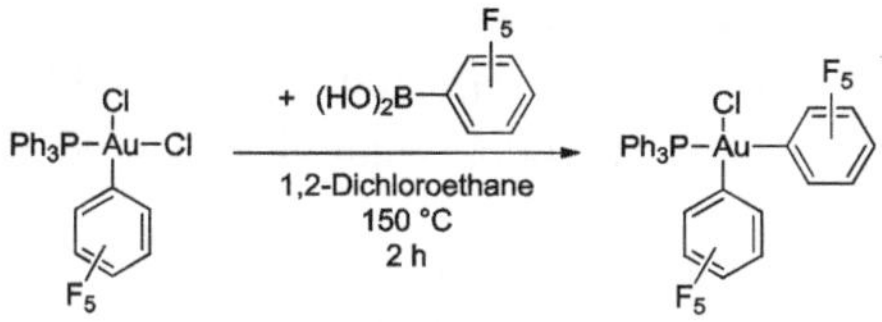

Scheme 3.57 A boron-to-gold(III) transmetallation

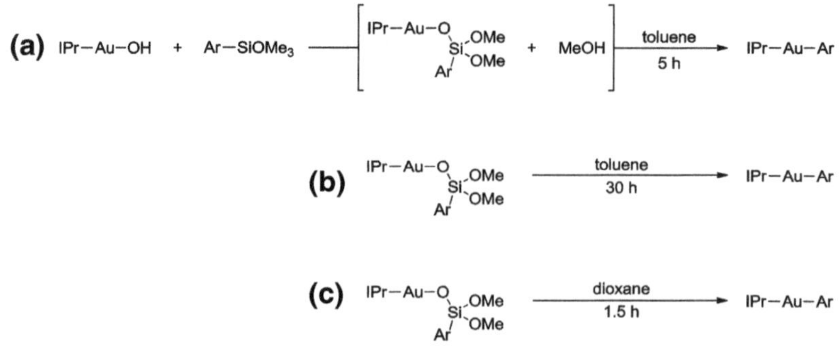

Scheme 3.58 Synthesis of arylgold complexes by Si-to-Au transmetallation. **a** starting from aryltrimethoxysilanes; **b** and **c** starting from isolated gold silanoate complexes

be discussed below (see Sect. 3.9.2). To the best of our knowledge, a single example of stoichiometric Si-to-Au transmetallation was reported. Nolan and coworkers [147] investigated the reactivity of an NHC gold hydroxide complex towards aryltrimethoxysilanes aiming at the transmetallation of the aryl group to give (IPr)arylgold complexes (Scheme 3.58).

The reaction occurred in two steps. First, a gold silanoate complex was formed which was identified as a key intermediate that subsequently reacted further to the arylgold complex. When the reaction was performed without isolation of the gold silanoate complex, the arylgold complex formed cleanly (Scheme 3.58, a)). However, starting from the isolated gold silanoate complex in nonpolar solvents, the transmetallation step was found to be sluggish and to proceed at much slower rate, indicating the important role of the eliminated methanol molecule in the direct reaction without isolation of the silanoate complex (Scheme 3.58, b)). Indeed, addition of methanol or the use of a polar, coordinating solvent such as 1,4-dioxane led to similar reaction rates for the arylgold complex formation based on the corresponding isolated silanoate complex (Scheme 3.58, c)). This finding was rationalized by formation of a pentacoordinate silicon species by coordination of either methanol or the coordinating solvent, thus facilitating the transmetallation step.

3.7.3 Sn-to-Au

Tin-to-gold transmetallations were used occasionally for the synthesis of organogold complexes. Usón et al. [148] synthesized early on a dibenzoaurol complex via transmetallation from the corresponding stannol to a gold(III) precursor complex (Scheme 3.59).

Mohr and coworkers [149, 150] studied the transmetallation of various arylstannanes to gold(I) chloride complexes. While aryl(tributyl)stannanes were found to be inert, the trimethyl counterparts afforded the corresponding arylgold(I) complexes (Scheme 3.60).

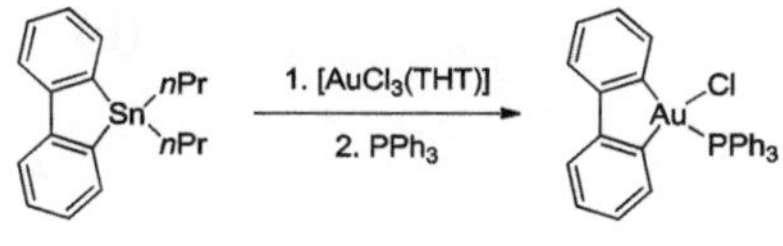

Scheme 3.59 Early example for a tin-to-gold transmetallation giving rise to a dibenzoaurol complex

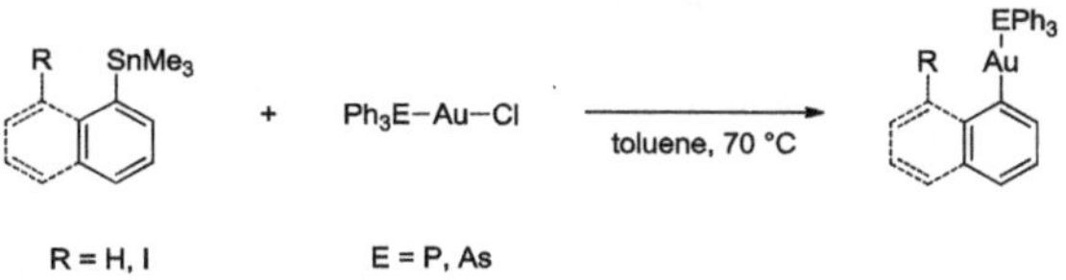

Scheme 3.60 Sn-to-Au transmetallations yielding gold(I) aryl complexes

Liu and coworkers [151] developed a gold-catalyzed cyclization/stannylation tandem reaction which presumably involves a tin-to-gold transmetallation step (Scheme 3.61). Stannylated naphthalenes were synthesized by gold-catalyzed reaction of 1,6-diyne-4-en-3-ols with tributylstannylfuran.

The proposed mechanism for the tandem process involves first cyclization of the diyne to give an arylgold complex. Stannylation of this intermediate by an in situ generated stannylium cation ($n\mathrm{Bu_3Sn^+}$) leads to formation of the product. $n\mathrm{Bu_3Sn^+}$ was generated from 2-tributylstannylfuran and proposed to be formed by an adjoint catalytic cycle in which the furanyl group is transferred to gold (Scheme 3.62, left cycle). A direct transfer of the $n\mathrm{Bu_3Sn}$ group from tributylstannylfuran to the arylgold complex seemed less likely but was not excluded by the authors.

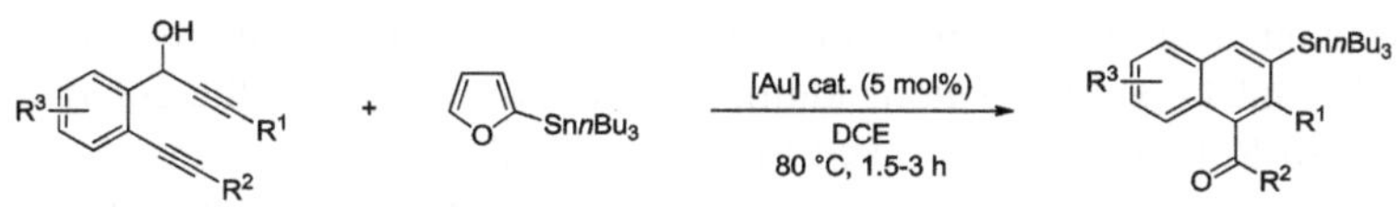

Scheme 3.61 Gold-catalyzed cyclization/stannylation tandem reaction affording stannylated naphthalenes

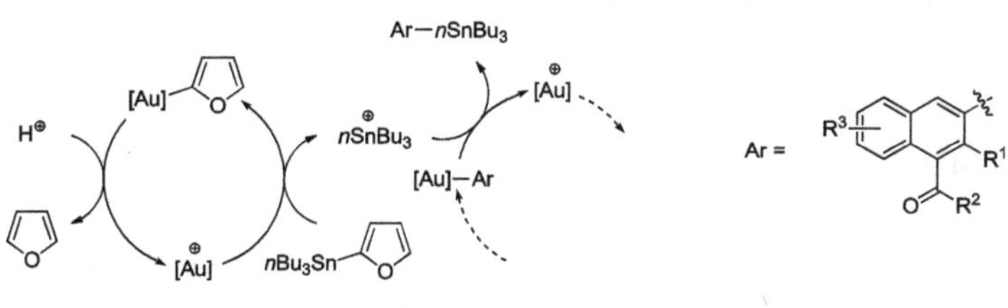

Scheme 3.62 Proposed mechanism for a gold-catalyzed stannylation involving a tin-to-gold transmetallation liberating a stannylium cation (left cycle)

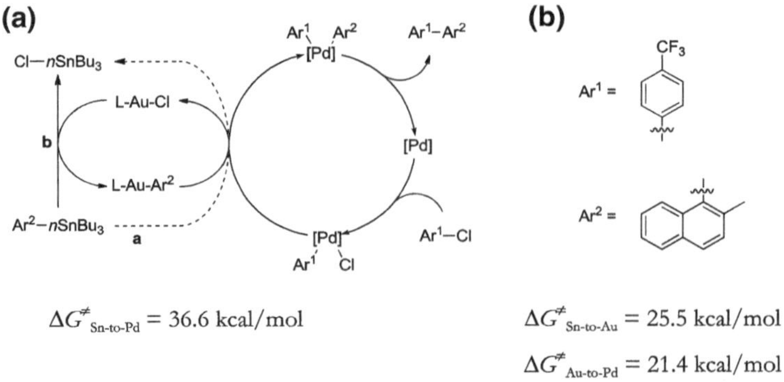

Scheme 3.63 Mechanistic pathways for the classical (palladium-only) Stille reaction (**a**), and the Au/Pd cocatalyzed reaction (**b**), together with the respective computed activation energies

Casares et al. [152] investigated the influence of gold(I) cocatalysts on the Stille reaction. In preliminary experiments, the thermodynamics of the tin-to-gold transmetallation of an aryl group was studied in detail and found to depend strongly on the halide involved. Formation of a strong SnCl bond (in comparison to SnI) was found to favor the transfer of an aryl group from $n\mathrm{Bu_3SnAr}$ to LAuX (L = PPh_3, $AsPh_3$; X = Cl, I). The gold/palladium cocatalyzed Stille reaction was examinated and proposed to proceed via tin-to-gold and then gold-to-palladium transmetallation steps. This double transmetallation pathway was studied computationally and allowed for the rationalization of the experimentally observed efficient reaction when bulky aryl groups where employed as nucleophilic cross-coupling partners. While direct transmetallation from Sn-to-Pd was hampered by steric repulsion between the stannane and ligands on Pd, the Sn-to-Au and Au-to-Pd transmetallation sequence was found to be kinetically more favorable due to the decreased steric hindrance of the linear arylgold(I) complex involved (Scheme 3.63).

Consistent with the finding that the transmetallation is thermodynamically favored by SnCl bond formation, addition of lithium chloride to the gold/palladium co-catalyzed Stille reactions proved to be beneficial.

All of these examples show that the transmetallation of organic fragments to gold complexes from p-block elements is possible and often facile. However, compared to the chemistry of palladium, where many studies contributed to the better understanding of transmetallation reactions, this field remains largely underdeveloped for gold complexes.

3.8 Arylation

A particularity of simple gold(III) salts is their ability to direct CH bond activation of simple arenes. This reactivity pattern is based on the high electrophilicity of these salts and is believed to proceed via an S_EAr-type (or Friedel-Crafts-type)

mechanism. Although not a veritable elementary step, this arylation reaction is discussed here, as it represents an important aspect of organogold chemistry.

Because these reactions were recently summarized (in the context of gold catalysis), only the general concept is presented here, and the reader may be referred to the recent reviews in the literature for further information [153–155].

The facile auration of benzene by $AuCl_3$ was reported early on by Kharasch and Isbell. However, this reactivity was subsequently only sparely examined which is likely due to the instability of the aurated products. A systematic study of this reactivity was undertaken by Fuchita and coworkers [156]. Electron-rich arenes reacted rapidly with $AuCl_3$ to form an unstable, uncharacterized intermediate; hydrogen chloride was eliminated. Addition of lutidine led to the isolation of corresponding, air- and moisture stable, arylated complexes (Scheme 3.64).

The observed regioselectivity for the auration of substituted arenes is in line with an S_EAr-type mechanism.

The auration of arenes employing gold(I) complexes was first achieved by Larrosa et al. [157]. A phosphine gold(I) complex reacted in the presence of a silver salt, a mild base and pivalic acid with electron-deficient arenes to give the corresponding arylated complexes (Scheme 3.65). Contrasting the reactivity for gold(III) salts, electron-rich arenes did not react with the gold(I) complexes. The reaction was found to be regioselective for the most electron-poor CH bond. The authors propose a concerted metallation-deprotonation pathway to account for the observed reactivity, but solid mechanistic evidence was not provided. An intermediate gold (III) hydride complex stemming from CH bond oxidative addition was not excluded.

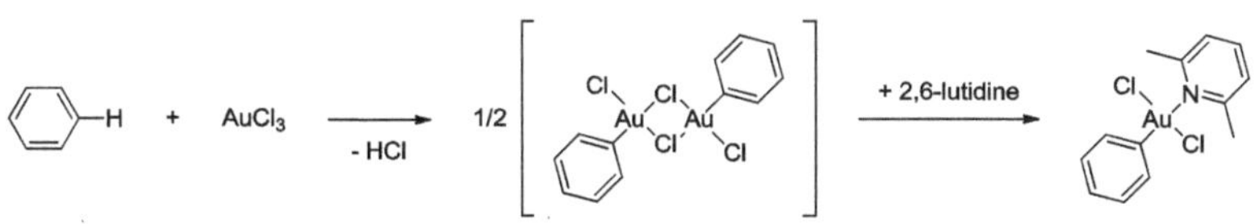

Scheme 3.64 Arylation of $AuCl_3$

Scheme 3.65 Arylation of a phosphine gold(I) complex

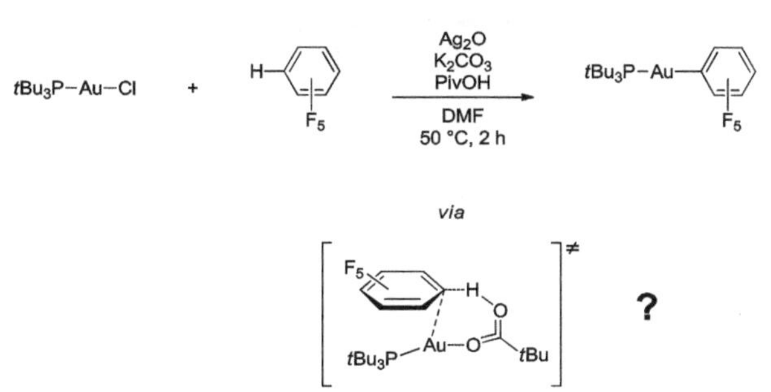

The combination of these two different arylation reactions, i.e. activation of electron-poor arenes by gold(I) and activation of electron-rich arenes by gold(III) complexes, is rather unusual for late transition metals and might be relevant in the further development of gold-mediated organic transformations.

3.9 Gold Catalysis Beyond π-Bond Activation

The vast majority of catalytic reactions in the field of homogeneous gold catalysis rely on the typical π-activation of unsaturated organic substrates by gold complexes [101]. However, this is not the only *modus operandi*; a notable exception is the gold-catalyzed dehydrogenative silylation of alcohols, involving gold-hydride species [157–160].

The following section will summarize the recently developed gold-catalyzed reactions that are based on one or more of the elementary steps presented in the previous sections.

3.9.1 Hydrodefluorination of Fluoroarenes

Zhang et al. [161] studied the hydrodefluorination of activated aryl fluorides with hydrosilanes catalyzed by Xanthphos-based gold(I) complexes (Scheme 3.66). Several mechanistic scenarios were studied by computational means. The authors proposed a mechanism involving oxidative addition of the CF bond of the aryl

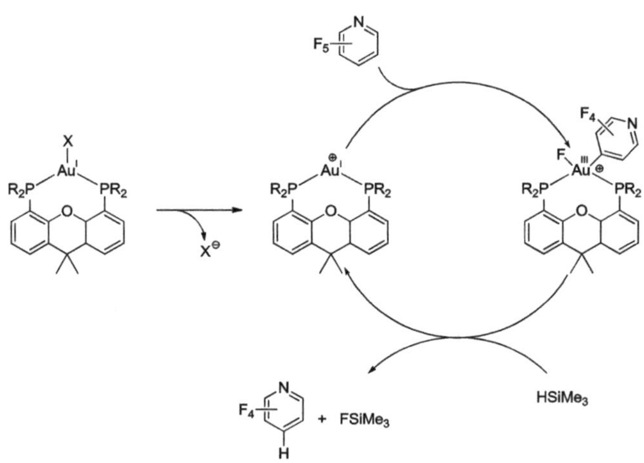

Scheme 3.66 Proposed mechanism for the gold-catalyzed hydrodefluorination of aryl fluorides with hydrosilanes

fluoride with subsequent concerted CH and SiF bond formation. An activation barrier of 30.4 kcal/mol was calculated for oxidative addition of the *para*-CF bond of pentafluoropyridine to a $[\kappa^2 P\text{-Xanthphos-Au}]^+$ fragment, somewhat lower than the value for the corresponding *ortho*-CF activation (34.6 kcal/mol), rationalizing the observed regioselectivity. Oxidative addition of the CF bond was suggested to be the rate determining step of the catalytic transformation. The gold(III) intermediate was neither isolated, nor observed experimentally.

The same group studied the reactivity of NHC gold(I) hydride complexes towards hydrodefluorination of pentafluoronitrobenzene in presence of a hydrosilane to yield the corresponding *para*-hydrodefluorinated arene. A similar mechanism involving a gold(III) species was proposed [162].

3.9.2 Catalytic Gold-Redox Chemistry with External Oxidants

Apart from the catalytic homocoupling of alkyl iodides to the corresponding higher homologues (see Sect. 3.3.2), redox-based gold-catalysis was dormant for a long period. A growing research activity in this field was stimulated by the rising interest in this metal for application in organic synthesis. The lack of reactivity of gold(I) complexes towards oxidative addition of aryl halides as usually employed in palladium-catalyzed coupling reactions led to the development of alternative protocols to achieve Au(I)/Au(III) redox catalytic cycles providing synthetically useful transformations. Hereafter, important achievements in this field will be summarized.

Tse and Beller [163, 164] reported in 2008 on the homocoupling of simple arenes with catalytic amounts of chloroauric acid and $PhI(OAc)_2$ as the stoichiometric oxidant (Scheme 3.67). The oxidative coupling was proposed to proceed via double CH activation, most likely at a gold(III) species, followed by reductive elimination, although direct evidence is not available.

Hashmi and coworkers [165] reported on the cyclization of allenyl carbinols and observed dimerized side-products in low yields when $AuCl_3$ was employed as the catalyst. A transmetallation/reductive elimination sequence was proposed to be operative. This hypothesis was in line with the observation that gold(I)-based catalysts yield exclusively the simple cyclization product.

Wegner et al. [166] developed a catalytic cyclization/dimerization of aryl propiolates in the presence of *tert*-butylhydroperoxide (Scheme 3.68). The reaction is

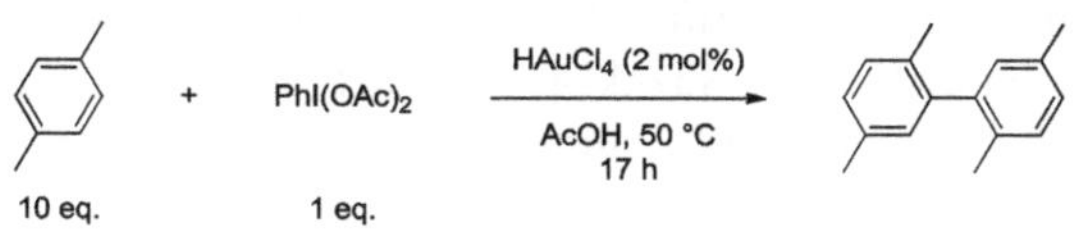

Scheme 3.67 Gold-catalyzed homocoupling of simple arenes

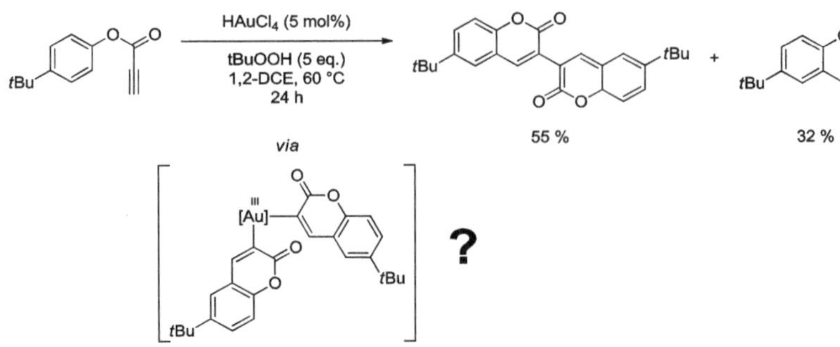

Scheme 3.68 A gold-catalyzed cyclization-oxidative coupling sequence giving rise to dicoumarins

proposed to proceed via cyclization of two substrate molecules at a gold(III) species, followed by reductive elimination to give the dimerized product. By employing stoichiometric amounts of *tert*-butylhydroperoxide as the sacrificial oxidant, the catalytically active gold(III) is regenerated. Without any oxidant present, only traces of the dimerized product were observed. Although a mechanism involving reductive elimination at a gold(III) species to form the homocoupling products is reasonable, no information about the putative key intermediate is available.

Employing alkynyl phenols as the starting material, bis(arylbenzofurans) were obtained via a similar cyclization/oxidative coupling process [167]. In all these examples, the protodeauration reaction giving rise to the simple cyclization products could not be completely suppressed, limiting therefore the yield of the homocoupling product.

Muñiz and Zhang [168, 169] further extended the scope of gold-catalysis employing sacrificial oxidants to oxidative CN and CO bond formations, respectively.

When Zhang and coworkers [170, 171] attempted the gold-catalyzed synthesis of α-fluoroenones from propargylic acetates employing Selectfluor as an electrophilic fluorinating reagent, important quantities of homocoupling products were observed instead of the expected fluorinated product.

This reactivity was rationalized by the oxidation of the gold(I) catalyst by the highly electrophilic F^{+}-species delivered by Selectfluor, giving rise to a tricoordinate gold(III) fluoride. This species reacts further via transmetallation of an intermediate vinylgold(I) complex to a bis(vinyl)gold(III) complex, followed by reductive elimination yielding the homocoupling product (Scheme 3.69, left cycle) and after decoordination of fluoride the initial catalytic species. This unexpected reactivity was extended to a Suzuki-type reaction: transmetallation was also effective with arylboronic acids and gave rise to cross-coupled products (Scheme 3.69, right cycle).

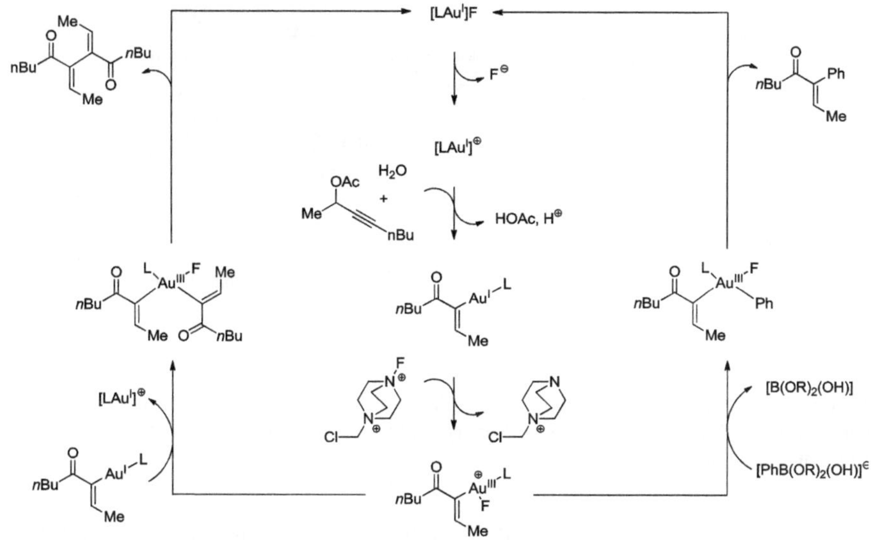

Scheme 3.69 Proposed mechanistic cycles for the gold-catalyzed coupling reactions developed by Zhang and coworkers, employing Selectfluor as the sacrificial oxidant. *Left* homocoupling cycle. *Right* Cross-coupling cycle. L = PPh_3

Dynamic light scattering experiments excluded the formation of nanoparticles and were in favor of homogeneous catalysis. However, the postulated key intermediate gold(III) species were not observed.

Following this key discovery, oxy- and aminoarylations proceeding via Selectfluor-enabled similar Au(I)–Au(III) redox cycles employing an alkene, an O- or N-nucleophile and an arylboronic acid were developed (Scheme 3.70) [172–174].

Some additional information concerning the mechanism of this transformation has been gained by Toste and coworkers [174] in the course of their studies concerning gold-catalyzed aminoarylations. It was shown that Ph_3PAuPh is an active catalyst for the aminoarylation reaction. However, when this complex is employed in stoichiometric quantities and in the absence of an arylboronic acid, no phenylated product was formed (Scheme 3.71). Addition of arylboronic acid gives rise to the

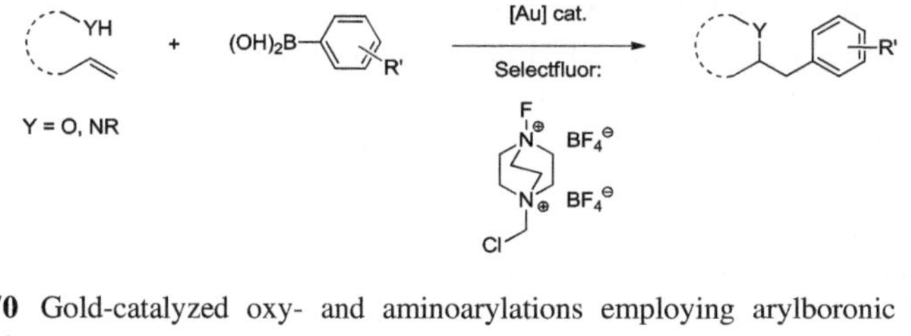

Scheme 3.70 Gold-catalyzed oxy- and aminoarylations employing arylboronic acids as the coupling partner

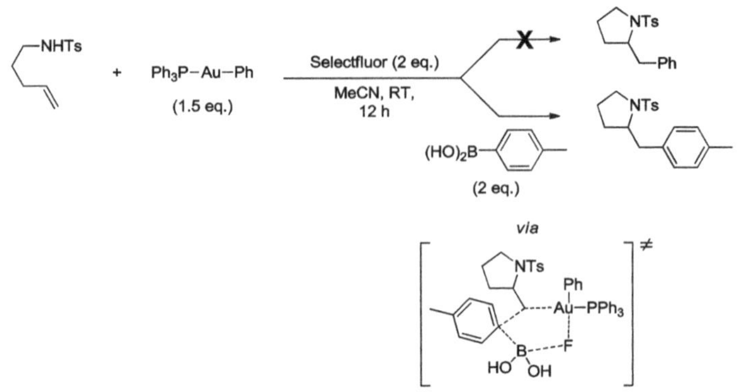

Scheme 3.71 Stoichiometric study of the gold-mediated aminoarylation developed by Toste et al.: [Au(Ph)(PPh_3)] alone is not capable to arylate the aminoalkene. Arylation proceeds however in presence of arylboronic acids

arylated product. These findings led the authors to the conclusion that the CC bond formation occurs likely via a bimolecular outer-sphere mechanism, instead of a transmetallation/reductive elimination pathway. A related study concerning the reactivity of an (NHC) methylgold(III) difluoride complex gives these assumptions further weight (see Sect. 3.4.2).

A subsequent experimental and computational analysis by Toste and Goddard [175] further corroborated this observation and helped to establish a more complete mechanistic picture. In contrast to the mechanistic proposal of Zhang et al. (Scheme 3.69), it was concluded that (i) activation of the alkene/alkyne occurs after oxidation of gold(I) to gold(III), (ii) CC and BF bond formation proceeds indeed via a concerted, asynchronous mechanism, and (iii) Au(II)–Au(II) intermediates lower significantly the barriers for all of the elementary steps involved in the catalytic cycle and explain the superior catalytic activity of bimolecular complexes in this transformation.

The aromatic coupling reagent is not limited to boronic acid derivatives as shown by the groups of Lloyd-Jones/Russell and Toste [176, 177] who employed arylsilanes or arylsiloxanes together with Selectfluor in the gold-catalyzed oxy-arylation of alkenes. Iodine(III) reagents were found as well to act as the sacrificial oxidant to promote this transformation and allowed for the methoxyarylation of *gem*-disubstituted olefins and styrenes [178].

Consequently, Lloyd-Jones and Russell further enhanced the gold-based redox catalysis by combining catalytic CH activation/homocoupling reported by Tse and Beller (Scheme 3.67) with the activation of arylsilanes. A gold-catalyzed reaction for the coupling of simple arenes with aryltrimethylsilanes was developed (Scheme 3.72) [179].

The mechanism of this reaction was meticulously studied by means of a kinetic analysis and a detailed catalytic cycle was established [180]. A rather simplified representation is represented in Scheme 3.73. It was shown that the active catalytic

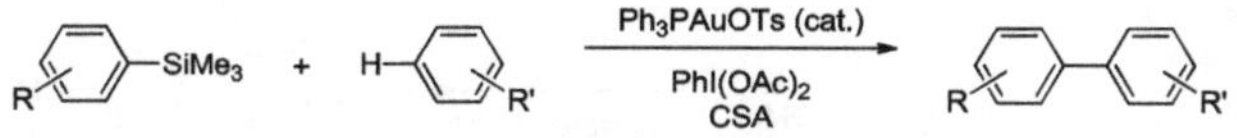

Scheme 3.72 Gold-catalyzed direct arylation

metal species can be considered as ligand-less (i.e. coordinated by the solvent (methanol), camphorsulfonic acid and the respective conjugate bases). The cycle is initiated by electrophilic *ipso*-arylation of the arylsilane ("transmetallation", step A in Scheme 3.73). The rate-limiting step corresponds to the π-coordination of the arene to gold (not explicitly shown below), which is followed by auration (step B). The diarylgold(III) complex subsequently undergoes reductive elimination, liberating the cross-coupled product (step C). Re-oxidation of the gold(I) species regenerates the catalyst (step D).

Heterocoupling is highly favored over homocoupling. This chemoselectivity is rationalized by a pronounced discrimination of the two electrophilic auration steps for CSi and CH activation, respectively. Essential is the lower barrier for auration of the arylsilane (step A) and irreversible formation the π-complex during step B. Arene coordination is favored over arylsilane coordination for the employed arenes. As the π-complex formation in step B is turnover-limiting, the reaction is cross-coupling selective.

A related cross-coupling reaction of two different arenes was established by Larrosa and coworkers [181], though stoichiometric in gold. Based on the orthogonal reactivity of Au(I)/(Au(III) complexes towards CH activation of electron-poor/electron-rich arenes, respectively (see Sect. 3.8) [154, 155, 157], an oxidative cross-coupling protocol was developed (Scheme 3.74). A gold(I) complex in the presence of a base underwent selective CH activation of an electron-poor arene, yielding an arylgold(I) complex. Because the second arylation is not compatible with bases, the reaction media had to be filtered through kieselguhr prior to oxidation of the gold(I) complex with hydroxy(phenyl)iodonium tosylate. Subsequent selective CH activation of the electron-rich arene was proposed to give a bis(aryl)gold(III) complex which undergoes rapidly reductive elimination to yield the biaryl. Virtually no homocoupling products were observed.

Scheme 3.73 Simplified catalytic cycle for the gold-catalyzed direct arylation of arylsilanes with arenes

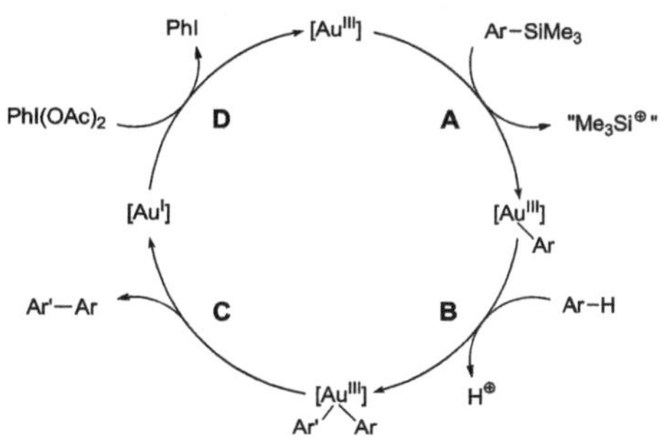

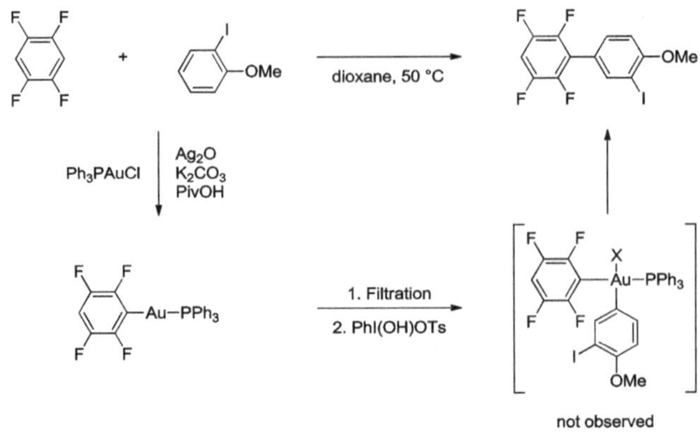

Scheme 3.74 Sequential Au(I)–Au(III) CH bond activation, followed by reductive elimination yielding biphenyls

3.9.3 C(sp^3)–C(sp^2) Cross-Coupling

Toste and coworkers developed very recently a gold-catalyzed C(sp^2)–C(sp^3) cross-coupling reaction which does not require the use of external oxidants. Allyl bromides and arylboronic acids were reacted in the presence of mononuclear or dinuclear gold(I) chloride complexes to give allylated arenes (Scheme 3.75) [182].

Mono- and dinuclear phosphine gold(I) chlorides were found to be efficient catalysts for this cross-coupling reaction. A dinuclear gold complex based on a bis (diphenylphosphino)amine ligand (see Scheme 3.75) proved to be considerably more active than mononuclear catalysts. The precise origin for the beneficial influence of the bimetallic catalyst on the reaction remained unclear.

The proposed catalytic cycle involves first transmetallation of the aryl moiety from the boronic acid to a gold(I) species (Scheme 3.76, step A). The formed arylgold(I) complex undergoes subsequently (S_N2-type) oxidative addition of allyl bromide affording a gold(III) complex (step B). Reductive elimination closes the cycle with formation of the allylated arene and regeneration of the initial gold(I) species.

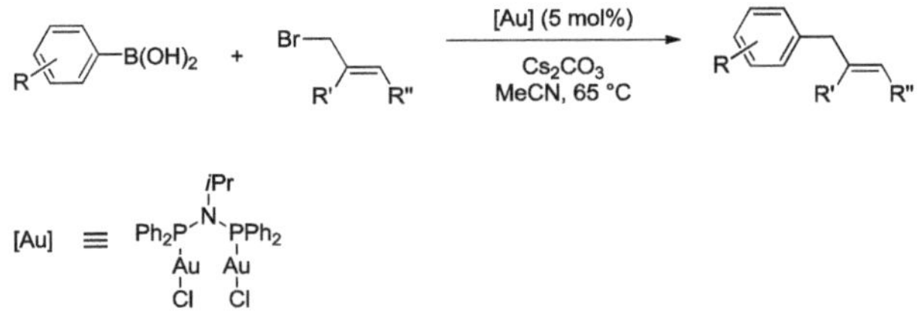

Scheme 3.75 Gold-catalyzed C(sp^3)–C(sp^2) cross-coupling

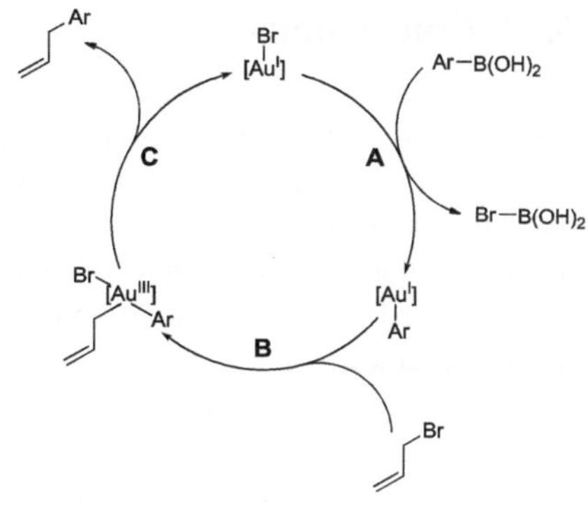

Scheme 3.76 Proposed catalytic cycle for the gold-catalyzed cross-coupling of arylboronic acids with allyl bromides

To corroborate the viability of the proposed mechanism, stoichiometric studies were conducted with a test substrate featuring both the allyl bromide and the arylboronic acid functionality (Scheme 3.77). The synthesis of the arylgold(I) intermediate complex was achieved by reaction of an NHC-gold(I) hydroxide with the substrate, supporting a mechanism in which transmetallation occurs before oxidative addition. Heating the arylgold(I) complex in acetonitrile solution led to the formation of the (aryl)allylgold(III) compound via oxidative addition of the CBr bond. The formed auracycle did not undergo spontaneously CC bond forming reductive elimination, unlike for the catalytic reaction. This step had to be initiated by abstraction of the bromide on gold by addition of $AgSbF_6$, pointing at more facile reductive elimination from a three coordinated intermediate.

The presented reaction proved to be fully selective for the oxidative addition of allyl halides. Oxidative addition of C(sp^2)-I bonds in substrates bearing an aryl halide was not observed.

The developed catalytic protocol demonstrates that gold-catalyzed C(sp^2)–C (sp^3) cross-coupling reactions are feasible without the need of sacrificial oxidants to achieve the gold(I)–gold(III) transition.

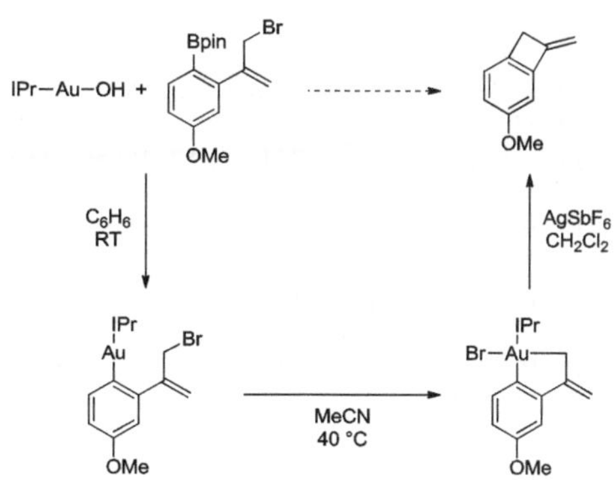

Scheme 3.77 Intramolecular gold-mediated transmetallation/oxidative addition/reductive elimination sequence

3.10 Conclusion

The here presented bibliographic overview shows that gold differs in many ways from other transition metals, especially from the neighboring group-10 metals. Despite sometimes unusual conditions or reagents having to be employed, many of the elementary steps commonly encountered in organometallic chemistry were demonstrated (i.e. oxidative addition, reductive elimination, arylation and transmetallation).

Migratory insertion and oxidative addition—although demonstrated in principle—still lack answers to many fundamental questions. For example, an unambiguous example for the migratory *syn* insertion at gold complexes is still missing and the oxidative addition of unpolar reagents or aryl halides to gold(I) merits further investigation.

The reader may be referred to the following chapters where it will be shown that gold complexes offer indeed still latent and unexplored reactivities concerning these two elementary steps.

References

1. A.S.K. Hashmi, Chem. Rev. **107**, 3180–3211 (2007)
2. A.S.K. Hashmi, Angew. Chem. Int. Ed. **49**, 5232–5241 (2010)
3. A.M. Echavarren, C. Obradors, Chem. Commun. **50**, 16–28 (2014)
4. L.-P. Liu, G.B. Hammond, Chem. Soc. Rev. **41**, 3129 (2012)
5. R.E.M. Brooner, R.A. Widenhoefer, Angew. Chem. Int. Ed. **52**, 11714–11724 (2013)
6. A.S.K. Hashmi, Acc. Chem. Res. **47**, 864–876 (2014)
7. D.S. Weinberger, M. Melaimi, C.E. Moore, A.L. Rheingold, G. Frenking, P. Jerabek, G. Bertrand, Angew. Chem. Int. Ed. **52**, 8964–8967 (2013)
8. N. Mézailles, N. Avarvari, N. Maigrot, L. Ricard, F. Mathey, P. Le Floch, L. Cataldo, T. Berclaz, M. Geoffroy, Angew. Chem. Int. Ed. **38**, 3194–3197 (1999)
9. K. Leary, A. Zalkin, N. Bartlett, Inorg. Chem. **13**, 775–779 (1974)
10. R.V. Parish, Gold Bull. **15**, 51–63 (1982)
11. S.H. Elder, G.M. Lucier, F.J. Hollander, N. Bartlett, J. Am. Chem. Soc. **119**, 1020–1026 (1997)
12. S. Seidel, K. Seppelt, Science **290**, 117–118 (2000)
13. M.A. Carvajal, J.J. Novoa, S. Alvarez, J. Am. Chem. Soc. **126**, 1465–1477 (2004)
14. A.F. Holleman, E. Wiberg, N. Wiberg, *Lehrbuch der anorganischen Chemie* (De Gruyter, Berlin, 1985)
15. W.T. Robinson, E. Sinn, J. Chem. Soc. Dalton Trans. 726–731 (1975)
16. C.J. O'Connor, E. Sinn, Inorg. Chem. **17**, 2067–2071 (1978)
17. J. Vicente, M.T. Chicote, M.D. Bermúdez, P. G. Jones, C. Fittschen, G. M. Sheldrick, J. Chem. Soc. Dalton Trans. 2361–2366 (1986)
18. P. Pyykko, J.P. Desclaux, Acc. Chem. Res. **12**, 276–281 (1979)
19. P. Pyykkö, Angew. Chem. Int. Ed. **43**, 4412–4456 (2004)
20. D.J. Gorin, F.D. Toste, Nature **446**, 395–403 (2007)
21. F. Mohr, *Gold Chemistry: Applications and Future Directions in the Life Sciences* (Wiley-VCH, Weinheim, 2009)
22. H.-C. Tai, I. Krossing, M. Seth, D.V. Deubel, Organometallics **23**, 2343–2349 (2004)

23. S.G. Bratsch, J. Phys. Chem. Ref. Data **18**, 1–21 (1989)
24. J. Barrett, *Structure and Bonding* (Royal Society Of Chemistry, Cambridge, 2001)
25. B. Cordero, V. Gómez, A.E. Platero-Prats, M. Revés, J. Echeverría, E. Cremades, F. Barragán, S. Alvarez, Dalton Trans. 2832–2838 (2008)
26. A.L. Allred, J. Inorg. Nucl. Chem. **17**, 215–221 (1961)
27. P. Schwerdtfeger, J. Am. Chem. Soc. **111**, 7261–7262 (1989)
28. P. Schwerdtfeger, P.D.W. Boyd, S. Brienne, A.K. Burrell, Inorg. Chem. **31**, 3411–3422 (1992)
29. W. Nakanishi, M. Yamanaka, E. Nakamura, J. Am. Chem. Soc. **127**, 1446–1453 (2005)
30. A.S.K. Hashmi, C. Lothschütz, R. Döpp, M. Ackermann, J. De Buck Becker, M. Rudolph, C. Scholz, F. Rominger, Adv. Synth. Catal. **354**, 133–147 (2012)
31. J.J. Hirner, Y. Shi, S.A. Blum, Acc. Chem. Res. **44**, 603–613 (2011)
32. F.G. Mann, D. Purdie, J. Chem. Soc. Resumed 1235 (1940)
33. D. Schneider, A. Schier, H. Schmidbaur, Dalton Trans. 1995–2005 (2004)
34. D. Schneider, O. Schuster, H. Schmidbaur, Dalton Trans. 1940 (2005)
35. D. Schneider, O. Schuster, H. Schmidbaur, Organometallics **24**, 3547–3551 (2005)
36. P. de Frémont, R. Singh, E.D. Stevens, J.L. Petersen, S.P. Nolan, Organometallics **26**, 1376–1385 (2007)
37. J. Vicente, A. Arcas, M. Mora, X. Solans, M. Font-Altaba, J. Organomet. Chem. **309**, 369–378 (1986)
38. V.J. Scott, J.A. Labinger, J.E. Bercaw, Organometallics **29**, 4090–4096 (2010)
39. H. Schmidbaur, R. Franke, Angew. Chem. Int. Ed. Engl. **12**, 416–417 (1973)
40. H. Schmidbaur, R. Franke, Inorganica Chim. Acta **13**, 85–89 (1975)
41. A. Tamaki, J.K. Kochi, J. Organomet. Chem. **40**, C81–C84 (1972)
42. A. Shiotani, H. Schmidbaur, J. Organomet. Chem. **37**, C24–C26 (1972)
43. A. Tamaki, J.K. Kochi, J. Organomet. Chem. **64**, 411–425 (1974)
44. A. Johnson, R. Puddephatt, Inorg. Nucl. Chem. Lett. **9**, 1175–1177 (1973)
45. A. Johnson, R.J. Puddephatt, J. Organomet. Chem. **85**, 115–121 (1975)
46. J.F. Hartwig, *Organotransition Metal Chemistry: From Bonding to Catalysis* (University Science Books, Sausalito, 2009)
47. A. Laguna, M. Laguna, Coord. Chem. Rev. **193–195**, 837–856 (1999)
48. J.P. Fackler, Inorg. Chem. **41**, 6959–6972 (2002)
49. J.P. Fackler, Polyhedron **16**, 1–17 (1997)
50. P. Jandik, U. Schubert, H. Schmidbaur, Angew. Chem. Int. Ed. Engl. **21**, 73 (1982)
51. M.S. Winston, W.J. Wolf, F.D. Toste, J. Am. Chem. Soc. **136**, 7777–7782 (2014)
52. A. Davison, D.V. Howe, E.T. Shawl, Inorg. Chem. **6**, 458–463 (1967)
53. J.H. Enemark, J.A. Ibers, Inorg. Chem. **7**, 2636–2642 (1968)
54. R.E. Bachman, S.A. Bodolosky-Bettis, C.J. Pyle, M.A. Gray, J. Am. Chem. Soc. **130**, 14303–14310 (2008)
55. P. Gualco, S. Ladeira, K. Miqueu, A. Amgoune, D. Bourissou, Angew. Chem. Int. Ed. **50**, 8320–8324 (2011)
56. N. Lassauque, P. Gualco, S. Mallet-Ladeira, K. Miqueu, A. Amgoune, D. Bourissou, J. Am. Chem. Soc. **135**, 13827–13834 (2013)
57. P. Gualco, S. Ladeira, K. Miqueu, A. Amgoune, D. Bourissou, Organometallics **31**, 6001–6004 (2012)
58. C. González-Arellano, A. Corma, M. Iglesias, F. Sánchez, J. Catal. **238**, 497–501 (2006)
59. S. Carrettin, J. Guzman, A. Corma, Angew. Chem. Int. Ed. **44**, 2242–2245 (2005)
60. C. González-Arellano, A. Corma, M. Iglesias, F. Sánchez, Chem. Commun. 1990–1992 (2005)
61. C. González-Arellano, A. Abad, A. Corma, H. García, M. Iglesias, F. Sánchez, Angew. Chem. Int. Ed. **46**, 1536–1538 (2007)
62. P. Li, L. Wang, M. Wang, F. You, Eur. J. Org. Chem. **2008**, 5946–5951 (2008)
63. T. Lauterbach, M. Livendahl, A. Rosellón, P. Espinet, A.M. Echavarren, Org. Lett. **12**, 3006–3009 (2010)

64. A. Corma, R. Juárez, M. Boronat, F. Sánchez, M. Iglesias, H. García, Chem. Commun. **47**, 1446 (2011)
65. M. Boronat, A. Leyva-Pérez, A. Corma, Acc. Chem. Res. **47**, 834–844 (2014)
66. P.S.D. Robinson, G.N. Khairallah, G. da Silva, H. Lioe, R.A.J. O'Hair, Angew. Chem. Int. Ed. **51**, 3812–3817 (2012)
67. J. Guenther, S. Mallet-Ladeira, L. Estevez, K. Miqueu, A. Amgoune, D. Bourissou, J. Am. Chem. Soc. **136**, 1778–1781 (2014)
68. A.M. Echavarren, M. Livendahl, C. Goehry, F. Maseras, Chem. Commun. **50**, 1533–1536 (2014)
69. A. Burawoy, C.S. Gibson, J. Chem. Soc. Resumed 860 (1934)
70. A. Burawoy, C.S. Gibson, J. Chem. Soc. Resumed 219 (1935)
71. N.P. Mankad, F.D. Toste, Chem. Sci. **3**, 72–76 (2012)
72. M. Aresta, G. Vasapollo, J. Organomet. Chem. **50**, C51–C53 (1973)
73. R. Usón, A. Laguna, J. Pardo, Synth. React. Inorg. Met.-Org. Chem. **4**, 499–513 (1974)
74. R. Usón, A. Laguna, J. Vicente, J. Organomet. Chem. **86**, 415–421 (1975)
75. M.J. Ghidiu, A.J. Pistner, G.P.A. Yap, D.A. Lutterman, J. Rosenthal, Organometallics **32**, 5026–5029 (2013)
76. M.J. Ghidiu, A.J. Pistner, G.P.A. Yap, D.A. Lutterman, J. Rosenthal, Organometallics **32**, 6628 (2013)
77. S. Lavy, J.J. Miller, M. Pažický, A.-S. Rodrigues, F. Rominger, C. Jäkel, D. Serra, N. Vinokurov, M. Limbach, Adv. Synth. Catal. **352**, 2993–3000 (2010)
78. H. Schmidbaur, C. Hartmann, J. Riede, B. Huber, G. Mueller, Organometallics **5**, 1652–1656 (1986)
79. M.A. Bennett, S.K. Bhargava, K.D. Griffiths, G.B. Robertson, Angew. Chem. Int. Ed. Engl. **26**, 260–261 (1987)
80. M.A. Bennett, S.K. Bhargava, D.C.R. Hockless, L.L. Welling, A.C. Willis, J. Am. Chem. Soc. **118**, 10469–10478 (1996)
81. M.A. Bennett, D.C.R. Hockless, A.D. Rae, L.L. Welling, A.C. Willis, Organometallics **20**, 79–87 (2001)
82. F.H. Brain, C.S. Gibson, J. Chem. Soc. Resumed 762–767 (1939)
83. S. Komiya, J.K. Kochi, J. Am. Chem. Soc. **98**, 7599–7607 (1976)
84. S. Komiya, T.A. Albright, R. Hoffmann, J.K. Kochi, J. Am. Chem. Soc. **98**, 7255–7265 (1976)
85. L.P. Kuch, T.R. Stuart, J. Organomet. Chem. **122**, 429–446 (1976)
86. S. Komiya, A. Shibue, Organometallics **4**, 684–687 (1985)
87. S. Komiya, S. Ozaki, A. Shibue, J. Chem. Soc. Chem. Commun. 1555–1556 (1986)
88. N.P. Mankad, F.D. Toste, J. Am. Chem. Soc. **132**, 12859–12861 (2010)
89. E.C. Constable, L.R. Sousa, J. Organomet. Chem. **427**, 125–139 (1992)
90. E.C. Constable, R.P.G. Henney, D.A. Tocher, J. Chem. Soc. Dalton Trans. 2467–2474 (1992)
91. Y. Fuchita, H. Ieda, S. Wada, S. Kameda, M. Mikuriya, J. Chem. Soc. Dalton Trans. 4431–4435 (1999)
92. F. Zamora, P. Amo-Ochoa, B. Fischer, A. Schimanski, B. Lippert, Angew. Chem. Int. Ed. **38**, 2274–2275 (1999)
93. A.K. Sahoo, Y. Nakamura, N. Aratani, K.S. Kim, S.B. Noh, H. Shinokubo, D. Kim, A. Osuka, Org. Lett. **8**, 4141–4144 (2006)
94. M.T. Johnson, J.M.J. van Rensburg, M. Axelsson, M.S.G. Ahlquist, O.F. Wendt, Chem. Sci. **2**, 2373–2377 (2011)
95. M. Hofer, C. Nevado, Tetrahedron **69**, 5751–5757 (2013)
96. M. Hofer, E. Gomez-Bengoa, C. Nevado, Organometallics **33**, 1328–1332 (2014)
97. J. Vicente, M. Dolores Bermudez, J. Escribano, Organometallics **10**, 3380–3384 (1991)
98. J. Vicente, M.D. Bermúdez, F.J. Carrión, Inorganica Chim. Acta **220**, 1–3 (1994)
99. J. Vicente, M.D. Bermúdez, J. Escribano, M.P. Carrillo, P.G. Jones, J. Chem. Soc. Dalton Trans. 3083 (1990)

100. W.J. Wolf, M.S. Winston, F.D. Toste, Nat. Chem. **6**, 159–164 (2014)
101. A.S.K. Hashmi, D.F. Toste, *Modern Gold Catalyzed Synthesis* (Wiley-VCH, Weinheim, 2012)
102. E. Mizushima, T. Hayashi, M. Tanaka, Org. Lett. **5**, 3349–3352 (2003)
103. V. Lavallo, G.D. Frey, B. Donnadieu, M. Soleilhavoup, G. Bertrand, Angew. Chem. Int. Ed. **47**, 5224–5228 (2008)
104. R. Kinjo, B. Donnadieu, G. Bertrand, Angew. Chem. Int. Ed. **50**, 5560–5563 (2011)
105. K.D. Hesp, M. Stradiotto, J. Am. Chem. Soc. **132**, 18026–18029 (2010)
106. N. Nishina, Y. Yamamoto, Angew. Chem. Int. Ed. **45**, 3314–3317 (2006)
107. N. Nishina, Y. Yamamoto, Synlett **2007**, 1767–1770 (2007)
108. X. Zeng, M. Soleilhavoup, G. Bertrand, Org. Lett. **11**, 3166–3169 (2009)
109. J.H. Teles, S. Brode, M. Chabanas, Angew. Chem. Int. Ed. **37**, 1415–1418 (1998)
110. C.M. Krauter, A.S.K. Hashmi, M. Pernpointner, ChemCatChem **2**, 1226–1230 (2010)
111. A. Johnson, R.J. Puddephatt, J.L. Quirk, J. Chem. Soc. Chem. Commun. 938b–939 (1972)
112. J.A.J. Jarvis, A. Johnson, R.J. Puddephatt, J. Chem. Soc. Chem. Commun. 373–374 (1973)
113. M.A. Bennett, K. Hoskins, W.R. Kneen, R.S. Nyholm, P.B. Hitchcock, R. Mason, G.B. Robertson, A.D.C. Towl, J. Am. Chem. Soc. **93**, 4591–4592 (1971)
114. M.A. Bennett, J. Organomet. Chem. **300**, 7–19 (1986)
115. J.A. Akana, K.X. Bhattacharyya, P. Müller, J.P. Sadighi, J. Am. Chem. Soc. **129**, 7736–7737 (2007)
116. E.Y. Tsui, P. Müller, J.P. Sadighi, Angew. Chem. Int. Ed. **47**, 8937–8940 (2008)
117. G. Klatt, R. Xu, M. Pernpointner, L. Molinari, T. Quang Hung, F. Rominger, A.S.K. Hashmi, H. Köppel, Chem.—Eur. J. **19**, 3954–3961 (2013)
118. Z.J. Wang, D. Benitez, E. Tkatchouk, W.A. Goddard III, F.D. Toste, J. Am. Chem. Soc. **132**, 13064–13071 (2010)
119. R.L. LaLonde, J. William E. Brenzovich, D. Benitez, E. Tkatchouk, K. Kelley, I. I. I. William A. Goddard, F. D. Toste, Chem. Sci., **1**, 226–233 (2010)
120. M.W. Johnson, S.L. Shevick, F.D. Toste, R.G. Bergman, Chem. Sci. **4**, 1023–1027 (2013)
121. G. Kovács, A. Lledós, G. Ujaque, Angew. Chem. Int. Ed. **50**, 11147–11151 (2011)
122. D.-A. Roşca, D.A. Smith, D.L. Hughes, M. Bochmann, Angew. Chem. Int. Ed. **51**, 10643–10646 (2012)
123. M.A. Cinellu, G. Minghetti, F. Cocco, S. Stoccoro, A. Zucca, M. Manassero, Angew. Chem. Int. Ed. **44**, 6892–6895 (2005)
124. F. Mohr, L.R. Falvello, M. Laguna, Eur. J. Inorg. Chem. **2006**, 833–838 (2006)
125. M. Peña-López, M. Ayán-Varela, L.A. Sarandeses, J.P. Sestelo, Org. Biomol. Chem. **10**, 1686–1694 (2012)
126. A.S.K. Hashmi, T.D. Ramamurthi, F. Rominger, J. Organomet. Chem. **694**, 592–597 (2009)
127. Y. Shi, S.A. Blum, Organometallics **30**, 1776–1779 (2011)
128. Y. Shi, S.D. Ramgren, S.A. Blum, Organometallics **28**, 1275–1277 (2009)
129. T.P. Cornell, Y. Shi, S.A. Blum, Organometallics **31**, 5990–5993 (2012)
130. H. Ye, Z. Lu, D. You, Z. Chen, Z.H. Li, H. Wang, Angew. Chem. Int. Ed. **51**, 12047–12050 (2012)
131. J.J. Hirner, S.A. Blum, Organometallics **30**, 1299–1302 (2011)
132. A. Corma, A. Leyva-Pérez, M.J. Sabater, Chem. Rev. **111**, 1657–1712 (2011)
133. H. Schmidbaur, H.G. Raubenheimer, L. Dobrzańska, Chem. Soc. Rev. **43**, 345–380 (2013)
134. A. Sladek, S. Hofreiter, M. Paul, H. Schmidbaur, J. Organomet. Chem. **501**, 47–51 (1995)
135. J.M. Forward, J.P. Fackler, R.J. Staples, Organometallics **14**, 4194–4198 (1995)
136. S.G. Weber, D. Zahner, F. Rominger, B.F. Straub, Chem. Commun. **48**, 11325–11327 (2012)
137. D.V. Partyka, M. Zeller, A.D. Hunter, T.G. Gray, Angew. Chem. Int. Ed. **45**, 8188–8191 (2006)
138. D.V. Partyka, M. Zeller, A.D. Hunter, T.G. Gray, Inorg. Chem. **51**, 8394–8401 (2012)
139. H.K. Lenker, T.G. Gray, R.A. Stockland, Dalton Trans. **41**, 13274 (2012)

140. D.V. Partyka, A.J. Esswein, M. Zeller, A.D. Hunter, T.G. Gray, Organometallics **26**, 3279–3282 (2007)
141. D.V. Partyka, J.B. Updegraff, M. Zeller, A.D. Hunter, T.G. Gray, Organometallics **28**, 1666–1674 (2009)
142. L. Gao, M.A. Peay, D.V. Partyka, J.B. Updegraff, T.S. Teets, A.J. Esswein, M. Zeller, A.D. Hunter, T.G. Gray, Organometallics **28**, 5669–5681 (2009)
143. D.Y. Melgarejo, G.M. Chiarella, J.P. Fackler, Acta Crystallogr. Sect. C **65**, m299–m301 (2009)
144. S. Dupuy, L. Crawford, M. Bühl, A.M.Z. Slawin, S.P. Nolan, Adv. Synth. Catal. **354**, 2380–2386 (2012)
145. G. Seidel, C.W. Lehmann, A. Fürstner, Angew. Chem. Int. Ed. **49**, 8466–8470 (2010)
146. J.E. Heckler, M. Zeller, A.D. Hunter, T.G. Gray, Angew. Chem. Int. Ed. **51**, 5924–5928 (2012)
147. S. Dupuy, A.M.Z. Slawin, S.P. Nolan, Chem.—Eur. J. **18**, 14923–14928 (2012)
148. R. Usón, J. Vicente, J.A. Cirac, M.T. Chicote, J. Organomet. Chem. **198**, 105–112 (1980)
149. N. Meyer, C.W. Lehmann, T.K.-M. Lee, J. Rust, V.W.-W. Yam, F. Mohr, Organometallics **28**, 2931–2934 (2009)
150. N. Meyer, S. Sivanathan, F. Mohr, J. Organomet. Chem. **696**, 1244–1247 (2011)
151. Y. Chen, M. Chen, Y. Liu, Angew. Chem. Int. Ed. **51**, 6181–6186 (2012)
152. J. delPozo, D. Carrasco, M.H. Pérez-Temprano, M. García-Melchor, R. Álvarez, J.A. Casares, P. Espinet. Angew, Chem. Int. Ed. **52**, 2189–2193 (2013)
153. Z. Li, C. Brouwer, C. He, Chem. Rev. **108**, 3239–3265 (2008)
154. T.C. Boorman, I. Larrosa, Chem. Soc. Rev. **40**, 1910–1925 (2011)
155. T. de Haro, C. Nevado, Synthesis **2011**, 2530–2539 (2011)
156. Y. Fuchita, Y. Utsunomiya, M. Yasutake, J. Chem. Soc. Dalton Trans. 2330–2334 (2001)
157. P. Lu, T.C. Boorman, A.M.Z. Slawin, I. Larrosa, J. Am. Chem. Soc. **132**, 5580–5581 (2010)
158. H. Ito, K. Takagi, T. Miyahara, M. Sawamura, Org. Lett. **7**, 3001–3004 (2005)
159. H. Ito, T. Saito, T. Miyahara, C. Zhong, M. Sawamura, Organometallics **28**, 4829–4840 (2009)
160. S. Labouille, A. Escalle-Lewis, Y. Jean, N. Mézailles, P. Le Floch, Chem.—Eur. J. **17**, 2256–2265 (2011)
161. J.-H. Zhan, H. Lv, Y. Yu, J.-L. Zhang, Adv. Synth. Catal. **354**, 1529–1541 (2012)
162. H. Lv, J.-H. Zhan, Y.-B. Cai, Y. Yu, B. Wang, J.-L. Zhang, J. Am. Chem. Soc. **134**, 16216–16227 (2012)
163. A. Kar, N. Mangu, H.M. Kaiser, M. Beller, M.K. Tse, Chem. Commun. 386–388 (2008)
164. A. Kar, N. Mangu, H.M. Kaiser, M.K. Tse, J. Organomet. Chem. **694**, 524–537 (2009)
165. A.S.K. Hashmi, M.C. Blanco, D. Fischer, J.W. Bats, Eur. J. Org. Chem. **2006**, 1387–1389 (2006)
166. H.A. Wegner, S. Ahles, M. Neuburger, Chem.—Eur. J. **14**, 11310–11313 (2008)
167. M. Auzias, M. Neuburger, H. Wegner, Synlett 2443–2448 (2010)
168. A. Iglesias, K. Muñiz, Chem.—Eur. J. **15**, 10563–10569 (2009)
169. Y. Peng, L. Cui, G. Zhang, L. Zhang, J. Am. Chem. Soc. **131**, 5062–5063 (2009)
170. L. Cui, G. Zhang, L. Zhang, Bioorg. Med. Chem. Lett. **19**, 3884–3887 (2009)
171. G. Zhang, Y. Peng, L. Cui, L. Zhang, Angew. Chem. Int. Ed. **48**, 3112–3115 (2009)
172. G. Zhang, L. Cui, Y. Wang, L. Zhang, J. Am. Chem. Soc. **132**, 1474–1475 (2010)
173. A.D. Melhado, W.E. Brenzovich, A.D. Lackner, F.D. Toste, J. Am. Chem. Soc. **132**, 8885–8887 (2010)
174. W.E. Brenzovich, D. Benitez, A.D. Lackner, H.P. Shunatona, E. Tkatchouk, W.A. Goddard, F.D. Toste, Angew. Chem. Int. Ed. **49**, 5519–5522 (2010)
175. E. Tkatchouk, N.P. Mankad, D. Benitez, W.A. Goddard, F.D. Toste, J. Am. Chem. Soc. **133**, 14293–14300 (2011)

176. L.T. Ball, M. Green, G.C. Lloyd-Jones, C.A. Russell, Org. Lett. **12**, 4724–4727 (2010)
177. W.E. Brenzovich, J.-F. Brazeau, F.D. Toste, Org. Lett. **12**, 4728–4731 (2010)
178. L.T. Ball, G.C. Lloyd-Jones, C.A. Russell, Chem.—Eur. J. **18**, 2931–2937 (2012)
179. L.T. Ball, G.C. Lloyd-Jones, C.A. Russell, Science **337**, 1644–1648 (2012)
180. L.T. Ball, G.C. Lloyd-Jones, C.A. Russell, J. Am. Chem. Soc. **136**, 254–264 (2014)
181. X.C. Cambeiro, T.C. Boorman, P. Lu, I. Larrosa, Angew. Chem. Int. Ed. **52**, 1781–1784 (2013)
182. M.D. Levin, F.D. Toste, Angew. Chem. Int. Ed. **53**, 6211–6215 (2014)

Chapter 4
Gold: Migratory Insertion at Au(I)

It has been established that organogold compounds are mostly inert towards migratory insertion reactions with olefins and alkynes (see Sect. 3.5). By contrast, the analogous reactivity of silylgold complexes has not been explored. In this chapter, the experimental results concerning our studies on the reactivity of AuSi bonds towards alkynes and allenes, as well as a theoretical analysis will be discussed.

4.1 Reactions of Alkynes with Silylgold(I) Complexes

Since significant experience with silylgold complexes stemming from the activation of disilanes was gained in our group [1], we decided to investigate the reactivity of AuSi bonds toward alkynes and to explore the possibility of insertion reactions at Au(I). Silylgold(I) complexes have been synthesized and structurally characterized early on by Schubert [2], but very little is known about the reactivity of the AuSi bond. Klinkhammer studied recently the reactivity of AuSi bonds in a dihypersilylaurate complex towards chlorotrimethylsilane [3]. Particularly noteworthy is the recent contribution of Stratakis and coworkers, reporting on the dehydrogenative addition of dihydrodisilanes and dihydrodisiloxanes to alkynes catalyzed by gold nanoparticles [4, 5]. The postulated mechanistic cycle accounting for the formation of the observed product is shown in Scheme 4.1. A key step of this proposed mechanism is the insertion of an alkyne into a gold-silicon bond, leading to a (silyl)vinylgold intermediate (Step C).

4.1.1 Preliminary Test: Insertion of Methyl Propiolate

In order to survey the feasibility of an analogous insertion reaction employing molecular gold species, we synthesized phosphine silylgold(I) complexes. Sterically demanding substituents are usually necessary to impart reasonable stability to such complexes, and we thus worked on $(Ph_3P)Au\text{–}Si\textit{t}BuPh_2$ complex **8**.

M. Joost, *Synthesis and Original Reactivity of Copper and Gold Complexes*,
Springer Theses, DOI 10.1007/978-3-319-18690-0_4

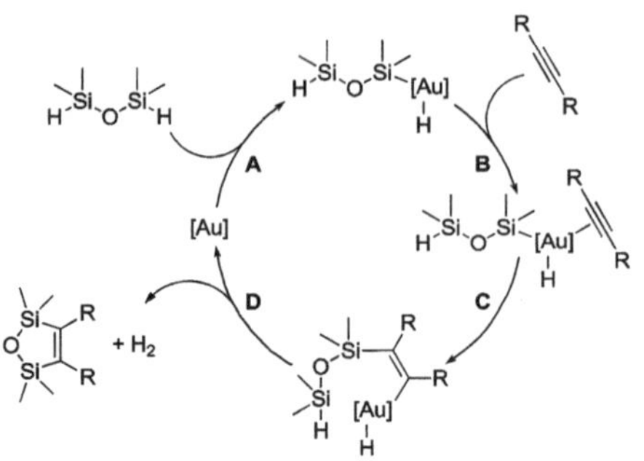

Scheme 4.1 Proposed mechanistic cycle for the dehydrogenative disilylation of alkynes at gold nanoparticles

Following the reported methodology [2, 6], that is coupling the corresponding phosphine gold chloride and lithiosilane, complex **8** was obtained in 91 % yield.

In a preliminary test, its ability to undergo an insertion reaction was evaluated with methyl propiolate (5 eq.) in toluene at 60 °C (Scheme 4.2). $^{31}P\{^{1}H\}$ NMR monitoring indicated progressive and clean transformation of **8** (δ = 56.5 ppm) into a single species **9** displaying a resonance signal at δ = 41.5 ppm, in the typical region for (triphenylphosphine)vinylgold complexes [7]. The reaction reached completion within 8 h. After work-up, complex **9** was isolated as an air-stable white solid (95 % yield) and fully characterized.

Several NMR signals are diagnostic of a (β-silyl)vinylgold complex: The vinylic proton appears in the ^{1}H NMR spectrum at δ = 6.79 ppm (d, $^{4}J_{HP}$ = 17.8 Hz). The ^{29}Si NMR resonance signal is shifted to higher field (δ = 35.4 ppm for **8**, vs. δ = −12.5 ppm for **9**) and the associated J_{SiP} coupling constant decreases drastically (from 164.8 Hz in **8** to 3.9 Hz in **9**). To unambiguously assess the structure of **9**, crystals were grown (from a pentane/toluene solution) and an X-ray diffraction study was performed (Fig. 4.1). Indeed, in agreement with the NMR spectroscopic data, complex **9** in the solid state was found to be a (β-silyl)vinylgold complex. Despite severe steric crowding, the central CC double bond adopts a perfectly planar geometry, with *cis* arrangement of the gold and silicon atoms, and *trans* arrangement of the silyl and CO_2Me moieties.

Complex **9** is formed in a selective manner. The observed regioselectivity is consistent with the polarity of the alkyne and the nucleophilic character of the silyl group at gold. The stereoselectivity indicates that the reaction takes place by *syn*

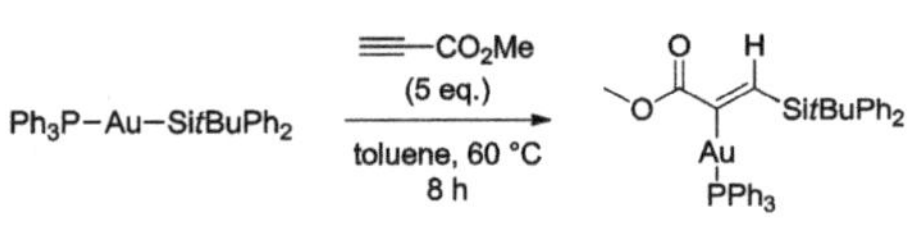

8 **9** (95 %)

Scheme 4.2 *syn* insertion of methyl propiolate in the AuSi bond of silylgold complex **8**

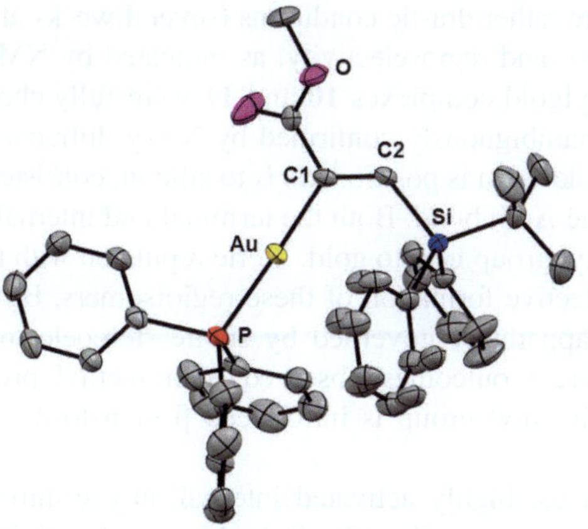

Fig. 4.1 Molecular structure of **9** determined by single crystal X-ray diffraction. Hydrogen atoms are omitted for clarity. Selected bond lengths [Å] and angles [°]: AuC 2.046(3), AuP 2.2878(8), C1C2 1.342(5), PAuC1 176.98(9), AuC1C2 124.8(2), SiC2C1 125.7(3), AuC1C2Si -2.6(4)

insertion of the alkyne into the AuSi bond. This pathway markedly contrasts with the outer-sphere mechanism normally observed upon functionalization of alkynes at gold (see Sect. 3.5).

4.1.2 Scope of the Insertion Reaction with Alkynes

The generality of the insertion process, as observed with methyl propiolate, was then evaluated by treating complex **8** with various substrates. No reaction occurred with ethylene, methyl acrylate or ethyl 2,3-butadienoate even under forcing conditions (solvent-free, pressures up to 10 bar, temperatures up to 80 °C). However, complex **8** was found to react with phenylacetylene and methyl phenylpropiolate (Scheme 4.3).

Scheme 4.3 *syn* insertion of phenylacetylene and methyl phenylpropiolate in the AuSi bond of silylgold complex **8**

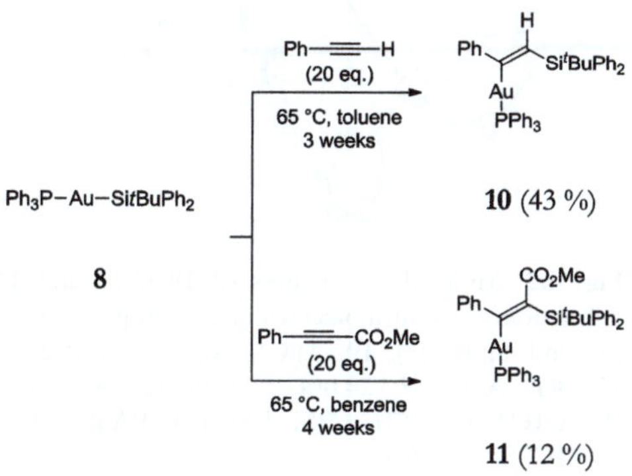

The reactions require rather drastic conditions (several weeks at 65 °C), but proceed with complete stereo- and regioselectivity, as indicated by NMR monitoring.

The ensuing vinylgold complexes **10** and **11** were fully characterized, and their connectivity was unambiguously confirmed by X-ray diffraction studies (Fig. 4.2). In both cases, the gold atom is positioned *cis* to silicon, consistent with *syn* insertion of the alkyne into the AuSi bond. Both the terminal and internal alkynes insert so as to position the phenyl group *gem* to gold. Steric repulsion with the bulky silyl group may explain the selective formation of these regioisomers. But the regioselectivity of the insertion is apparently governed by subtle stereoelectronic effects, as suggested by the different outcomes observed with methyl propiolate and methyl phenylpropiolate (the silyl group is introduced β or α to the CO_2Me substituent, respectively).

Curiously, when the highly activated internal alkyne dimethyl acetylenedicarboxylate (DMAD) was reacted with silylgold complex **8** in toluene at 60 °C, monitoring by $^{31}P\{^1H\}$ revealed the formation of two products. The two species, featuring singlet resonances at δ = 40.5 ppm (**12**) and δ = 16.1 ppm (**13**), respectively, were formed in about equal amounts. While the $^{31}P\{^1H\}$ NMR chemical shift of **12** corresponds well to other (triphenylphosphine)vinylgold complexes as expected for the insertion product of DMAD into the AuSi bond of **8**, the $^{31}P\{^1H\}$ NMR data of **13** are in the region of (vinyl)triphenylphosphonium compounds [8]. Formation of the latter could be explained by insertion of DMAD into the AuP instead of the AuSi bond. The two products were therefore tentatively attributed to the structures shown in Scheme 4.4, where **12** is the expected vinylgold complex stemming from insertion into the AuSi bond and **13** a zwitterionic phosphonium silylaurate complex. Isolation by selective precipitation and characterization by NMR spectroscopy corroborated this hypothesis. The most striking difference in the ^{29}Si NMR spectra of the two products is their chemical shift difference. While the

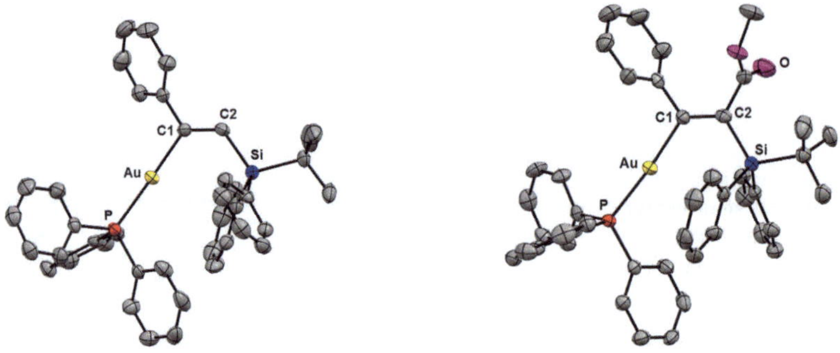

Fig. 4.2 Molecular structures of **10** (*left*) and **11** (*right*) determined by single crystal X-ray diffraction. Solvent molecules and hydrogen atoms are omitted for clarity. Selected bond lengths [Å] and angles [°]: **10**: AuC1 2.059(5), AuP 2.292(1), C1C2 1.332(8), C2Si 1.863(6), PAuC1 176.9(1), AuC1C2 124.0(4), C1C2Si 124.0(4), AuC1C2Si 7.3(7); **11**: AuC1 2.051(5), AuP 2.280 (1), C1C2 1.357(7), C2Si 1.895(5), PAuC1 174.1(1), AuC1C2 124.2(3), C1C2Si 119.3(3), AuC1C2Si -13.3(6)

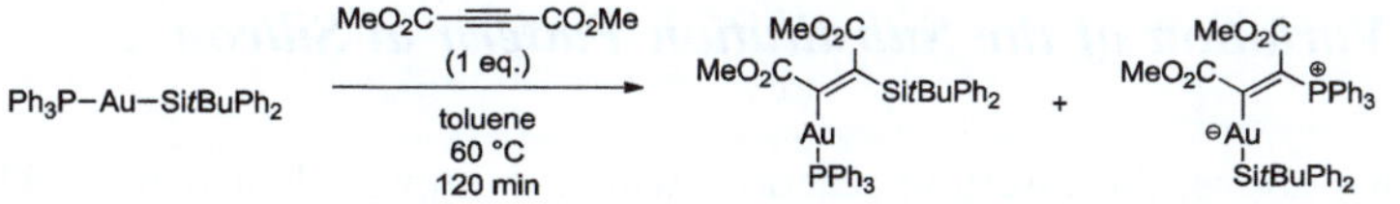

8 **12** (35 %) **13** (28 %)

Scheme 4.4 *syn* insertion of DMAD into the AuSi and AuP bond of complex **8**, respectively

expected highfield shift is observed for **12** (from δ = 35.4 ppm in **8** to δ = −7.7 ppm in **12**), the ^{29}Si NMR chemical shift is for **13** at δ = 31.4 ppm in the range of silylgold complexes, in accordance with the proposed zwitterionic structure, featuring a Si atom directly bound to gold.

In order to substantiate the attributed structures, single crystals of both products were grown by slow diffusion of a pentane/benzene mixture in saturated dichloromethane solutions of the respective compound and X-ray diffraction analyses were performed. The solid state structures of **12** and **13** (Fig. 4.3) are in agreement with the NMR spectroscopic data and confirm indeed the proposed connectivity.

A notable feature of these structures is the *cis* configuration at the central CC double bond adopted by both **12** and **13**. The geometry around gold is quasi-linear for the two complexes (**12**: PAuC1 176.86(7)°, **13**: SiAuC2 170.05(4)°). The AuSi distance in complex **13** (2.3487(4) Å) is comparable to those seen previously (2.23–2.36 Å) [2, 6, 9–11].

The peculiar reactivity of DMAD giving rise to **12** and **13** was investigated computationally (vide infra) and two competing mechanisms are proposed to account for the formation of the two products.

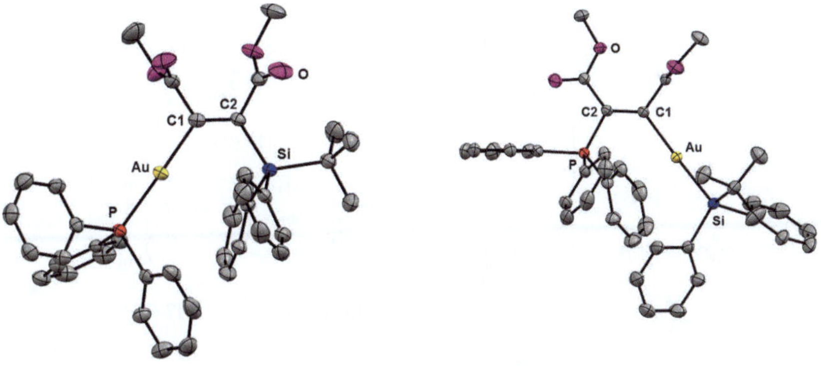

Fig. 4.3 Molecular structures of **12** (*left*) and **13** (*right*) determined by single crystal X-ray diffraction. Solvent molecules and hydrogen atoms are omitted for clarity. Selected bond lengths [Å] and angles [°]: **12**: AuC1 2.054(3), AuP 2.2811(7), C1C2 1.345(4), C2Si 1.900(3), PAuC1 176.86(7), AuC1C2 125.3(2), C1C2Si 120.5(2), AuC1C2Si 7.3(3); **13**: SiAu 2.3487(4), AuC1 2.119(1), C1C2 1.350(2), C2P 1.800(1), SiAuC2 170.05(4), AuC1C2 126.8(1), C1C2P 116.4(1), AuC1C2P -14.2(2)

4.1.3 Variation of the Substitution Pattern at Silicon

As shown above, the insertion reaction, which is unprecedented for gold(I) complexes, gives access readily to (β-silyl)vinylgold complexes in a highly regio- and stereochemically manner. However, the long reaction times observed for internal alkynes and the lack of reactivity towards allenes prompted us to vary the substitution pattern around Si with the objective of facilitating the insertion reaction by decrease of the steric bulk around gold.

Therefore, the Si*t*BuPh$_2$ group was replaced by SiPh$_3$. The corresponding gold silyl complex **14** [6] was also found to react readily with methyl propiolate, yielding selectively the product of *syn* insertion (**15**, 93 % isolated yield). Gratifyingly, complex **14** proved far more reactive than **8** towards methyl phenylpropiolate (Scheme 4.5).

At 60 °C, the insertion proceeded to completion within 5 days (instead of 3 weeks) and gave two vinylgold complexes **16** and **17** (~ 1:1 ratio according to $^{31}P\{^1H\}$ NMR) that were separated by column chromatography. Their respective structures were unequivocally assigned by multi-nuclear NMR spectroscopy and X-ray diffraction analysis (for **17**, see Fig. 4.4).

They both result from *syn* insertion of the internal alkyne into the Au–Si bond of **14** (*cis* arrangement of the Au and Si atoms) and correspond to the two conceivable regioisomers, with the Ph group *gem* to Au (**16**) or *gem* to Si (**17**). Finally, the increase in reactivity observed upon lowering the steric shielding around silicon prompted us to reconsider the insertion of allenes. Satisfyingly, complex **14** was found to react indeed with ethyl 2,3-butadienoate at 60 °C giving complex **18** as sole product. As shown by crystallography, the insertion involves the terminal C = C bond of the allene and the silyl group is introduced in γ position to the CO$_2$Et substituent (Fig. 4.5).

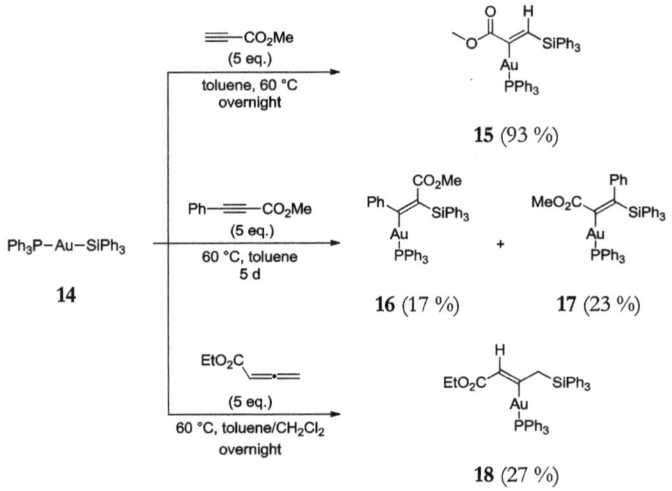

Scheme 4.5 Alkyne and allene *syn* insertion in the AuSi bond of complex **14**

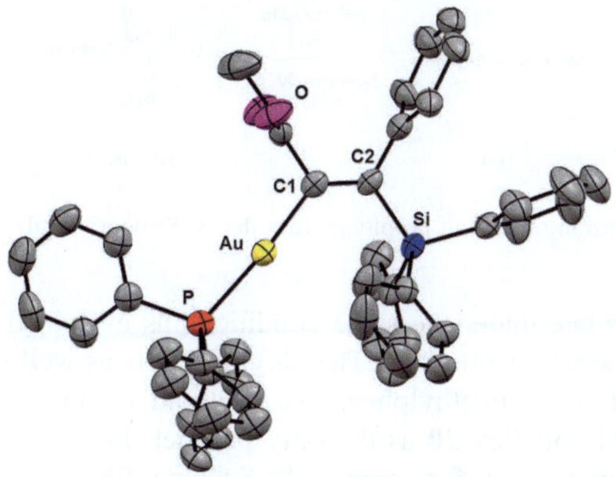

Fig. 4.4 Molecular structure of **17** determined by single crystal X-ray diffraction. Solvent molecules and hydrogen atoms are omitted for clarity. Selected bond lengths [Å] and angles [°]: AuC1 2.055(6), AuP 2.283(2), C1C2 1.33(1), C2Si 1.871(7), PAuC1 174.5(2), AuC1C2 128.6(5), C1C2Si 121.1(5), AuC1C2Si 0.3(8)

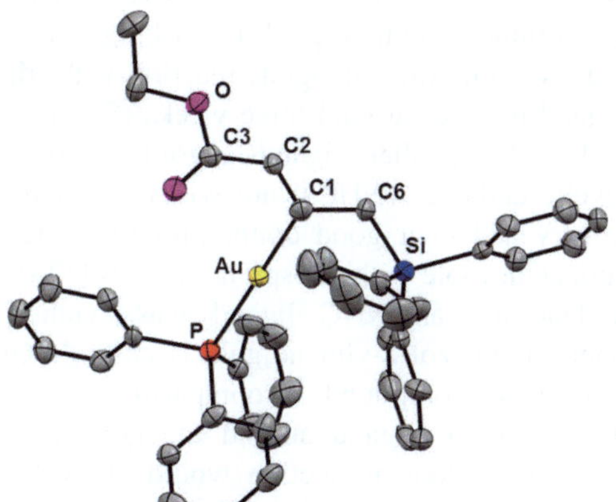

Fig. 4.5 Molecular structure of **18** determined by single crystal X-ray diffraction. Solvent molecules and hydrogen atoms are omitted for clarity. Selected bond lengths [Å] and angles [°]: AuC1 2.041(7), AuP 2.2978(3), C1C2 1.36(1), C2C3 1.48(1), C1C6 1.51(1), C6Si 1.891(8), PAuC1 179.4(2), AuC1C2 122.8(5), C1C2C3 123.6(6)

4.1.4 Influence of the Neutral Donor Ligand

Our attention was drawn as well to the influence of the neutral donor ligand at gold. We changed the phosphine ligand in order to check on the reactivity of the corresponding silylgold complexes. The trimethylphosphine complex **19**, $(Me_3P)Au$–Si*t*BuPh$_2$, was synthesized by the typical protocol (i.e. reaction of the lithiosilane with $(Me_3P)Au$–Cl) in 74 % yield. As a first test on its reactivity, **19** was reacted

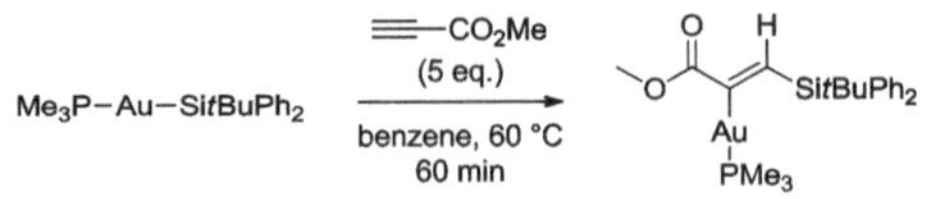

19 **20** (96 %)

Scheme 4.6 *syn* insertion of methyl propiolate into the AuSi bond of silylgold complex **19**

with methyl propiolate under the same conditions as evaluated for the triphenylphosphine complex **8** (Scheme 2.1). The alkyne inserts as well in *syn* fashion into the AuSi bond of the (trimethylphosphine)silylgold complex to give the corresponding vinylgold complex **20** as the only product. Full conversion is achieved within 1 h (compared to 8 h for complex **8**) Scheme 4.6.

Complex **20** was isolated in 96 % yield and fully characterized. The *cis* configuration of the Si and Au atoms is unambiguously established by the characteristic ^{1}H–^{31}P coupling constant of the vinylic proton at $\delta = 8.16$ ppm (d, $^{4}J_{HP} = 18.9$ Hz) in the ^{1}H NMR spectrum of **20**.

Furthermore, silylgold complex **19** was found to react also with phenylacetylene (Scheme 4.7), yielding the corresponding (β-silyl)vinylgold complex **21**. Again, the influence of the environment around gold is striking: full conversion is here achieved after two days, while the analogous reaction with triphenylphosphine as the neutral donor ligand requires around three weeks (Scheme 4.3).

Next, we explored if other ancillary ligands at gold promote as well the insertion reaction. *N*-heterocyclic carbenes (NHCs) are widely used in stoichiometric and catalytic gold chemistry and their good donor properties are known to stabilize unusual structures not achievable with phosphine ligands [12–16]. Therefore a new silylgold complex featuring an NHC ligand was synthesized. The [1,3-bis ((2,6-diisopropyl)phenyl)imidazol-2-ylidene]gold(I) chloride complex ((IPr)AuCl) was chosen as the precursor compound. In comparison to phosphine ligands, the steric shielding of the carbene ligand around the gold atom is increased. We expected that the increased steric protection would allow for a stable silylgold complex with a less bulky silyl moiety. Starting from chlorodimethylphenylsilane and lithium metal the corresponding lithiosilane was prepared and in situ reacted with (IPr)AuCl to give the corresponding (NHC)silylgold complex **22** (Scheme 4.8) which is to the best of our knowledge the first compound of this type.

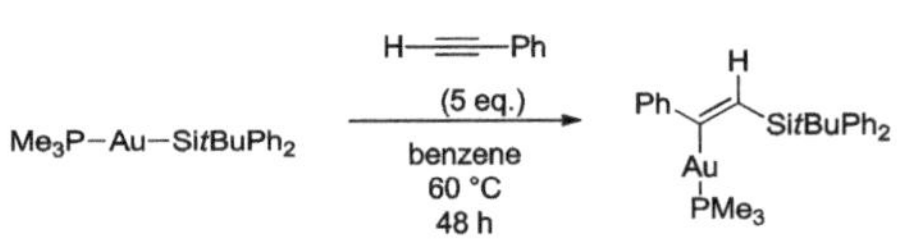

19 **21** (quant.)

Scheme 4.7 *syn* insertion of phenylacetylene into the AuSi bond of silylgold complex **19**

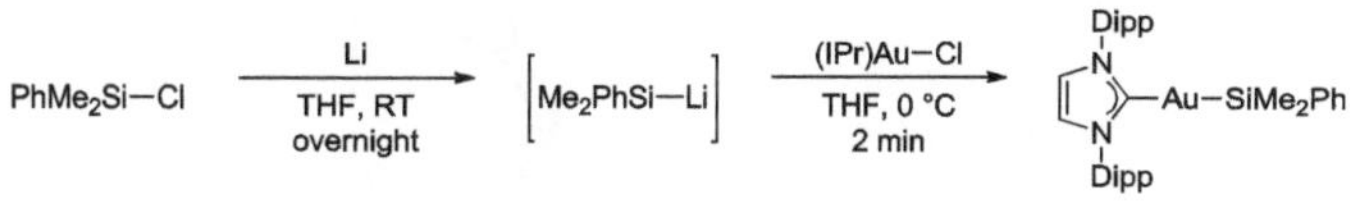

22 (75 %)

Scheme 4.8 Synthesis of (NHC)silylgold complex **22**. IPr = *N,N′*-Bis(2,6-diisopropylphenyl) imidazol-2-ylidene, Dipp = (2,6-diisopropyl)phenyl

The ^{29}Si NMR spectrum of **22** exhibits a singlet resonance at δ = 12.9 ppm for the silyl ligand. In the ^{13}C NMR spectrum, a singlet resonance at δ = 217.4 ppm corresponding to the carbene carbon is observed. The downfield shift of the carbene signal when going from (IPr)AuCl (δ = 175.1 ppm) [17] to **22**, is indicative of the presence of the silyl ligand at gold. This shift results from the important *trans* influence of the silyl group and can be compared to the ^{13}C NMR carbene resonance of Sadighi's (IPr)AuH complex (see Sect. 3.5.2), featuring a hydride ligand with known high *trans* influence. The hydride provokes a downfield shift of the carbene singlet resonance to δ = 204.9 ppm [13].

When **22** was reacted with methyl propiolate (10 equivalents) at 100 °C in toluene/dichloromethane (1:1), monitoring by ^{29}Si{^{1}H} NMR indicated the clean formation of a new species **23**, resonating at δ = −13.6 ppm (s). The observed highfield shift is indicative of a Si atom bound to carbon and this ^{29}Si chemical shift is indeed in accord with reported values for (vinyl)$SiMe_2Ph$ species [18, 19], pointing at the insertion of the alkyne into the gold-silicon bond. Full conversion was achieved after 12 h (Scheme 4.9).

Compound **23** was isolated (63 % yield) and completely characterized. High resolution mass spectroscopy confirms an adduct of methyl propiolate and the silylgold complex **22**. A peculiar feature in the ^{1}H NMR spectrum of **23** is the resonance at δ = 9.20 ppm, attributed to a vinylic proton. For vinylgold complexes with the proton sitting at β-position to gold, a shift at ca. δ = 8 ppm is usually observed (vide supra). The unusual low field resonance signal observed for **23**, might be indicative of a vinylic proton in α-position to gold, instead of the commonly observed β-position. In order to elucidate the regio- and stereo-chemistry of complex **23**, single crystals were grown and an X-ray diffraction analysis was performed. The molecular structure of the (NHC) vinylgold complex is shown in Fig. 4.6.

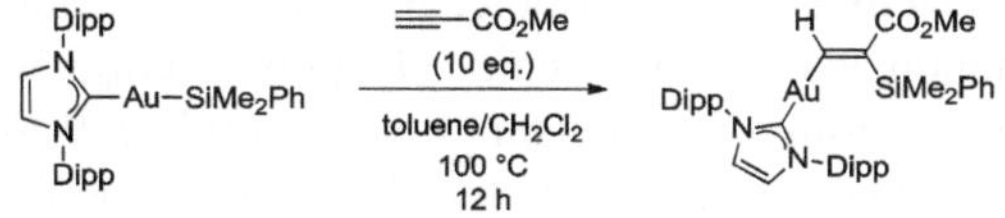

22 **23** (63 %)

Scheme 4.9 syn insertion of methyl propiolate into the AuSi bond of (NHC)silylgold(I) complex **23**. Dipp = (2,6-diisopropyl)phenyl

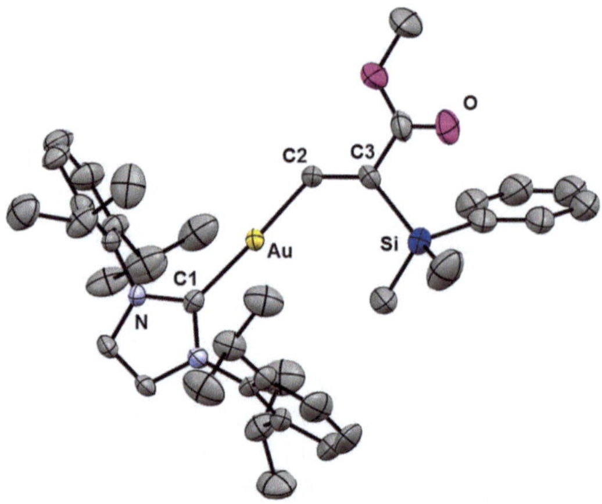

Fig. 4.6 Molecular structure of **23** determined by single crystal X-ray diffraction. Hydrogen atoms are omitted for clarity. Selected bond lengths [Å] and angles [°]: C1Au 2.03(1), AuC2 2.03(1), C2C3 1.35(1), C1AuC2 174.3(3), AuC2C3 132.1(7), C2C3Si 126.5(7), AuC2C3Si 5(1)

According to the determined structure, the insertion reaction giving rise to **23** is *syn* selective, yielding a vinylgold complex with Au and Si in a *cis* arrangement. However, the regioselectivity is inverted as compared to the insertion reaction of (phosphine)silylgold(I) complexes with methyl propiolate stated above. As suggested by the ^{1}H NMR data of complex **23**, the vinylic proton and the gold atom are positioned in a *gem* arrangement, such as the Si atom and the carboxylate group. A reasonable explanation is furnished by the important steric bulk of the NHC (diisopropyl)phenyl wing groups. The formed regioisomer is the one featuring less steric repulsion between the NHC ligand and the methyl carboxylate group of the alkyne. This reaction outcome demonstrates again that the insertion of alkynes into the AuSi bond is not simply governed by a Michael-type reactivity, but also to a large extent by steric factors.

4.1.5 Application of Vinylgold Complexes in Synthesis

In order to capitalize on the direct access to the (β-silyl)vinylgold complexes, we then explored its reactivity for intermolecular functionalization reactions. First, Au/Sn transmetallation was readily achieved upon reaction with nBu$_3$SnOTf at room temperature, affording the α-stannyl-β-silylacrylate **9-Sn** in 81 % yield. Complex **9** was also engaged in Pd-catalyzed CC cross-coupling reactions in the presence of 5 mo % of [PdCl$_2$(cod)] at 60 °C, using allyl bromide and *p*-nitroiodobenzene as electrophiles. The resulting allyl and aryl functionalized products **9-Allyl** and **9-Ar** were thereby obtained in good yields (73 and 84 %, respectively).

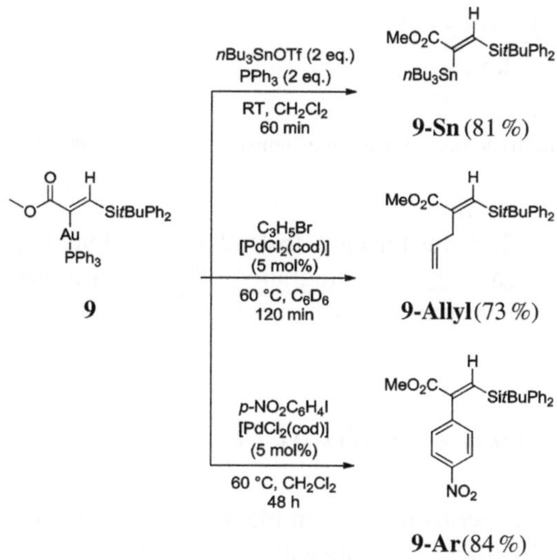

Scheme 4.10 Functionalization of vinylgold complex **9** by transmetallation and cross-coupling

Complex **9** is highly functional and sterically hindered, yet it reacts smoothly and cleanly affording efficient access to trisubstituted olefins with complete regio- and stereo-control (Scheme 4.10). The isolated yields are comparable to those reported by Blum for similar reactions with less substituted vinylgold(I) complexes [20].

Given the fact that the silyl group of those olefins presents a useful handle for subsequent cross-coupling via Hiyama reactions, (β-silyl)vinylgold complexes may be interesting building blocks for the selective synthesis of highly functionalized *Z*-olefins [21].

4.2 Mechanistic Study of the Insertion Reaction

In order to gain insight into the mechanism of the insertion reaction, a combined experimental and theoretical study was performed.

4.2.1 Working Hypothesis

In the course of the experimental work on the insertion of alkynes and allenes in the AuSi bond of silylgold(I) complexes to give (β-silyl)vinylgold complexes, we observed that the reactions give rise exclusively to *Z*-configurated products, i.e. Au and Si atoms are in a *cis* arrangement. Based on this observation, the coordination/ *anti* addition mechanistic scheme which is commonly observed for gold(I)

Scheme 4.11 Working hypothesis for the mechanism of the *syn* insertion of alkynes in the AuSi bond

complexes (see Sect. 3.5) was rejected. Instead, our working hypothesis consists of a *syn* insertion reaction with direct migration of the silyl moiety from gold to the alkyne in an inner-sphere process (Scheme 4.11).

4.2.2 Kinetic Studies with Methyl Propiolate

We turned to kinetic experiments to examine the *syn*-insertion process. The reaction of 10 equivalents of methyl propiolate with complex **8** in benzene-d_6 was monitored using ^{1}H NMR spectroscopy (Scheme 4.12). The progress of the reaction over time was followed by integration of characteristic resonances of reactants (alkyne: H_{OCH3}, δ = 3.12 ppm; **8**: H_{tBu}, δ = 1.50 ppm) and product; (**9**: $H_{C=CH}$, δ = 8.40 ppm) relative to mesitylene (δ = 2.15 ppm) as internal standard. The insertion reaction afforded complex **9** as the exclusive product without any intermediate being detected.

According to the kinetic plots, the consumption of silylgold(I) complex **8** follows a first-order decay (Fig. 4.7a). The reaction obeys overall a second-order dependence (Fig. 4.7b).

Thus, the reaction shows a first-order dependence for both the silylgold complex and the alkyne (rate law: $\upsilon = k_{obs}\cdot[\mathbf{8}]\cdot[\text{alkyne}]$), in agreement with a bimolecular process. Rate constants for the formation of the vinylgold complex **9** were determined at several temperatures (25, 35, 45, 55 and 65 °C).

From the temperature dependence of the observed rate constants, the activation parameters for the insertion process were calculated via an Eyring plot (Fig. 4.8). The following values have been obtained: $\Delta H^{\neq}$ = 13.7 (±1.6) kcal/mol and $\Delta S^{\neq}$ = −32.0 (±5.0) cal/(mol K).

In the same way, the formation of vinylgold complex **20** from the reaction of methyl propiolate with silylgold complex **19** bearing a PMe_3 ligand instead of PPh_3 was monitored by ^{1}H NMR spectroscopy at several temperatures (Scheme 4.13).

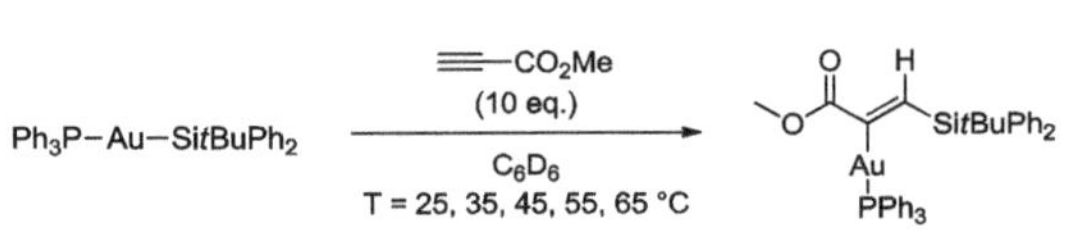

8 **9**

Scheme 4.12 ^{1}H NMR-spectroscopic monitoring of the *syn* insertion of methyl propiolate in the AuSi bond of silylgold complex **8** at different temperatures

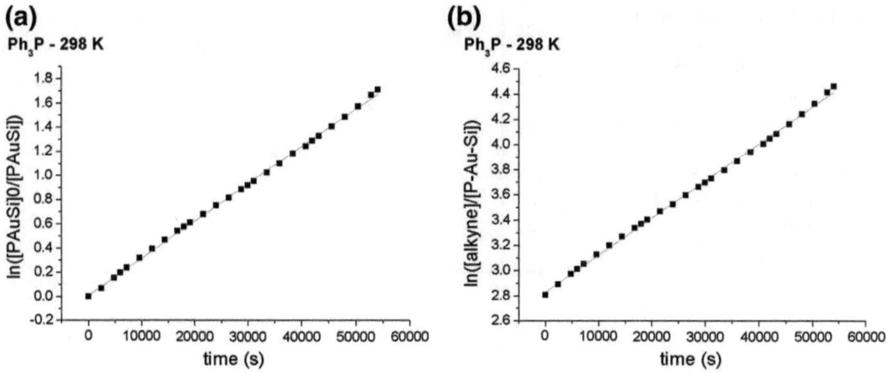

Fig. 4.7 Kinetic plots for the reaction of **8** with methyl propiolate at 25 °C. $[\mathbf{8}]_0 = 0.0483$ mol/L, $[\text{alkyne}]_0 = 0.483$ mol/L. **a** Plot of $\ln([\mathbf{8}]_0/[\mathbf{8}])$ versus time **b** plot of $\ln([\text{alkyne}]/[\mathbf{8}])$ versus time

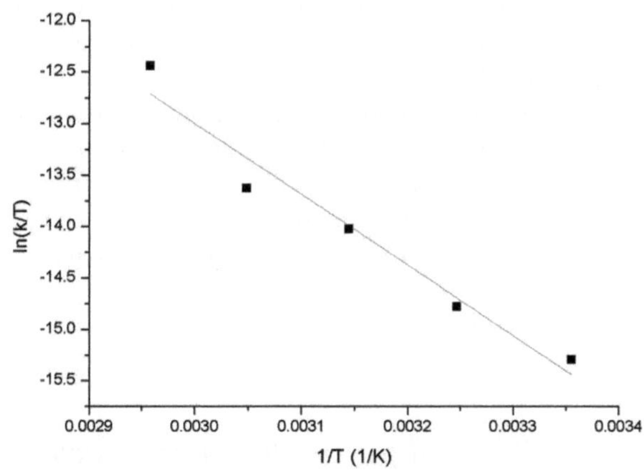

Fig. 4.8 Eyring plot for the *syn* insertion of methyl propiolate into the AuSi bond of silylgold complex **8**. Temperature range: 25–65 °C. Slope: (−6874.075 ± 804.316) K; intercept: 7.624 ± 2.537. $R^2 = 0.950$

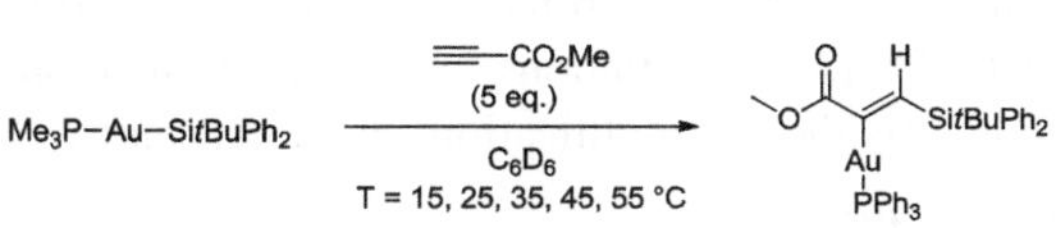

19 **20**

Scheme 4.13 ^{1}H NMR-spectroscopic monitoring of the *syn* insertion of methyl propiolate in the AuSi bond of silylgold complex **19** at different temperatures

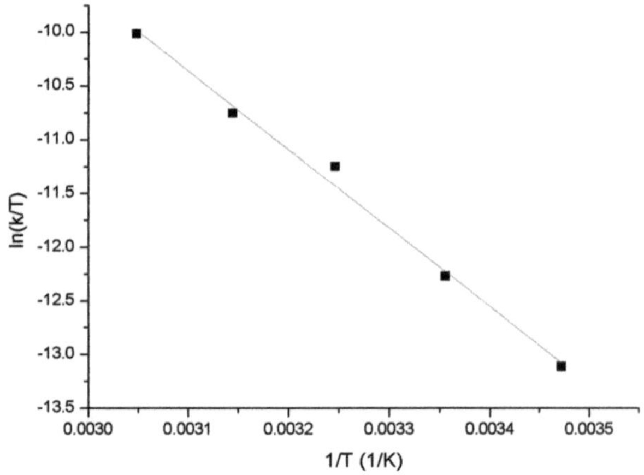

Fig. 4.9 Eyring plot for the *syn* insertion of methyl propiolate into the AuSi bond of silylgold complex **19**. Temperature range: 15 °C to 55 °C. Slope: (−7302.783 ± 355.088) K; intercept: 12.278 ± 1.157. $R^2 = 0.989$

As for the formation of vinylgold complex **9**, a first-order dependence for each of the reactants was observed. The activation parameters were determined via an Eyring plot (Fig. 4.9), and are as follows: $\Delta H^{\neq}$ = 14.5 (±0.7) kcal/mol and $\Delta S^{\neq}$ = −21.4 (±2.3) cal/mol K).

The negative value of $\Delta S^{\neq}$ for both reactions is consistent with associative mechanisms. When comparing the activation parameters for the insertion reactions of complexes **8** and **19** respectively, the main difference stems from the entropic term which is lower for the reaction employing trimethylphosphine complex **19**, resulting in a decrease of the experimental Gibbs free energy of activation by about 2 kcal/mol ($\Delta G^{\neq}$ = 24.4 (±2.3) kcal/mol for complex **8** and $\Delta G^{\neq}$ = 21.8 (±1.0) kcal/mol for complex **19** at T = 60 °C).

4.2.3 Mechanistic Proposal for the Insertion Reaction

Two plausible mechanistic pathways leading to the *syn* insertion products may be considered (Scheme 4.14): (i) a concerted 1,2-addition of the Au–Si fragment to the alkyne or (ii) a 2-step pathway involving first the coordination of the alkyne to the gold complex followed by a migratory insertion process. Both mechanistic schemes were investigated in collaboration with the group of Dr. K. Miqueu (Université de Pau) using DFT (B97D) [22] calculations on the reaction of silylgold complex **8** with methyl propiolate without simplification of the reactants.

(i) **Concerted 1,2-Addition**

The potential energy surface (PES) was carefully examined aiming at the localization of a transition state (TS) accounting for a direct 1,2-addition of the AuSi bond to the alkyne. Optimizations from diverse starting geometries using a broad range of AuC1 and SiC2 bond distances (from 1.900 to 2.350 Å) did not

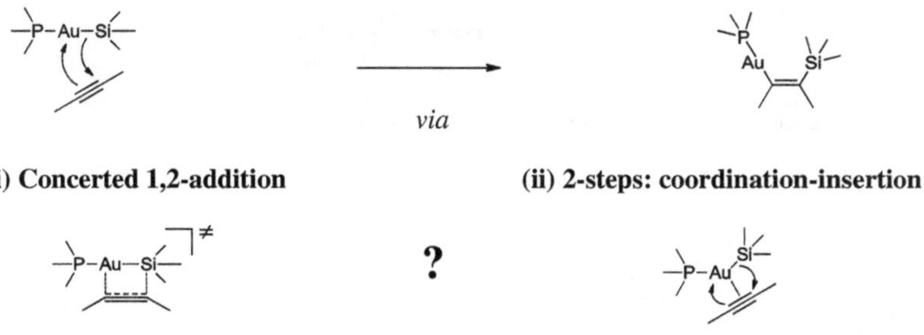

Scheme 4.14 Two possible mechanistic pathways for the insertion of alkynes into the AuSi bond

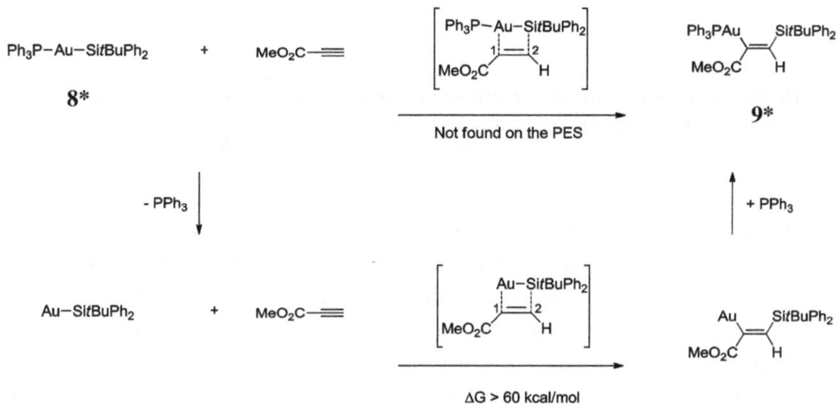

Scheme 4.15 Computational analysis of a hypothetical 1,2-addition pathway for the reaction of methyl propiolate with silylgold complex **8***

allow for the localization of the expected in-plane (Au, Si, C1, C2) four-center transition state structure (Scheme 4.15). However, a TS could be localized if a prior dissociation of the PPh_3 ligand from complex **8*** was taken into account. Phosphine dissociation is predicted to be endothermic by 22.0 kcal/mol. However, the 1,2-addition of the ensuing phosphine-free silylgold complex to the alkyne is found to proceed via a transition state prohibitively high in energy ($\Delta G^{\neq}$ = 64.6 kcal/mol, $\Delta H^{\neq}$ = 63.0 kcal/mol), ruling out the formation of complex **9*** via this pathway.

(ii) **Coordination-Insertion Pathway**

While for the reaction of **8*** with methyl propiolate no in-plane TS was found on the PES, without preceding phosphine dissociation, an out-of-plane distorted four-center TS ($\mathbf{TS_{8^* \to 9^*}}$, Fig. 4.10) corresponding to the insertion of the alkyne into the AuSi bond was localized. Intrinsic Reaction Coordinate (IRC) calculations indicate that $\mathbf{TS_{8^* \to 9^*}}$ is connected to the final product **9*** and to an adduct $\mathbf{8^* - \pi}$ between the alkyne and the gold complex. The adduct $\mathbf{8^* - \pi}$ is slightly higher in energy than the free reactants (ΔG = 5.2 kcal/mol), but only a very low activation

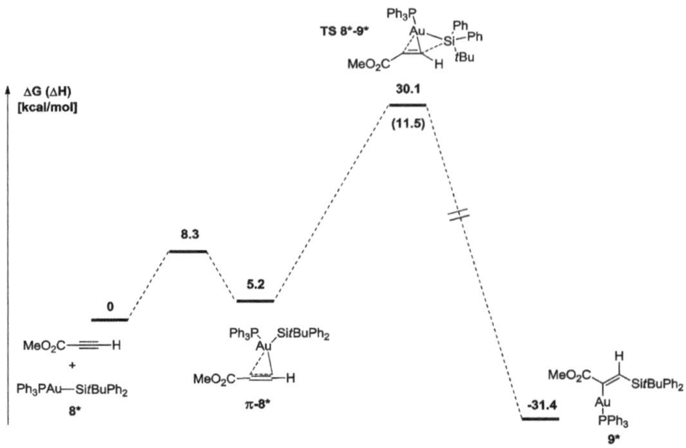

Fig. 4.10 Reaction profile for the insertion of methyl propiolate into the AuSi bond of **8***. ΔG values and, enthalpy of activation (in parentheses) in kcal/mol. Computed at the B97D/SDD + f(Au)-6-31G** level of theory

barrier ($\Delta G^{\neq}$ = 8.3 kcal/mol) is involved in its formation. The rate determining step for the formation of the thermodynamically stable vinyl complex **9*** is therefore the insertion of the alkyne into the AuSi bond $\left(\Delta G^{\neq}{}_{\mathbf{8*}\rightarrow\mathbf{9*}} = 30.1\ \mathrm{kcal/mol}\right)$.

The BD97 functional which takes dispersion effects into account was found to give enthalpy values that fit well with the experimental data: An enthalpy value $\Delta H^{\neq}{}_{\mathbf{8*}\rightarrow\mathbf{9*}}$ for the insertion of the alkyne of about 11.5 kcal/mol was predicted, close to that determined experimentally $(\Delta H^{\neq}{}_{8\rightarrow 9} = 13.7(\pm 1.6)\ \mathrm{kcal/mol})$. However, the computed entropy differs significantly from the experimentally obtained values. The accurate determination of entropic contributions is non-trivial, especially for reactions which involve a change of molecularity [23]. The overestimation of the entropic term is at the origin of the differences between experimental and theoretical Gibbs energy values.

Overall, the 2-step pathway is a probable mechanistic scheme for the insertion of alkynes into the AuSi bond of silylgold complexes.

4.2.4 Regioselectivity of the Insertion Reaction

To gain insight into the factors controlling the reaction, in particular its regioselectivity, the structures of the key intermediate $\mathbf{8*} - \boldsymbol{\pi}$ and the transition state $\mathbf{TS}_{\mathbf{8*}\rightarrow\mathbf{9*}}$ were analyzed in more detail. The structure of the adduct $\mathbf{8*} - \boldsymbol{\pi}$ reveals that the alkyne is coordinated to the gold center in a side-on fashion and lies in the same plane as the phosphorus and silicon atoms (Fig. 4.11, left). Upon coordination of the alkyne to gold, the C1C2 bond distance is elongated by 0.07 Å and the PAuSi

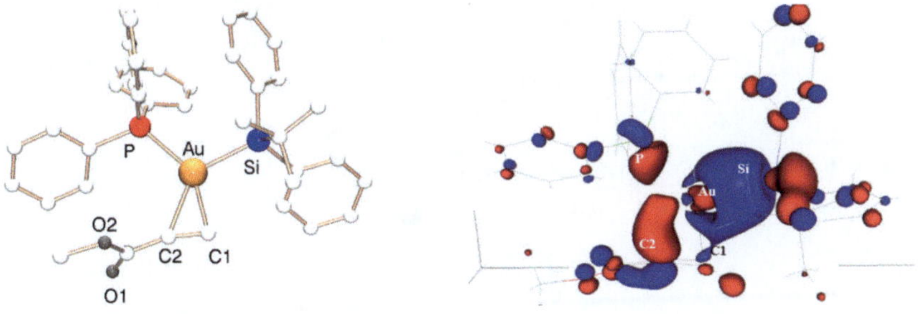

8*-π **8*-π**

Fig. 4.11 *Left* Calculated intermediate (**8*** − **π**) of the insertion of methyl propiolate into the AuSi bond of **8*** (B97D/SDD + f(Au)-6-31G** level of theory). Selected bond distances (Å) and angles (°): 8* − π: AuC1: 2.108, AuC2: 2.267, C1C2: 1.280, AuSi: 2.405, AuP: 2.395, PAuSi: 104.7. Right: Plot of ± 0.04 au isosurfaces for the HOMO of **8***−

bond angle is significantly bent (104.7°). A non-symmetric coordination of C1 and C2 to gold is observed, with the AuC1 bond being significantly shorter than the AuC2 bond (2.108 Å vs. 2.267 Å).

Analysis of the key frontier molecular orbitals in the adduct **8*** − **π** (Fig. 4.11, right) indicates that the prominent interaction between **8*** and methyl propiolate involves the HOMO of the silylgold complex (a combination of σ_{AuSi} and d_{yz}(Au) orbitals) and the LUMO of the methyl propiolate ($\pi^*_{C1\equiv C2}$). NPA charges calculated by means of an NBO analysis also revealed a net charge transfer of 0.47 e^- from complex **1** to methyl propiolate.

This bonding scheme differs from that typically observed for the coordination of alkynes to LAu^+ fragments (for example π-complexes involved in external nucleophilic additions), in which the dominant interaction involves donation of the occupied $\pi_{C\,\equiv\,C}$ orbital of the alkyne to gold [24, 25].

The structure of the transition state $\mathbf{TS_{8*\rightarrow 9*}}$ corresponding to the insertion step was then studied (Fig. 4.12). Starting from **8*** − **π**, the reaction involves migration of the silyl group from Au to C1 with concomitant shift of gold towards C2, resulting overall in the *syn* insertion of methyl propiolate into the AuSi bond. The most important geometrical modifications from **8*** − **π** to $\mathbf{TS_{8*\rightarrow 9*}}$ are (i) the displacement of the silyl group out of the (AuC1C2) plane (C1C2AuSi = −34.5° in $\mathbf{TS_{8*\rightarrow 9*}}$ vs. −7.0° in **8*** − **π**) and (ii) the widening of the PAuSi bond angle (130.8°), which altogether bring the Si atom close to C1 (SiC1 = 2.357 Å). It is also interesting to note that, from an electronic standpoint, the charge transfer from **8** to methyl propiolate increases further by 0.33 e^- in $\mathbf{TS_{8*\rightarrow 9*}}$.

For comparative purposes, the formation of the experimentally unobserved regioisomer of vinylgold complex **9**, featuring the silyl group and the CO_2Me moiety in a *gem* arrangement, was computationally investigated: The reaction profile corresponds as well to a 2-step coordination-insertion process. However, the TS involving nucleophilic attack of the silyl group to the carbon atom in α-position

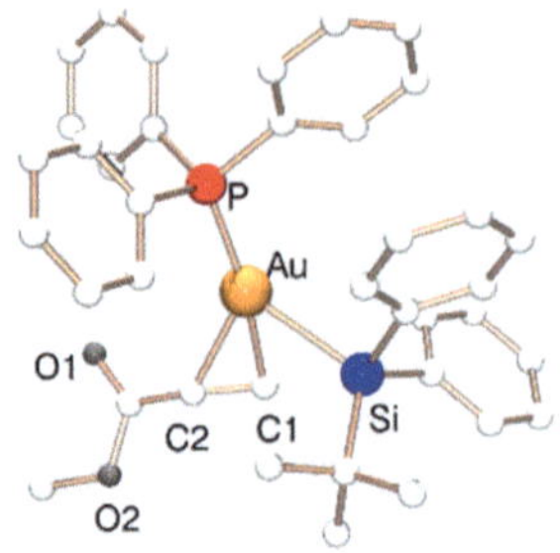

$TS_{8^* \rightarrow 9^*}$

Fig. 4.12 Calculated transition state ($TS_{8^* \rightarrow 9^*}$) of the insertion of methyl propiolate into the AuSi bond of **8*** (B97D/SDD + f(Au)-6-31G** level of theory). Selected bond distances (Å) and angles (°): AuC1: 2.125, AuC2: 2.318, C1C2: 1.323, AuSi: 2.587, AuP: 2.273, C1Si: 2.357, PAuSi: 130.8

to the CO_2Me group is 6–7 kcal/mol higher in energy compared to $\mathbf{TS_{8^* \rightarrow 9^*}}$, in agreement with the experimentally observed selective formation of regioisomer **9**.

The computed reaction profile for (trimethylphosphine)silylgold complex **19*** is very similar to that found with complex **8*** with comparable activation barriers. The enthalpy value for the rate determining insertion step was computationally well reproduced using the B97D functional ($\Delta H^{\neq}{}_{19* \rightarrow 20*} = 13.3\ \text{kcal/mol}$). The electronic situation in the $\mathbf{TS_{19^* \rightarrow 20^*}}$ is also very similar indicating a charge transfer from complex **19*** to methyl propiolate mainly involving Si and C1 (0.54 e^- from computed NBO charges).

According to this analysis, the likely mechanistic scheme among the conceivable scenarios for the insertion reaction is the 2-step coordination-insertion process. Coordination of methyl propiolate to silylgold complexes **8** and **19** involves the HOMO of the complexes and the LUMO of the alkyne. The rate determining insertion step takes place via an asynchronous concerted mechanism. The regioselectivity of the reaction with methyl propiolate is in line with the polarity of the alkyne (Michael-type addition), and originates from the charge transfer from Si to C1 in the transition state.

4.2.5 Insertion of DMAD

In order to investigate the origin of the unexpected zwitterionic product (**13**) stemming from the insertion of DMAD into the AuP bond of silylgold complex **8** (see Scheme 4.4), this reaction was probed computationally as well.

The *syn* insertion of DMAD into the AuSi bond of complex **8*** follows in general the same coordination-insertion pathway than the one found for methyl propiolate (vide supra). In agreement with the experimental observations (the reaction proceeds smoothly with equimolar reactant concentrations in contrast to the insertion with methyl propiolate where an excess of alkyne is employed), the intermediate

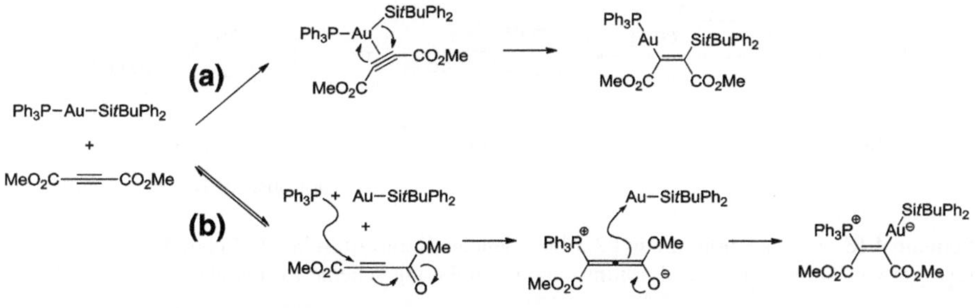

Scheme 4.16 Mechanistic scheme accounting for the insertion of DMAD in the AuSi bond (*pathway a*)) or in the AuP bond (*pathway b*)) of silylgold complex **8** to yield **12** and **13**, respectively

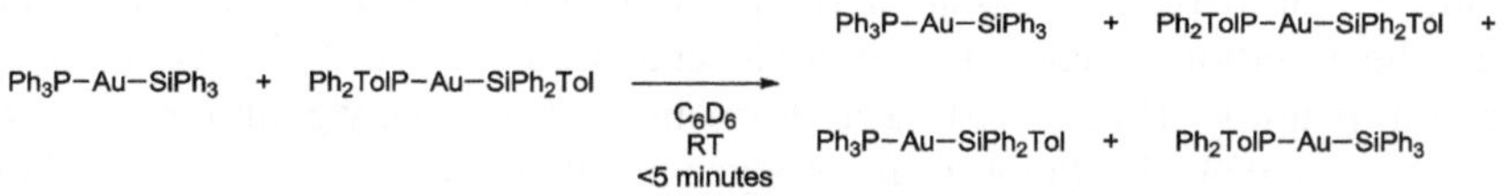

Scheme 4.17 Ligand scrambling upon mixing of two different (phosphine)silylgold(I) complexes

$\mathbf{8^* - \pi_{DMAD}}$ and the transition state $\mathbf{TS_{8^*\rightarrow 12^*}}$ are found at lower energy on the PES. $\mathbf{8^* - \pi_{DMAD}}$ is located only 3.3 kcal/mol above the reactants and the transition state $\mathbf{TS_{8^*\rightarrow 12^*}}$ for the rate determining insertion is calculated to be 8 kcal/mol lower in energy than $\mathbf{TS_{8^*\rightarrow 9^*}}$($\Delta H^{\neq}{}_{8^*\rightarrow 12^*} = 3.2$ kcal/mol; $\Delta G^{\neq}{}_{8^*\rightarrow 12^*} = 22.4$ kcal/mol).

Calculations indicate that neither a 1,2-addition nor a 2-step coordination-insertion pathways can account for the insertion of DMAD into the AuP bond. Instead, complex **13*** was predicted to result from a conjugate addition of PPh_3 to DMAD to form a phosphonium enolate species [26]. Finally, nucleophilic addition of the vinyl anion to the silylgold complex may afford complex **13*** (Scheme 4.16). The dissociation of the phosphine from complex **8*** ($\Delta H_{\text{P-diss}}$ = 33.0 kcal/mol; $\Delta G_{\text{P-diss}}$ = 22.0 kcal/mol) is facilitated by the strong *trans* influence of the silyl ligand, thus rendering it competitive with the rate determining step of the coordination-insertion mechanism for the insertion of DMAD into the AuSi bond of **8***.

In line with the proposed mechanism, the lability of the phosphine ligand was observed experimentally by ^{31}P NMR spectroscopy: rapid ligand scrambling occurs upon mixing of two different (phosphine)silylgold complexes (Scheme 4.17).

4.2.6 Insertion of Ethyl 2,3-Butadienoate

The reaction of silylgold complex **14** with ethyl 2,3-butadienoate led to the formation of vinylgold complex **18** with a rather unexpected regioselectivity. The insertion reaction involved the terminal, less-activated double bond of the allene, and the silyl group is introduced in γ-position to the CO_2Et substituent

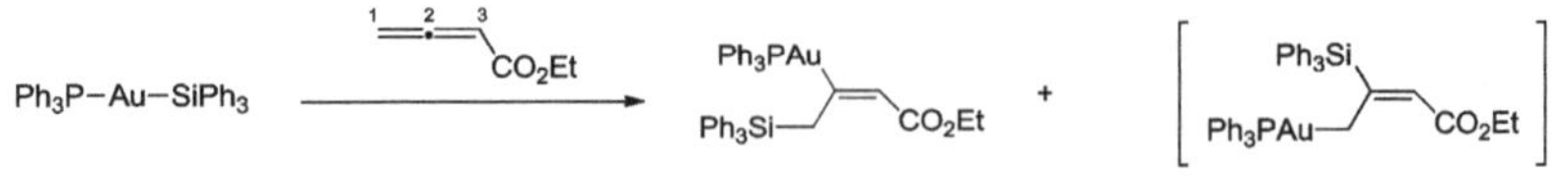

Scheme 4.18 *syn* Insertion of ethyl 2,3-butadienoate in the AuSi bond of complex **14**. The main text refers to the here indicated arbitrary atom numbering scheme for the allene

(Scheme 4.18). To gain insight into the origin for the observed selectivity, the mechanism for this reaction was investigated computationally.

First, we studied the reaction of complex **14*** with the terminal C1 = C2 bond of the allene, considering the formation of the two regioisomers **18*** and **18b***. In both cases, the insertion reaction is exergonic and occurs via a 2-step coordination-insertion pathway (Fig. 4.13). The formation of the vinylgold complex **18*** ($\Delta G_{14^* \rightarrow 18^*} = -23.1$ kcal/mol) is thermodynamically favored by ca. 3 kcal/mol over

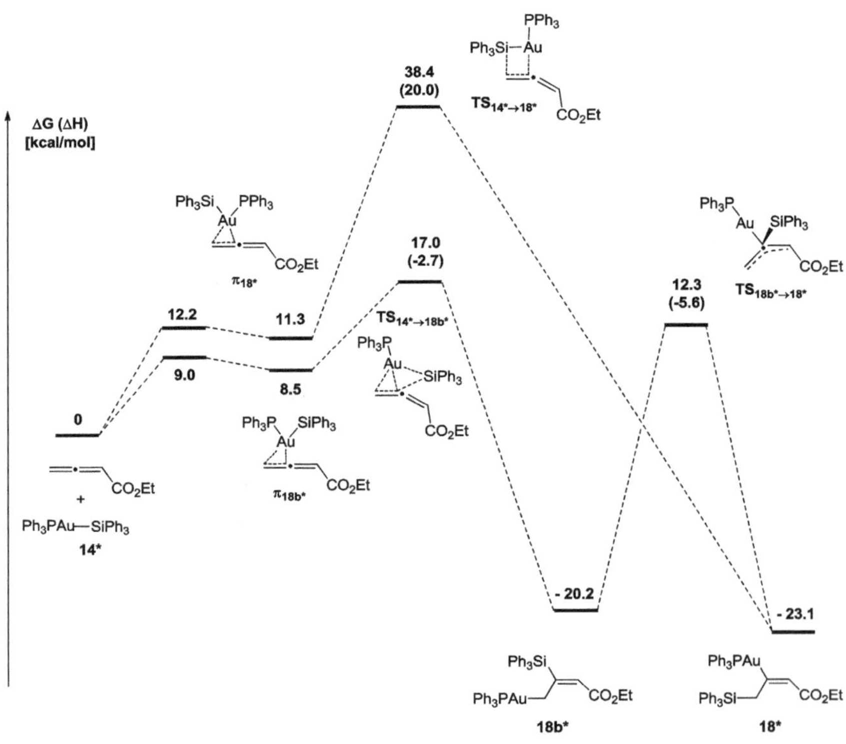

Fig. 4.13 Reaction profile for the insertion of 2,3-butadienoate in the AuSi bond of silylgold complex **14**. ΔG values (in kcal/mol) computed at the B97D/SDD + f(Au)-6-31G** level of theory. Enthalpies of activation $\Delta H^{\neq}$ (in kcal/mol) in parentheses

the formation of the alkylgold complex **18b*** ($\Delta G_{14*\rightarrow18b*}$ = −20.2 kcal/mol), in agreement with experimental observations. Both pathways are initiated by coordination of the terminal C1 = C2 bond of the allene to gold, which is endergonic by about 10 kcal/mol. The resulting complexes π_{18*} and π_{18b*} adopt η^2-allene structures and their relative stability is opposite to that of the insertion products **18*** and **18b*** (π_{18*} leading to the more stable insertion product, is less stable than π_{18b*} by about 3 kcal/mol).

However, the activation barrier for the pathway leading to complex **18*** is much higher than the one leading to **18b*** ($\Delta G^{\neq}_{14*\rightarrow18*} = 38.4$ kcal/mol vs. $\Delta G^{\neq}_{14*\rightarrow18b*} = 17.0$ kcal/mol), suggesting that the formation of **18b*** is kinetically much more favorable than of **18***. This energy profile is not consistent with the ready and selective formation of **18***, as observed experimentally.

A possibility to reconcile the profile predicted computationally and the selective formation of regioisomer **18*** observed experimentally would be that the kinetically favored insertion product **18b*** subsequently rearranges into the thermodynamically favored vinyl gold complex **18***. Consistently, a transition state ($\mathbf{TS}_{18b*\rightarrow18*}$) linking both products could be located on the PES. This transition state is located only 12.3 kcal/mol higher in energy than the initial reactants (and thus lower in energy than the insertion transition states $\mathbf{TS}_{14*\rightarrow18*}$ and $\mathbf{TS}_{14*\rightarrow18b*}$) and can certainly be reached under the experimental conditions (heating at 60 °C overnight). Interconversion of the alkyl- and vinylgold complexes **18b*** and **18*** is a rather unusual process. $\mathbf{TS}_{18b*\rightarrow18*}$ corresponds to the exchange of the relative positions of gold and silicon at C1/C2.

For sake of comprehensiveness, *syn* insertion of the C2 = C3 bond of ethyl 2,3-butadienoate into the AuSi bond of complex **14** was also considered and found to be less favored thermodynamically and to require higher activation barriers for both conceivable regioisomers. Thus, insertion of the activated C2 = C3 double bond of ethyl 2,3-butadienoate is both thermodynamically and kinetically disfavored over the insertion of the terminal C1 = C2 bond.

Altogether, the computational study carried out on the reaction between complex **14** and ethyl 2,3-butadienoate provides mechanistic insights and rationalizes the rather unexpected selectivity observed experimentally. Insertion of the terminal, less-activated C = C double bond of the allene into the AuSi bond of **14** is favored and first gives the alkyl gold complex **18b***, which subsequently rearranges into the vinyl gold complex **18*** via an Au/Si exchange process.

4.3 Conclusion

In conclusion, the experimental and computational results reported here provide evidence for a basic elementary step at gold, namely *syn* insertion. Terminal and internal alkynes, as well as allenes are shown to insert readily into AuSi bonds. The reaction is broad in scope and gives access to a variety of stable (β-silyl) vinylgold compounds in a stereoselective and regioselective manner. These vinylgold complexes can be employed in transmetallations or palladium-catalyzed cross-coupling reactions for further functionalization, giving access selectively to highly substituted *Z*-olefins.

By means of a combined experimental and computational study a likely mechanism for the insertion reaction was established: coordination of the alkyne to the silylgold complex gives rise to a three-coordinate intermediate which then forms the vinylgold complex by migration of the silyl moiety onto the alkyne with concomitant AuC bond formation.

These studies substantiate unequivocally the ability of gold to promote the functionalization of unsaturated organic substrates through an inner-sphere mechanism (alternatively to the commonly encountered outer-sphere nucleophilic addition). The here presented results give credit to the *syn* insertion pathway proposed occasionally for homogeneous [27] as well as heterogeneous [4, 5, 28] gold catalysts. The observed reactivity may have its origin in the particular electronic situation encountered in the silylgold complexes which differs remarkably compared to organogold complexes. Key factors rationalizing the observed selectivities have been identified.

This work may open the door to gold-mediated synthetic applications that are beyond today's scope for reactions with this metal.

4.4 Experimental Part

4.4.1 General Remarks

The silyl gold complexes **8** and **14** were prepared by slight modification of reported procedures [2, 6]. In order to exclude potential palladium-catalyzed processes, a palladium trace analysis of one of the gold(I) precursor complexes ([AuCl(PPh_3)]) was performed. The palladium content of this complex was shown to be lower than the detection limit of 2 ppm.

4.4.2 (Tert-*Butyldiphenylsilyl)(Triphenylphosphine)-Gold(I) Complex 8*

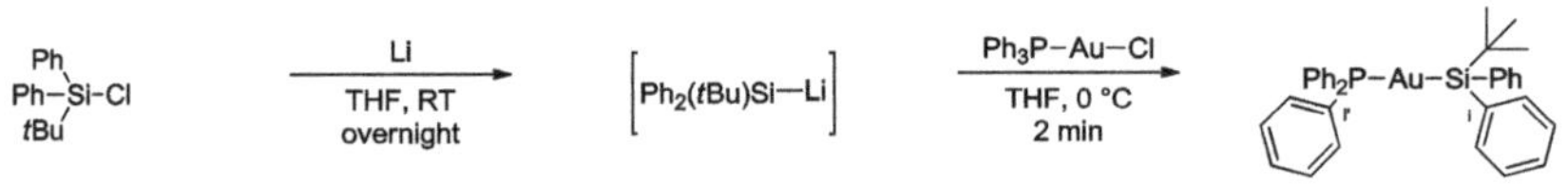

8

A solution of $Ph_2tBuSiCl$ (500 mg, 523 μL, 2 mmol, 1 eq) in tetrahydrofuran (6 mL) was added to finely cut, shiny lithium (276 mg, 40 mmol, 20 eq) in tetrahydrofuran (6 mL) placed in a Schlenk flask. The reaction mixture was stirred overnight at room temperature. The resulting dark brown solution was cooled down to 0 °C and then quickly added by cannula transfer to a dispersion of Ph_3PAuCl (1 g, 2 mmol, 1 eq) in tetrahydrofuran (20 mL) at 0 °C. The resulting orange-brown solution was stirred for 2 min, before a pea-sized piece of solid CO_2 was added. After stirring for 1 min at 0 °C, volatiles were removed under vacuum. The dark brown residue was re-dissolved in dichloromethane (3 mL), filtered over oven-dried alumina using 150 mL of a pentane/dichloromethane (9:1) mixture as the eluent. Volatiles of the bright orange filtrate were removed under vacuum and the solid residue was triturated twice with pentane. Upon scraping with the magnetic stirbar on the walls of the flask, the waxy residue turned into a light red to pink amorphous solid that was washed with pentane (5 × 3 mL) and then dried under vacuum to give silylgold complex **8** as an off-white powder. Yield: 1.28 g (91 %). Crystals suitable for X-ray crystallography were grown at −40 °C from a saturated toluene solution. **Mp**: 125 °C (decomposition); **^{1}H NMR** (300 MHz, $CDCl_3$): δ 8.20 (m, 2H, H_{ar}), 7.41 (m, 1H, H_{ar}), 7.38 (m, 4H, H_{ar}), 7.31 (m, 8H, H_{ar}), 7.28 (m, 2H, H_{ar}), 7.24 (m, 1H, H_{ar}), 7.21 (m, 1H, H_{ar}), 6.95 (m, 6H, H_{ar}), 1.53 (s, 9H, H_{tBu}); **^{31}P{^{1}H} NMR** (121 MHz, $CDCl_3$): δ 56.5 (s); **^{29}Si{^{1}H} NMR** (79 MHz, CD_2Cl_2, 193 K): δ 35.4 (d, $^2J_{SiP}$ = 164.8 Hz); **^{13}C{^{1}H}** NMR (75 MHz, $CDCl_3$): δ 144.9 (s, C_i), 137.9 (s, C_{ar}), 134.5 (d, J_{CP} = 14.2 Hz, C_{ar}), 132.1 (d, $^1J_{CP}$ = 39.7 Hz, $C_{i'}$), 130.7 (d, J_{CP} = 2.4 Hz, C_{ar}), 129.1 (d, J_{CP} = 9.8 Hz, C_{ar}), 127.7 (s, C_{ar}), 127.5 (s, C_{ar}), 30.6 (s, $C(CH_3)_3$), 22.6 (s, $C(CH_3)_3$); **HRMS (ESI+)**: calcd for $[M + H]^+$ = $C_{34}H_{35}PSiAu^+$: 699.1911. Found: 699.1936.

4.4.3 Reaction of 8 with Methyl Propiolate (9)

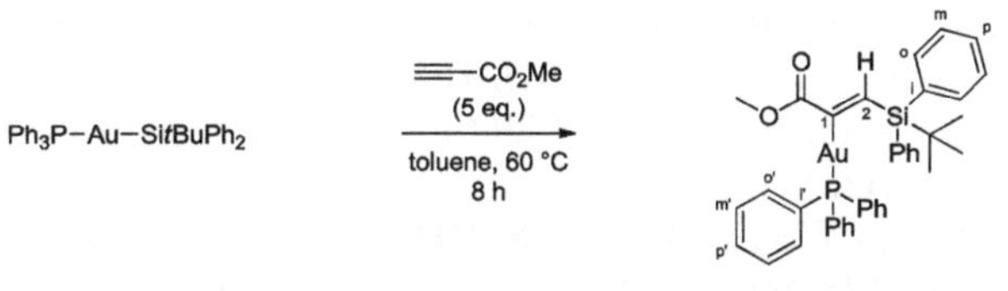

8 **9**

Methyl propiolate (256 μL, 2.88 mmol, 5 eq) was added to complex **8** (400 mg, 0.57 mmol, 1 eq) solubilized in toluene (4 mL) in a Schlenk tube. The Schlenk tube was sealed and heated for 8 h at 60 °C. After complete conversion, volatiles were removed in vacuo. The brownish residue was purified by column chromatography (silica, eluent: pentane/ethyl acetate, 9:1, R_f 0.41) to give complex **9** as a white solid. Yield: 425 mg (95 %). Crystals suitable for X-ray crystallography were grown at 4 °C from a saturated toluene solution layered with pentane. **Mp**: 153 °C; **^{1}H NMR** (300 MHz, C_6D_6): δ 8.47

(d, $^4J_{HP}$ = 17.8 Hz, 1H, H_{vinyl}), 8.03–7.95 (m, 4H, H_{ar}), 7.06–6.81 (m, 21H, H_{ar}), 3.68 (s, 3H, OCH_3), 1.15 (s, 9H, H_{tBu}); **^{31}P{^{1}H} NMR** (200 MHz, $CDCl_3$): δ 41.5 (s); **^{29}Si {^{1}H} NMR** (60 MHz, $CDCl_3$): δ −12.5 (d, $^4J_{SiP}$ = 3.9 Hz); **^{13}C{^{1}H} NMR** (75 MHz, C_6D_6) δ 192.4 (d, $^2J_{CP}$ = 115.9 Hz, C_1), 175.3 (d, $^3J_{CP}$ = 5.8 Hz, C = O), 144.0 (d, $^3J_{CP}$ = 2.8 Hz, C_2), 137.0 (s, C_o or C_m), 136.9 (s, C_i), 134.2 (d, J_{CP} = 13.9 Hz, $C_{o'}$ or $C_{m'}$), 130.7 (d, $^1J_{CP}$ = 49.8 Hz, $C_{i'}$), 130.3 (d, $^4J_{CP}$ = 2.6 Hz, $C_{p'}$), 128.5 (d, J_{CP} = 11.0 Hz, $C_{o'}$ or $C_{m'}$), 128.4 (s, C_p), 127.2 (s, C_o or C_m), 50.9 (s, OCH_3), 27.6 (s, C(CH$_3$)$_3$), 18.0 (s, C(CH$_3$)$_3$); **HRMS (ESI+)**: calcd for $[M + H]^+$ = $C_{38}H_{39}O_2SiPAu^+$: 783.2122. Found: 783.2131; **Elt. Anal.**: calcd for $C_{38}H_{38}O_2SiPAu$ + C_7H_8: C, 61.78; H, 5.30. Found: C, 61.16; H, 5.60.

4.4.4 Reaction of 9 with nBu$_3$SnOTf (9-Sn)

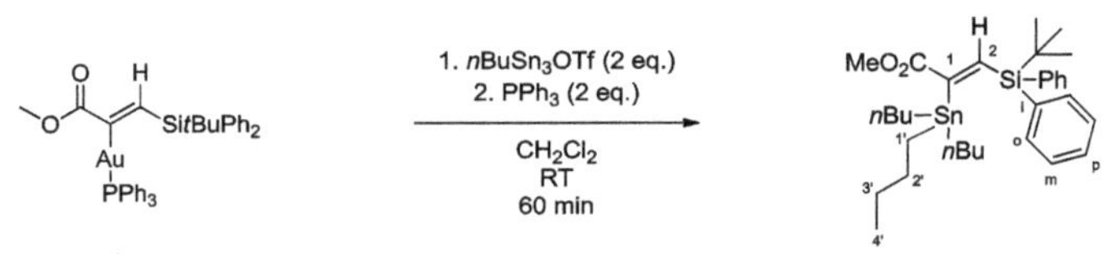

9 **9-Sn**

A solution of nBu$_3$SnOTf (78.5 mg, 0.18 mmol, 2 eq) in dichloromethane (1 mL) was added to complex **9** (70 mg, 0.089 mmol, 1 eq) placed in a Schlenk tube. The colorless mixture was stirred for about 3 min before a solution of triphenylphosphine (46.9 mg, 0.18 mmol, 2 eq) in dichloromethane (1 mL) was added. After 60 min of vigorous stirring, volatiles were removed in vacuo. The residue was purified by column chromatography (silica gel, eluent: pentane/dichloromethane, 9:1, R_f 0.15) to give the silyl-stannylated product **9-Sn** as a colorless viscous oil. Yield: 63.7 mg (81 %). **^{1}H NMR** (300 MHz, C_6D_6): δ 8.42 (s, Sn satellites: $^3J^1_{H17Sn}$ = 136 Hz, $^3J_{H119Sn}$ = 142 Hz, 1H, H_{vinyl}), 7.08–6.98 (m, 8H, H_{ar}), 6.88–6.82 (m, 2H, H_{ar}), 3.36 (s, 3H, OCH_3), 1.29 (m, 12H, $H_{nBu\text{-}CH2}$), 0.81 (s, 9H, H_{tBu}), 0.65 (t, $^3J_{HH}$ = 7.2 Hz, 9H, $H_{nBu\text{-}CH2}$), 0.54–0.45 (m, 6H, $H_{nBu\text{-}CH2}$); **^{29}Si{^{1}H} NMR** (60 MHz, C_6D_6): δ −13.44 (s); **^{119}Sn{^{1}H} NMR** (112 MHz, C_6D_6): δ −48.4 (s); **^{13}C{^{1}H} NMR** (125 MHz, C_6D_6): δ 170.7 (s, C = O), 164.0 (s, C_1), 153.1 (s, C_2), 136.3 (s, C_o), 133.0 (s, C_i), 129.5 (s, C_p), 128.1 (s, C_m), 51.7 (s, OCH_3), 29.2 (s, $C_{2'}$), 27.4 (s, C(CH$_3$)$_3$), 26.6 (s, $C_{3'}$), 18.0 (s, C(CH$_3$)$_3$), 13.6 (s, $C_{4'}$), 12.1 (s, $C_{1'}$); **HRMS (ESI+)**: calcd for $[M + Na]^+$ = $C_{32}H_{50}O_2SiSnNa$ $^+$: 637.2505. Found: 637.2502.

4.4.5 Cross-Coupling of 9 with 1-Iodo-4-Nitrobenzene (9-Ar)

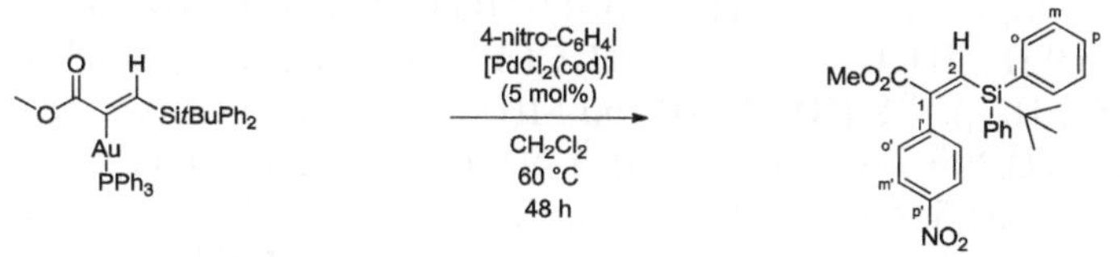

9 9-Ar

Complex **9** (96 mg, 0.123 mmol, 1 eq), 1-iodo-4-nitrobenzene (30.5 mg, 0.123 mmol, 1 eq) and [$PdCl_2$(cod)] (1.8 mg, 0.006 mmol, 0.05 eq) were transferred into a Schlenk tube. Dichloromethane (5 mL) was added. The mixture was stirred for 48 h at 60 °C. Volatiles were removed under vacuum and the remaining brown residue was purified by column chromatography (silica gel, eluent: pentane/ethyl acetate, 9:1, R_f 0.40) to give **9-Ar** as a colorless oil. Yield: 46 mg (84 %). **^{1}H NMR** (300 MHz, C_6D_6): δ 8.06 (s, 1H, H_{vinyl}), 7.32–6.86 (m, 14H, H_{ar}), 3.40 (s, 3H, OCH_3), 0.97 (s, 9H, H_{tBu}); **^{29}Si{^{1}H} NMR** (60 MHz, C_6D_6): δ −11.6 (s); **^{13}C{^{1}H} NMR** (75 MHz, C_6D_6): δ 165.6 (s, C = O), 149.2 (s, $C_{i'}$), 147.0 (s, $C_{p'}$), 143.0 (s, C_1), 140.7 (s, C_2), 135.8 (s, C_o or C_m), 132.5 (s, C_i), 129.9 (s, $C_{o'}$ or $C_{m'}$), 129.3 (s, $C_{o'}$ or $C_{m'}$), 127.6 (s, C_o or C_m), 121.9 (s, C_p), 51.9 (s, OCH_3), 27.3 (s, C(CH_3)$_3$), 18.3 (s, *C*(CH_3)$_3$); **HRMS (ESI+)**: calcd for $[M + H]^+ = C_{26}H_{28}NO_4Si^+$: 446.1788. Found: 446.1790.

4.4.6 Cross-Coupling of 9 with Allyl Bromide (9-Allyl)

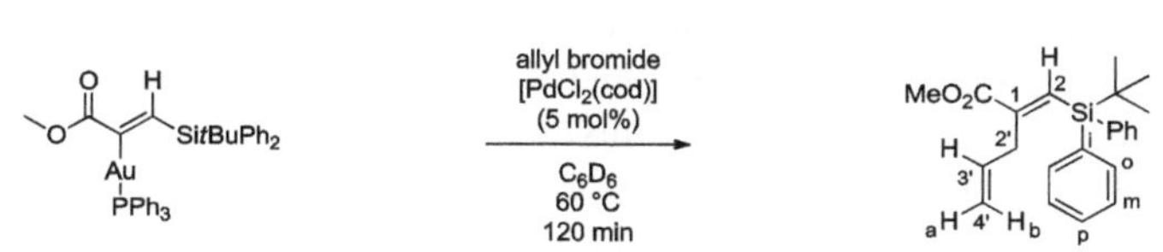

9 9-Allyl

Complex **9** (30 mg, 0.038 mmol, 1 eq) and [$PdCl_2$(cod)] (0.2 mg, 1.9 μmol, 0.05 eq) were introduced in a J. Young NMR tube and solubilized in 0.5 mL of benzene-d_6. Allyl bromide (39 μL, 0.19 mmol, 5 eq) was then added using a microsyringe. The tube was closed and heated for two hours in an oil bath at 60 °C. The ^{1}H–NMR spectra indicated quantitative conversion of **9**. The mixture was transferred into a Schlenk tube and volatiles were removed in vacuo. The brownish residue was purified by preparative thin layer chromatography (silica gel, eluent pentane/dichloromethane, 8:2, R_f 0.4) to give the diene **9-Allyl** as a colorless oil. Yield: 10.1 mg (73 %). **^{1}H NMR** (300 MHz, C_6D_6): δ 7.81–7.66 (m, 4H, H_{ar}), 7.63

(s, 1H, H_{vinyl}), 7.19–7.10 (m, 6H, H_{ar}), 5.56 (tdd, $^3J_{HH2'}$ = 6.5 Hz, $^3J_{HH4'a}$ = 10.4 Hz, $^3J_{HH4'b}$ = 16.9 Hz, 1H, $H_{3'}$), 4.75 (tdd, $^4J_{HH2'}$ = 1.3 Hz, $^2J_{HH4'b}$ = 1.7 Hz, $^3J_{HH3'}$ = 10.4 Hz, 1H, $H_{4'a}$), 4.62 (tdd, $^4J_{HH2'}$ = 1.6 Hz, $^2J_{HH4'a}$ = 1.7 Hz, $^3J_{HH3'}$ = 16.9 Hz, 1H, $H_{4'b}$), 3.43 (s, 3H, OCH_3), 3.05 (m, $^3J_{HH3'}$ = 6.5 Hz, 2H, $H_{2'}$), 1.04 (s, 9H, H_{tBu}); **^{29}Si{^{1}H} NMR** (60 MHz, C_6D_6): δ −12.7 (s); **^{13}C{^{1}H} NMR** (75 MHz, C_6D_6): δ 166.8 (s, C = O), 149.9 (s, C_1), 136.1 (s, C_o), 134.7 (s, $C_{3'}$), 134.3 (s, C_2), 133.5 (s, C_i), 129.5 (s, C_p), 127.8 (s, C_m), 116.2 ($C_{4'}$), 51.3 (s, OCH_3), 37.2 (s, $C_{2'}$), 27.1 (s, $C(CH_3)_3$), 18.0 (s, $C(CH_3)_3$); **HRMS (ESI+)**: calcd for $[M + Na]^+ = C_{23}H_{28}O_2SiNa^+$: 387.1756. Found: 387.1753.

4.4.7 Reaction of 8 with Phenylacetylene (10)

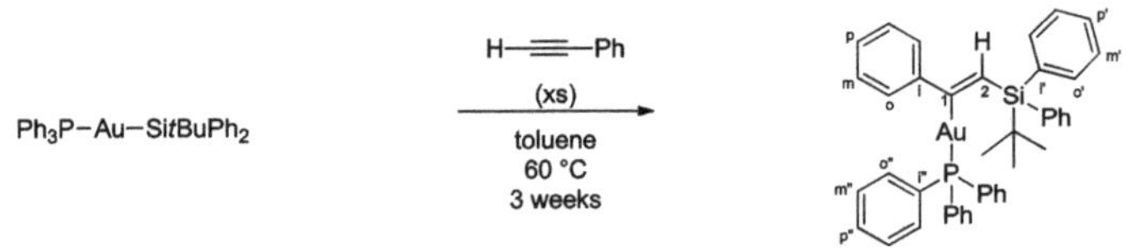

8 **10**

Phenylacetylene (0.5 mL, 4.5 mmol, 22 eq) was added to a dispersion of complex **8** (150 mg, 0.21 mmol, 1 eq) in toluene (1 mL) placed in a Schlenk tube. The Schlenk tube was sealed and heated for three weeks at 60 °C. Volatiles were removed in vacuo. The brownish residue was dispersed in dichloromethane (2 mL) and filtered over Celite to give a limpid solution. After concentrating, layering with pentane and storage at −40 °C overnight, a colorless solid precipitated. The solid residue was separated from the supernatant, washed with pentane (2 × 2 mL) and dried under vacuum to give **10** as a white solid. Yield: 72.3 mg (43 %). Although the ^{31}P{^{1}H} NMR spectrum of the crude reaction mixture suggests quantitative conversion, the isolated yield was low, which is probably due to the instability of **10** under the conditions of the purification by column chromatography. Crystals suitable for X-ray diffraction were grown at −40 °C from a saturated dichloromethane solution layered with pentane. **Mp**: 155 °C; **^{1}H NMR** (300 MHz, C_6D_6): δ 8.17–8.09 (m, 4H, H_{ar}), 8.07–7.98 (m, 2H, H_{ar}), 7.66 (d, $^4J_{HP}$ = 19.5 Hz, 1H, H_{vinyl}), 7.39–7.30 (m, 2H, H_{ar}), 7.10–7.02 (m, 6H, H_{ar}), 7.00–6.82 (m, 15H, H_{ar}), 1.23 (s, 9H, H_{tBu}); **^{31}P{^{1}H} NMR** (121 MHz, C_6D_6): δ 42.0 (s); **^{29}Si{^{1}H} NMR** (60 MHz, C_6D_6): δ −14.2 (d, $^4J_{SiP}$ = 4.2 Hz); **^{13}C{^{1}H} NMR** (75 MHz, C_6D_6): δ 204.6 (d, $^2J_{CP}$ = 112.2 Hz, C_1), 154.0 (d, $^3J_{CP}$ = 4.5 Hz, C_i), 138.4 (s, $C_{i'}$), 137.3 (s, $C_{p''}$),

134.2 (d, J_{CP} = 13.8 Hz, $C_{o''}$ or $C_{m''}$), 131.7 (d, $^3J_{CP}$ = 4.8 Hz, C_2), 130.9 (d, $^1J_{CP}$ = 47.7 Hz, $C_{i''}$), 130.3 (d, J_{CP} = 2.3 Hz, C_o or C_m), 128.5 (d, J_{CP} = 10.6 Hz, $C_{o''}$ or $C_{m''}$), 128.2 (s, $C_{o'}$ or $C_{m'}$), 128.0 (s, C_p), 127.9 (d, J_{CP} = 2.1 Hz, C_o or C_m), 127.1 (s, $C_{o'}$ or $C_{m'}$), 125.7 (s, $C_{p'}$), 27.8 (s, $C(CH_3)_3$), 18.3 (s, $C(CH_3)_3$); **HRMS (ESI+)**: calcd for $[M + H]^+ = C_{42}H_{41}SiPAu^+$: 801.2381. Found: 801.2402.

4.4.8 Reaction of 8 with Methyl 3-Phenylpropiolate (11)

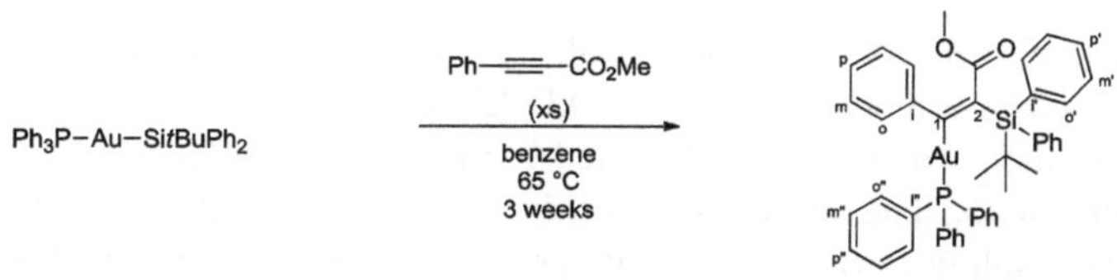

8 **11**

Methyl phenylpropiolate (1 mL, 6.8 mmol, 23 eq) was added to a dispersion of complex **8** (200 mg, 0.29 mmol, 1 eq) in benzene (1 mL) in a Schlenk tube. The Schlenk tube was sealed and heated at 65 °C. After four weeks, the mixture was filtered over Celite and volatiles were removed in vacuo. The brownish oily residue was purified by column chromatography (Neutral alumina, eluent: pentane/ethyl acetate (99:1), R_f (for pentane/ethyl acetate (8:2)): 0.23) to give **11** as a white solid. Yield: 31 mg (12 %). Although the $^{31}P\{^1H\}$NMR spectrum of the crude reaction mixture suggests quantitative conversion, the isolated yield was low, which is probably due to the instability of **11** under the conditions of the purification by column chromatography. Crystals suitable for X-ray diffraction were grown at −40 °C from a fluorobenzene solution layered with pentane. **Mp**: 152 °C (decomposition); **^{1}H NMR** (300 MHz, C_6D_6): δ 8.20–8.12 (m, 4H, H_{ar}), 7.68–7.54 (m, 4H, H_{ar}), 7.31–6.76 (m, 22H, H_{ar}), 3.38 (s, 3H, OCH_3), 1.03 (s, 9H, H_{tBu}); **^{31}P {^{1}H} NMR** (121 MHz, C_6D_6): δ 40.7 (s); **^{29}Si{^{1}H} NMR** (60 MHz, C_6D_6): δ −9.2 (d, $^4J_{SiP}$ = 4.6 Hz); **^{13}C{^{1}H} NMR** (75 MHz, C_6D_6) : δ 189.3 (d, $^2J_{CP}$ = 109.7 Hz, C_1), 175.9 (d, $^4J_{CP}$ = 13.9 Hz, C = O), 147.1 (d, $^3J_{CP}$ = 1.9 Hz, C_i), 146.3 (d, $^3J_{CP}$ = 10.0 Hz, C_2), 138.1 (s, $C_{i'}$), 137.3 (s, $C_{p''}$), 134.3 (d, $^1J_{CP}$ = 38.0 Hz, $C_{i''}$), 134.2 (d, J_{CP} = 14.5 Hz, $C_{o''}$ or $C_{m''}$), 130.8 (s, $C_{o'}$ or $C_{m'}$), 130.5 (d, J_{CP} = 1.9 Hz, C_o or C_m), 129.3 (s, $C_{o'}$ or $C_{m'}$), 128.8 (d, J_{CP} = 10.9 Hz, $C_{o''}$ or $C_{m''}$), 125.4 (s, $C_{p'}$), 49.8 (s, OCH_3), 28.8 (s, $C(CH_3)_3$), 19.3 (s, $C(CH_3)_3$); **HRMS (ESI+)**: calcd for $[M + H]^+ = C_{44}H_{43}O_2SiPAu^+$: 859.2435. Found: 859.2474; **Elt. Anal.**: calcd for $C_{45}H_{43}O_2Cl_3SiPAu$: C, 55.25; H, 4.43. Found: C, 55.26; H, 4.68.

4.4.9 Reaction of 8 with DMAD (12, 13)

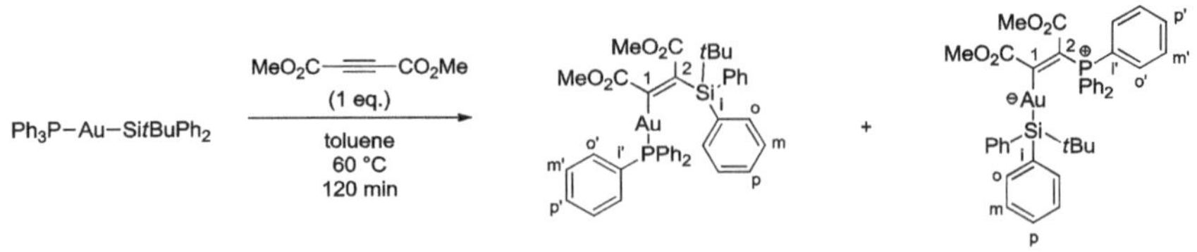

8 **12** **13**

Silylgold complex **8** (300 mg, 0.43 mmol, 1 eq) was transferred into a Schlenk tube and solubilized in toluene (4 mL). DMAD (53 μL, 0.43 mmol, 1 eq) was added. The Schlenk tube was sealed and heated for two hours at 60 °C. Volatiles were removed in vacuo. The orange-red residue was solubilized in dichloromethane (1.5 mL). This solution was layered with pentane (ca. 5 mL) and stored overnight at −60 °C to yield the zwitterionic complex **13** as a yellow precipitate, which was isolated by cannula filtration from the orange supernatant, washed with pentane (2 mL) and dried under vacuum. Volatiles of the supernatant containing vinylgold complex **12** were removed in vacuo and the orange-red residue was purified by column chromatography (silica, eluent: pentane/ethyl acetate, 8:2, R_f: 0.3) to give **12** as a white solid. Yield: **12**: 125 mg (35 %), **13**: 102 mg (28 %). Crystals suitable for X-ray crystallography of **12** and **13** were grown at 4 °C from saturated solutions of the respective compound in dichloromethane layered with pentane/benzene.

12: Mp: 223 °C; **^{1}H NMR** (300 MHz, C_6D_6): δ 8.12–8.01 (m, 4H, H_{ar}), 7.02–6.77 (m, 21H, H_{ar}), 3.66 (s, 3H, OCH_3), 3.60 (s, 3H, OCH_3), 1.24 (s, 9H, H_{tBu}); **^{31}P{^{1}H} NMR** (121 MHz, C_6D_6): δ 40.5 (s); **^{29}Si{^{1}H} NMR** (60 MHz, C_6D_6): δ −7.7 (d, $^4J_{SiP}$ = 4.1 Hz); **^{13}C{^{1}H} NMR** (126 MHz, C_6D_6): δ 194.3 (d, $^2J_{CP}$ = 111.5 Hz, C_1), 175.5 (d, J_{CP} = 3.0 Hz, C = O), 172.6 (d, J_{CP} = 11.8 Hz, C = O), 142.6 (d, $^3J_{CP}$ = 2.6 Hz, C_2), 137.2 (s, C_i), 136.9 (s, C_o or C_m), 134.2 (d, J_{CP} = 13.3 Hz, $C_{o'}$ or $C_{m'}$), 130.5 (d, $^4J_{CP}$ = 2.3 Hz, $C_{p'}$), 130.2 (d, $^1J_{CP}$ = 51.8 Hz, $C_{i'}$), 128.6 (s, C_p), 128.5 (d, J_{CP} = 10.8 Hz, $C_{o'}$ or $C_{m'}$), 127.3 (s, C_o or C_m), 50.9 (s, OCH_3), 50.8 (s, OCH_3), 28.5 (s, $C(CH_3)_3$), 18.9 (s, C($CH_3)_3$); **HRMS (ESI+)**: calcd for $[M + H]^+$ = $C_{40}H_{41}O_4SiPAu^+$: 841.2177. Found: 841.2181**; Elt. Anal.**: calcd for $C_{40}H_{40}O_4SiPAu$: C, 57.14; H, 4.80. Found: C, 58.45; H, 4.72.

13: Mp: 152 °C (decomposition); **^{1}H NMR** (300 MHz, $CDCl_3$): δ 7.81–7.63 (m, 6H, H_{ar}), 7.53–7.42 (m, 3H, H_{ar}), 7.41–7.29 (m, 6H, H_{ar}), 7.27–7.16 (m, 4H, H_{ar}), 7.11–6.98 (m, 6H, H_{ar}), 3.81 (s, 3H, OCH_3), 3.41 (s, 3H, OCH_3), 0.74 (s, 9H, H_{tBu}); **^{31}P{^{1}H} NMR** (121 MHz, $CDCl_3$): δ 16.1 (s); **^{29}Si{^{1}H} NMR** (60 MHz, $CDCl_3$): δ 31.4 (d, $^4J_{SiP}$ = 6.0 Hz); **^{13}C{^{1}H} NMR** (126 MHz, $CDCl_3$,): δ 244.36 (d, $^2J_{CP}$ = 38.3 Hz, C_1), 173.7 (d, J_{CP} = 31.3 Hz, C = O), 164.3 (d, J_{CP} = 38.2 Hz, C = O), 146.4 (s, C_i), 137.4 (s, C_o or C_m), 134.1 (d, J_{CP} = 9.6 Hz, $C_{o'}$ or $C_{m'}$), 133.6

(d, $^4J_{CP}$ = 3.2 Hz, $C_{p'}$), 129.3 (d, J_{CP} = 12.7 Hz, $C_{o'}$ or $C_{m'}$), 126.3 (s, C_p), 126.2 (s, C_o or C_m), 121.5 (d, $^1J_{CP}$ = 89.7 Hz, $C_{i'}$), 115.2 (d, $^1J_{CP}$ = 109.6 Hz, C_2), 52.2 (s, OCH_3), 51.4 (s, OCH_3), 30.0 (s, $C(CH_3)_3$), 21.1 (s, $C(CH_3)_3$); **HRMS (ESI+)**: calcd for $[M + H]^+ = C_{40}H_{41}O_4SiPAu^+$: 841.2177. Found: 841.2192.

4.4.10 (Triphenylsilyl)(Triphenylphosphine)Gold(I) Complex 14

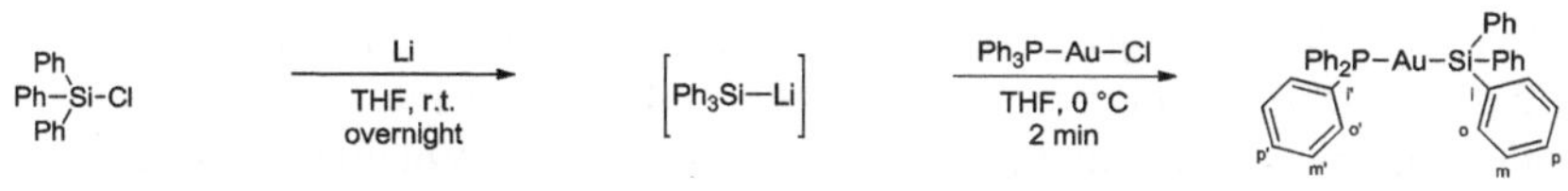

14

Finely cut, shiny lithium (278 mg, 40 mmol, 20 eq) was added to a solution of Ph_3SiCl (594.8 mg, 2 mmol, 1 eq) in tetrahydrofuran (20 mL) placed in a Schlenk flask. The reaction mixture was stirred overnight at room temperature. The dark brown-greenish solution was cooled down to 0 °C and added quickly by cannula transfer at 0 °C to a solution of Ph_3PAuCl (1 g, 2 mmol, 1 eq) in tetrahydrofuran (20 mL). The resulting brown solution was stirred for 2 min, before a pea-sized piece of solid CO_2 was added and volatiles were removed under vacuum. The dark brown residue was re-dissolved in dichloromethane (3 mL), filtered over oven-dried alumina using dichloromethane (50 mL) as the eluent. The bright red-orange filtrate was evaporated and the solid residue triturated twice with pentane. Upon scraping with the magnetic stir bar on the walls of the flask the oily residue turned into a brown-yellowish powder that was washed with pentane (3 × 3 mL) and then dried under vacuum to give the silylgold complex **14** as a brown-yellow solid. Yield: 1.18 g (81 %). **Mp**: The compound decomposes gradually without melting; **^{1}H NMR** (300 MHz, CD_2Cl_2): δ 7.70–7.10 (m); **^{31}P{^{1}H} NMR** (162 MHz, CD_2Cl_2): δ 57.1 (s); **^{29}Si{^{1}H} NMR** (79 MHz, CD_2Cl_2, 193 K): δ 22.7 (d, $^2J_{SiP}$ = 171.9 Hz); **^{13}C{^{1}H}** NMR (75 MHz, CD_2Cl_2): δ 144.3 (s, C^i), 136.8 (s, C^o or C^m), 134.3 (d, J_{CP} = 14.7 Hz, $C^{o'}$ or $C^{m'}$), 131.5 (d, $^1J_{CP}$ = 40.1 Hz, $C^{i'}$), 131.1 (d, $^4J_{CP}$ = 2.0 Hz, $C^{p'}$), 129.1 (d, J_{CP} = 10.5 Hz, $C^{o'}$ or $C^{m'}$), 127.7 (s, C^p), 127.4 (s, C^o or C^m). Compound **14** is stable over several days in dichloromethane. However, as previously reported [6], we observed rapid formation of the ion pair $[Au(PPh_3)_2][Au(SiPh_3)_2]$ in chloroform. **Elt. Anal.**: calcd for $C_{36}H_{30}PSiAu$: C, 60.17; H, 4.21. Found: C, 59.90; H, 4.07.

4.4.11 Reaction of 14 with Methyl Propiolate (15)

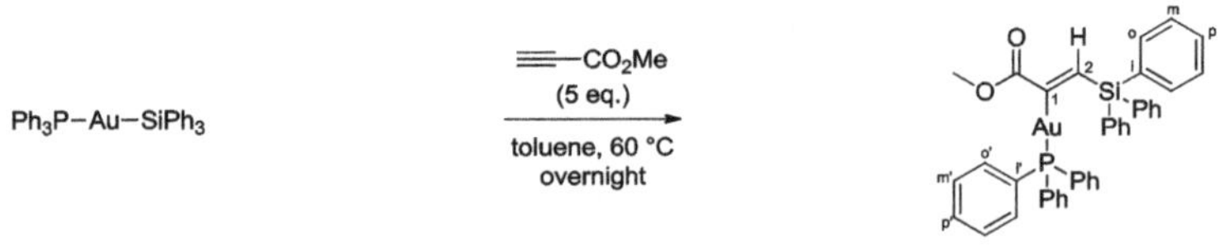

14 **15**

Methyl propiolate (125 μL, 1.40 mmol, 5 eq) was added to a solution of complex **14** (200 mg, 0.28 mmol, 1 eq) in toluene (5 mL) placed into a Schlenk tube. The Schlenk tube was sealed and heated overnight at 60 °C. Volatiles were removed in vacuo. The pale-yellow residue was re-dissolved in dichloromethane (2 mL) and filtered over neutral alumina. After removal of volatiles under vacuum, complex **15** was obtained as a white powder. Yield: 209 mg (93 %). **Mp**: 158 °C; **^{1}H NMR** (300 MHz, C_6D_6): δ 8.45 (d, $^{4}J_{HP}$ = 17.9 Hz, 1H, H_{vinyl}), 7.80–7.70 (m, 6H, H_{ar}), 7.15–7.72 (m, 24H, H_{ar}), 3.58 (s, 3H, OCH_3); **^{31}P{^{1}H} NMR** (121 MHz, C_6D_6): δ 41.6 (s); **^{29}Si{^{1}H} NMR** (60 MHz, C_6D_6): δ −21.5 (d, $^{4}J_{SiP}$ = 4.5 Hz); **^{13}C{^{1}H} NMR** (75 MHz, C_6D_6): δ 191.3 (d, $^{2}J_{CP}$ = 114.0 Hz, C_1), 175.2 (d, $^{3}J_{CP}$ = 5.5 Hz, C = O), 144.9 (d, $^{3}J_{CP}$ = 2.8 Hz, C_2), 137.0 (s, C_i), 136.6 (s, C_o), 134.3 (d, $^{2}J_{CP}$ = 14.2 Hz, $C_{o'}$), 130.7 (d, $^{1}J_{CP}$ = 50.0 Hz, $C_{i'}$), 130.5 (d, $^{4}J_{CP}$ = 2.4 Hz, $C_{p'}$), 128.8 (s, C_p), 128.6 (d, $^{3}J_{CP}$ = 11.0 Hz, $C_{m'}$), 127.5 (s, C_m), 51.0 (s, OCH_3); **HRMS (ESI +)**: calcd for $[M + H]^+$ = $C_{40}H_{35}O_2SiPAu^+$: 803.1809. Found: 803.1827; **Elt. Anal.**: calcd for $C_{40}H_{34}O_2SiPAu$: C, 59.85; H, 4.27. Found: C, 59.75; H, 3.87.

4.4.12 Reaction of 14 with Methyl 3-Phenylpropiolate (16, 17)

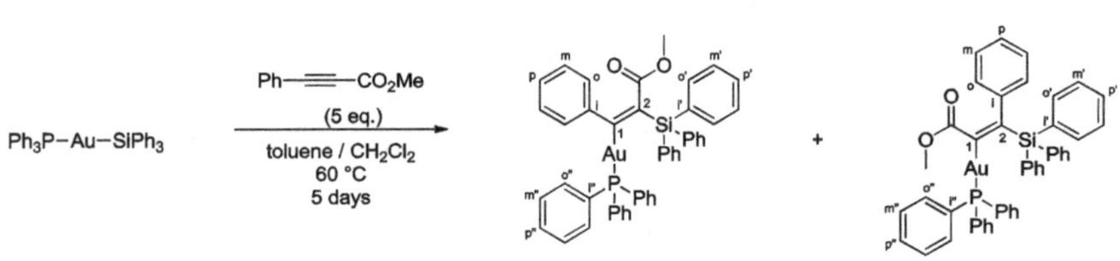

14 **16** **17**

Methyl 3-phenylpropiolate (205 μL, 1.40 mmol, 5 eq) was added to a solution of silylgold complex **14** (200 mg, 0.28 mmol, 1 eq) in toluene (2 mL) as well as dichloromethane (1 mL) placed into a Schlenk tube. The Schlenk tube was sealed and heated for five days at 60 °C. Volatiles were removed in vacuo. The residue was purified by column chromatography (neutral alumina, eluent: pentane, pentane/ethyl acetate, 9:1) to give the vinylgold complexes **16** (R_f 0.26) and **17** (R_f 0.21) as white solids. Yield: **16**, 42 mg (17 %); **17**, 57 mg (23 %). Although the $^{31}P\{^1H\}$NMR spectrum of the crude reaction mixture suggests quantitative conversion, isolated yields were low, which is probably due to the instability of **16** and **17** under the conditions of the purification by column chromatography. Crystals suitable for X-ray diffraction were grown at −40 °C from a fluorobenzene solution of **17** layered with pentane.

16: Mp: 251 °C (decomposition); **^{1}H NMR** (400 MHz, CD_2Cl_2): δ 7.76–6.87 (m, 35H, H_{ar}), 3.06 (s, 3H, OCH_3); **$^{31}P\{^1H\}$ NMR** (162 MHz, CD_2Cl_2): δ 40.7 (s); **$^{29}Si\{^1H\}$ NMR** (80 MHz, CD_2Cl_2): δ −23.1 (d, $^4J_{SiP}$ = 3.9 Hz); **$^{13}C\{^1H\}$ NMR** (100 MHz, CD_2Cl_2): δ 202.9 (d, $^2J_{CP}$ = 109.5 Hz, C_1), 172.7 (d, $^4J_{CP}$ = 13.6 Hz, C = O), 151.4 (d, $^3J_{CP}$ = 1.8 Hz, C_i), 147.8 (d, $^3J_{CP}$ = 8.0 Hz, C_2), 136.8 (s, C_p), 136.6 (s, $C_{i'}$), 136.5 (s, $C_{o'}$ or $C_{m'}$), 134.0 (d, J_{CP} = 13.9 Hz, $C_{o''}$ or $C_{m''}$), 130.8 (d, $^4J_{CP}$ = 2.6 Hz, $C_{p''}$), 130.2 (d, $^1J_{CP}$ = 50.6 Hz, $C_{i''}$), 129.0 (s, $C_{p'}$), 128.7 (d, J_{CP} = 10.9 Hz, $C_{o''}$ or $C_{m''}$); 127.7 (s, C_m); 127.4 (s, $C_{o'}$ or $C_{m'}$); 126.3 (d, $^4J_{CP}$ = 1.9 Hz, C_o); 50.0 (s, OCH_3); **HRMS (ESI+)**: calcd for $[M + H]^+ = C_{46}H_{39}O_2SiPAu^+$: 879.2122. Found: 879.2131.

17: Mp: 229 °C (decomposition); **^{1}H NMR** (400 MHz, CD_2Cl_2): δ 7.75–7.71 (m, 1H, H_{ar}), 7.62–6.83 (m, 34H, H_{ar}), 3.45 (s, 3H, OCH_3); **$^{31}P\{^1H\}$ NMR** (121 MHz, CD_2Cl_2): δ 40.9 (s); **$^{29}Si\{^1H\}$ NMR** (75 MHz, CD_2Cl_2): δ −9.2 (d, $^4J_{SiP}$ = 4.5 Hz); **$^{13}C\{^1H\}$ NMR** (100 MHz, C_6D_6): δ 186.8 (d, $^2J_{CP}$ = 108.5 Hz, C_1), 176.6 (d, $^3J_{CP}$ = 3.8 Hz, C = O), 147.8 (d, $^3J_{CP}$ = 1.9 Hz, C_2), 144.9 (d, $^4J_{CP}$ = 9.8 Hz, C_i), 136.7 (s, $C_{o'}$ or $C_{m'}$), 136.5 (s, $C_{i'}$), 134.2 (d, J_{CP} = 14.2 Hz, $C_{o''}$ or $C_{m''}$), 131.0 (d, $^4J_{CP}$ = 2.3 Hz, $C_{p''}$), 129.9 (d, $^1J_{CP}$ = 52.0 Hz, $C_{i''}$), 128.9 (s, C_o or C_m), 128.7 (d, $^3J_{CP}$ = 10.9 Hz, $C_{o''}$ or $C_{m''}$), 128.1 (s, C_o or C_m), 127.3 (s, $C_{o'}$ or $C_{m'}$), 127.0 (s, $C_{p'}$), 125.0 (s, C_p), 50.3 (s, OCH_3); **HRMS (ESI+)**: calcd for $[M + H]^+ = C_{46}H_{39}O_2SiPAu^+$: 879.2122. Found: 879.2140.

4.4.13 Reaction of 14 with Ethyl Buta-2,3-Dienoate (18)

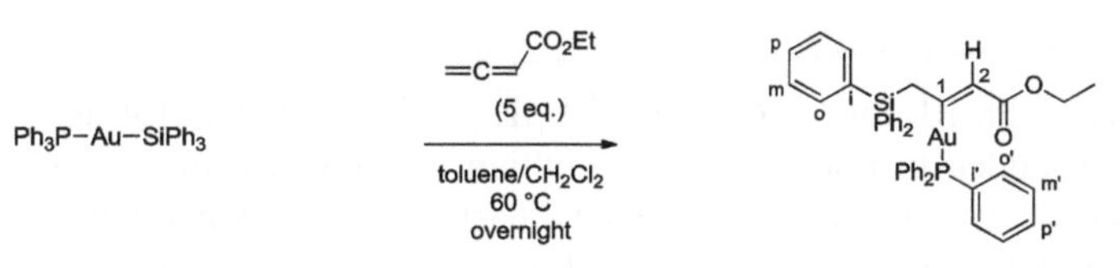

14 **18**

Ethyl buta-2,3-dienoate (163 μL, 1.40 mmol, 5 eq) was added to a solution of complex **5** (200 mg, 0.28 mmol, 1 eq) in toluene (3 mL) as well as dichloromethane (1 mL) placed into a Schlenk tube. The Schlenk tube was sealed and heated overnight at 60 °C under stirring. Volatiles were removed in vacuo. The residue was purified by column chromatography (neutral alumina, eluent: pentane, pentane/ethyl acetate, 9:1, R_f 0.50) to give gold complex **18** as a white powder. Yield: 63 mg (27 %). Although the $^{31}P\{^1H\}$NMR spectrum of the crude reaction mixture suggests quantitative conversion, the isolated yield was low, which is probably due to the instability of **18** under the conditions of the purification by column chromatography. Crystals suitable for X-ray diffraction were grown at −40 °C from a fluorobenzene solution of **18** layered with benzene and pentane. **Mp**: 133 °C; **^{1}H NMR** (300 MHz, CD_2Cl_2): δ 7.72 (m, 30H, H_{ar}), 6.44 (dt, $^4J_{HP}$ = 14.7 Hz, $^4J_{HH}$ = 1.0 Hz, 1H, H_{vinyl}), 4.12 (q, 2H, H_{CH2-O}), 3.21 (dd, $^4J_{HP}$ = 5.8 Hz, $^4J_{HH}$ = 1.0 Hz, 2H, H_{CH2-Si}), 1.22 (t, 3H, H_{CH3}); **$^{31}P\{^1H\}$ NMR** (121 MHz, CD_2Cl_2): δ 41.5 (s); **$^{29}Si\{^1H\}$ NMR** (60 MHz, CD_2Cl_2): δ −13.0 (s); **$^{13}C\{^1H\}$ NMR** (75 MHz, CD_2Cl_2): δ 199.2 (d, $^2J_{CP}$ = 109.3 Hz, C_1), 169.5 (d, $^4J_{CP}$ = 4.9 Hz, C = O), 136.8 (s, C_i), 136.2 (s, C_o or C_m), 134.4 (d, $^2J_{CP}$ = 14.0 Hz, $C_{o'}$), 131.0 (d, $^1J_{CP}$ = 49.8 Hz, $C_{i'}$) 130.9 (d, $^4J_{CP}$ = 2.1 Hz, $C_{p'}$), 129.2 (s, C_p), 128.8 (d, $^3J_{CP}$ = 10.8 Hz, $C_{m'}$), 127.5 (s, C_o or C_m), 127.2 (d, $^3J_{CP}$ = 1.3 Hz, C_2), 58.8 (s, C_{CH2-O}), 35.0 (d, $^3J_{CP}$ = 4.0 Hz, C_{CH2-Si}), 14.3 (s, C_{CH3}); **HRMS (ESI+)**: calcd for $[M + H]^+$ = $C_{42}H_{39}O_2SiPAu^+$: 831.2122. Found: 831.2147; **Elt. Anal.**: calcd for $C_{42}H_{38}O_2SiPAu$: C, 60.72; H, 4.61. Found: C, 60.65; H, 4.37.

4.4.14 (Tert-Butyldiphenylsilyl)(Trimethylphosphine)Gold(I) Complex 19

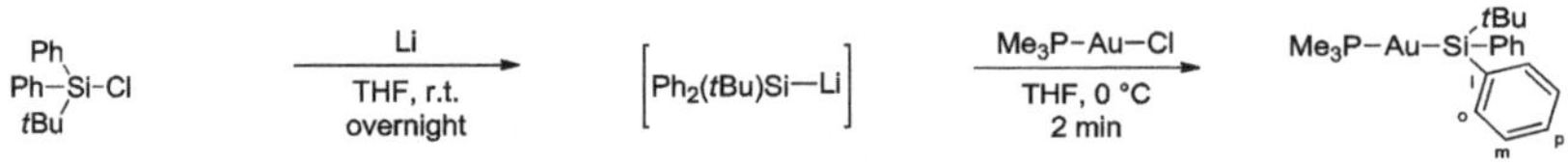

19

A solution of $Ph_2tBuSiCl$ (161 mg, 152 μL, 0.65 mmol, 1 eq) in tetrahydrofuran (2 mL) was added to finely cut, shiny lithium (90 mg, 13 mmol, 20 eq) in tetrahydrofuran (2 mL) placed in a Schlenk flask. The reaction mixture was stirred overnight at room temperature. The resulting dark brown solution was cooled down to −20 °C and then quickly added by cannula transfer to a solution of Me_3PAuCl (200 mg, 0.65 mmol, 1 eq) in tetrahydrofuran (3 mL) at −20 °C. The resulting red-brown solution was stirred for 2 min at −20 °C, then slowly warmed up to room temperature. The solution was concentrated by a gentle flux of argon gas to a volume of about 3 mL and then filtered over a pad of oven-dried alumina. The filter frit was additionally rinsed with dichloromethane (10 mL). The clear yellowish filtrate was concentrated by a gentle flux of argon gas to a volume of about 2 mL, filtered, layered with pentane (10 mL) and stored for 4 days at −60 °C. The formed colorless block-shaped crystals were isolated from the supernatant which was concentrated and stored at −60 °C for a second crop of product. Yield: 245 mg (74 %). **Mp**: 169–170 °C (decomposition); **^{1}H NMR** (300 MHz, C_6D_6): δ 8.14–8.04 (m, 4H, H_{ar}), 7.31–7.20 (m, 4H, H_{ar}), 7.18–7.04 (m, 2H, H_{ar}), 1.44 (s, 9H, H_{tBu}), 0.38 (d, $^2J_{HP}$ = 7.9 Hz, 9H, H_{PMe3}); **^{31}P{^{1}H} NMR** (121 MHz, C_6D_6): δ 31.3 (s); **^{29}Si{^{1}H} NMR** (79 MHz, CD_2Cl_2, 193 K): δ 35.4 (d, $^2J_{SiP}$ = 181.4 Hz); **^{13}C {^{1}H}** NMR (75 MHz, C_6D_6): δ 145.4 (s, C_i), 137.6 (s, C_o or C_m), 127.4 (s, C_p), 127.2 (s, C_o or C_m), 30.5 (s, C(CH$_3$)$_3$), 22.3 (s, C(CH$_3$)$_3$), 14.2 (d, $^1J_{CP}$ = 22.9 Hz, C_{PMe3}); **HRMS (ESI+)**: calcd for $[M + Na]^+$ = $C_{19}H_{28}PSiAuNa^+$: 535.1216. Found: 535.1261.

4.4.15 Reaction of 19 with Methyl Propiolate (20)

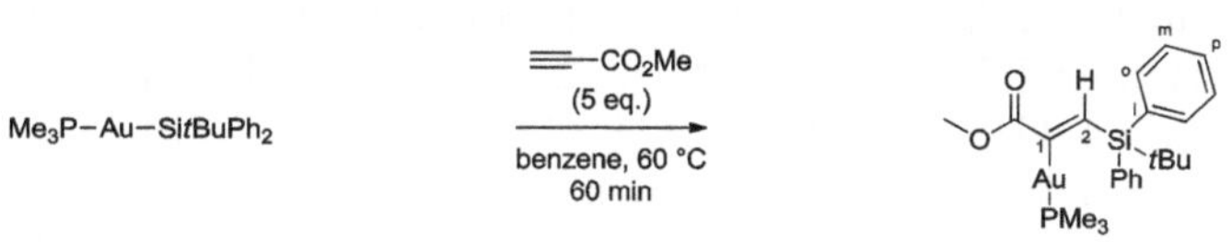

19 **20**

Methyl propiolate (87 μL, 0.98 mmol, 5 eq) was added to silylgold complex **19** (100 mg, 0.19 mmol, 1 eq) solubilized in benzene (4 mL) in a Schlenk tube. The Schlenk tube was sealed and heated for 60 min at 60 °C. After complete conversion, volatiles were removed in vacuo. The residue was re-dissolved in dichloromethane and filtered over a pad of silica gel. The solvent was removed and the residue triturated twice with pentane (2 × 2 mL) to give complex **20** as a white crystalline solid. Yield: 109 mg (96 %). **Mp**: 122–123 °C (decomposition); **^{1}H NMR** (300 MHz, C_6D_6): δ 8.16 (d, $^4J_{HP}$ = 18.9 Hz, 1H, H_{vinyl}), 8.00–7.93 (m, 4H, H_{ar}), 7.14–7.00 (m, 6H, H_{ar}), 3.59 (s, 3H, OCH_3), 1.10 (s, 9H, H_{tBu}), 0.11 (d, $^2J_{HP}$ = 9.5 Hz, 9H, H_{PMe3}); **^{31}P{^{1}H} NMR** (121 MHz, C_6D_6): δ 2.3 (s); **^{29}Si{^{1}H} NMR** (60 MHz, $CDCl_3$): δ −13.1 (d, $^4J_{SiP}$ = 4.2 Hz); **^{13}C{^{1}H} NMR** (75 MHz, C_6D_6) δ 193.7 (d, $^2J_{CP}$ = 119.6 Hz, C_1), 176.2 (d, $^3J_{CP}$ = 6.6 Hz, C = O), 142.6 (d, $^3J_{CP}$ = 2.5 Hz, C_2), 138.1 (s, C_i), 137.3 (s, C_o or C_m), 128.6 (s, C_p), 127.3 (s, C_o or C_m), 51.0 (s, OCH_3), 27.8 (s, C(CH$_3$)$_3$), 18.2 (s, C(CH$_3$)$_3$), 14.3 (d, $^1J_{CP}$ = 31.6 Hz, C_{PMe3}); **HRMS (ESI+)**: calcd for $[M + Na]^+$ = $C_{23}H_{32}O_2SiP$ $AuNa^+$: 619.1472. Found: 619.1469.

4.4.16 Reaction of 19 with Phenylacetylene (21)

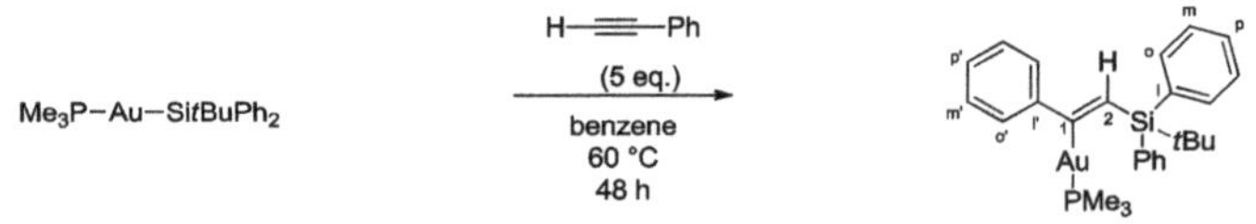

19 **21** (quant.)

Silylgold complex **19** (15 mg, 0.03 mmol, 1 eq.) in a J. Young NMR tube was solubilized in C_6D_6 (0.5 mL). Phenylacetylene (16.1 μL, 0.15 mmol, 5 eq.) was added. The tube was heated in an oilbath for 48 h at 60 °C. The reaction mixture was filtered over oven-dried alumina, eluted with dichloromethane (3 mL). Volatiles were removed in vacuo. The residue was triturated with pentane (3 × 2 mL) to give complex **21** as a waxy white solid. Yield: 18.5 mg (quant.). **^{1}H NMR** (300 MHz, C_6D_6): δ 8.15–8.05 (m, 4H, H_{ar}), 7.93–7.85 (m, 2H, H_{ar}), 7.45 (d, $^4J_{HP}$ = 20.7 Hz, 1H, H_{vinyl}), 7.34–7.23 (m, 3H, H_{ar}), 7.18–7.02 (m, 3H, H_{ar}), 6.85–6.76 (m, 3H, H_{ar}), 1.17 (s, 9H, H_{tBu}), 0.14 (d, $^2J_{HP}$ = 9.2 Hz, 9H, H_{PMe3}); **^{31}P{^{1}H} NMR** (121 MHz, C_6D_6): δ 4.0 (s); **^{29}Si{^{1}H} NMR** (60 MHz, $CDCl_3$): δ −15.2 (d, $^4J_{SiP}$ = 4.5 Hz); **^{13}C{^{1}H} NMR** (75 MHz, C_6D_6) δ 205.7 (d, $^2J_{CP}$ = 119.1 Hz, C_1), 153.9 (d, $^3J_{CP}$ = 4.9 Hz, $C_{i'}$), 139.4 (s, C_i), 137.3 (s, C_o or C_m), 132.0 (s, $C_{o'}$ or $C_{m'}$), 130.8 (d, $^3J_{CP}$ = 4.4 Hz, C_2), 128.2 (s, $C_{p'}$), 128.1 (s, C_p), 127.0 (s, C_o or C_m), 125.6 (s, C_o or C_m), 77.5 (s, *C*(CH_3)$_3$), 27.8 (s, C(*C*H_3)$_3$), 14.1 (d, $^1J_{CP}$ = 30.1 Hz, C_{PMe3}); **HRMS (ESI+)**: calcd for $[M + H]^+$ = $C_{27}H_{35}SiPAu^+$: 615.1911. Found: 615.1931.

4.4.17 (Dimethylphenylsilyl)(IPr)Gold(I) Complex 22

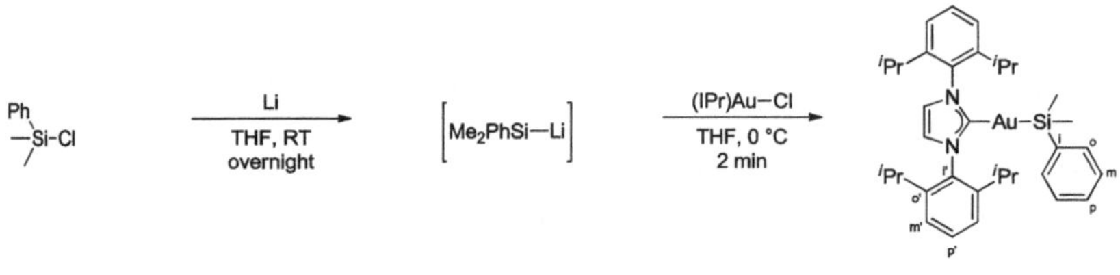

22

A solution of $Ph_2MeSiCl$ (137 mg, 135 μL, 0.81 mmol, 1 eq.) in tetrahydrofuran (2 mL) was added to finely cut, shiny lithium (167 mg, 24.15 mmol, 30 eq.) in tetrahydrofuran (2 mL) placed in a Schlenk flask. The reaction mixture was stirred overnight at room temperature. The resulting dark brown solution was cooled down to 0 °C and then quickly added by cannula transfer to a dispersion of (IPr)AuCl (500 mg, 0.81 mmol, 1 eq.) in tetrahydrofuran (15 mL) at 0 °C. The solution was stirred for 2 min at 0 °C and then warmed up to room temperature. Volatiles were removed under vacuum. The yellowish residue was re-dissolved in dichloromethane (2 mL), filtered over oven-dried alumina using dichloromethane (10 mL) as the eluent. The solvent of the clear-yellow filtrate was removed in vacuo. The off-white residue was washed twice with pentane (2 mL) and dried under vacuum to give silylgold complex **7** as a white powder. Yield: 440 mg (75 %). **Mp**: decomposition without melting at around 170 °C; **^{1}H NMR** (300 MHz, C_6D_6): δ 7.62–7.46 (m, 2H, H_{ar}), 7.36–6.97 (m, 9H, H_{ar}), 6.29 (s, 2H, $H_{NC(H)C(H)N}$), 2.62 (sept, $^{3}J_{HH}$ = 6.9 Hz, 4H, $H_{CH\text{-}iPr}$), 1.44 (d, $^{3}J_{HH}$ = 6.9 Hz, 12H, $H_{CH3\text{-}iPr}$), 1.09 (d, $^{3}J_{HH}$ = 6.9 Hz, 12H, $H_{CH3\text{-}iPr}$), 0.47 (s, 6H, H_{SiCH3}); **^{29}Si{^{1}H} NMR** (60 MHz, C_6D_6): δ 12.87 (s); **^{13}C{^{1}H} NMR** (75 MHz, $CDCl_3$): δ 217.4 (s, $C_{carbene}$), 152.2 (s, C_i), 145.8 (s, $C_{o'}$), 135.5 (s, C_o or C_m), 134.7 (s, $C_{i'}$), 130.3 (s, $C_{p'}$), 127.0 (s, C_o or C_m), 126.5 (s, C_p), 124.0 (s, $C_{m'}$), 122.4 (s, $C_{NC(H)C(H)N}$), 28.9 (s, $C_{iPr\text{-}CH}$), 24.7 (s, $C_{iPr\text{-}CH3}$), 23.7 (s, C $C_{iPr\text{-}CH3}$), 5.7 (s, C_{SiCH3}); **HRMS (ESI+)**: calcd for $[M + Na]^+$ = $C_{35}H_{47}N_2NaSiAu^+$: 743.3072. Found: 743.3083.

4.4.18 Reaction of 22 with Methyl Propiolate (23)

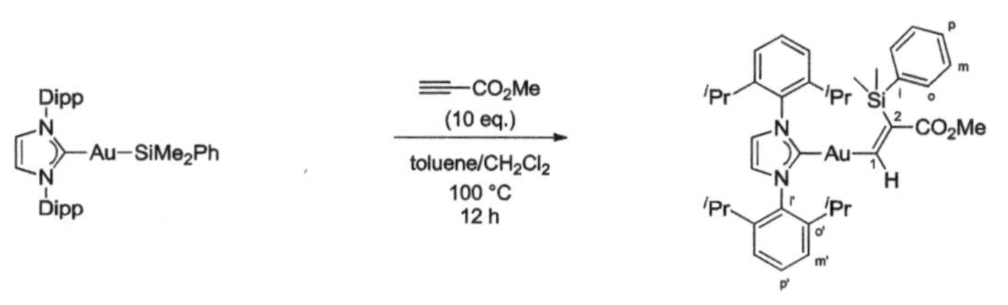

22 **23**

Methyl propiolate (125 μL, 0.98 mmol, 10 eq) was added to silylgold complex **22** (100 mg, 0.14 mmol, 1 eq) solubilized in a toluene/dichloromethane mixture (1:1, 4 mL) in a Schlenk tube. The Schlenk tube was sealed and heated for 12 h at 100 °C. After complete conversion, the reaction mixture was cooled down to room temperature, filtered over a Celite pad and eluted with dichloromethane (10 mL). Volatiles were removed in vacuo. The off-white residue was triturated twice with pentane (2 × 3 mL) and then resolubilized in a minimum volume of dichloromethane. Pentane (3 mL) was added. The solution was stored for 3 days at −80 °C to give complex **23** as a white crystalline solid. Yield: 71 mg (63 %). Crystals suitable for X-ray diffraction were grown at −80 °C from a saturated solution of **8** in a dichloromethane/pentane (1:3) mixture. **Mp**: 177–178 °C (decomposition); **^{1}H NMR** (300 MHz, $CDCl_3$): δ 9.20 (s, 1H, H_{vinyl}), 7.51–7.97 (m, 13H, H_{ar} & $H_{NC(H)C}$

$_{(H)N}$), 3.34 (s, 3H, OCH_3), 2.54 (sept, $^3J_{HH}$ = 6.9 Hz, 4H, $CH(CH_3)_2$), 1.23 (d, $^3J_{HH}$ = 6.9 Hz, 12H, $CH(CH_3)_2$), 1.14 (d, $^3J_{HH}$ = 6.9 Hz, 12H, $CH(CH_3)_2$), 0.01 (s, 6H, H_{SiMe2}); **^{29}Si{^{1}H} NMR** (60 MHz, $CDCl_3$): δ −13.6 (s); **^{13}C{^{1}H} NMR** (75 MHz, $CDCl_3$) δ 196.8 (s, C_2), 196.7 (s, $C_{carbene}$), 189.5 (s, C = O), 171.5 (s, C_1), 145.6 (s, $C_{o'}$), 145.0 (s, C_i), 134.6 (s, $C_{i'}$), 133.8 (s, C_o or C_m), 130.3 (s, $C_{p'}$), 127.6 (s, C_p), 127.0 (s, C_o or C_m), 124.1 ($C_{m'}$), 123.0 (s, $C_{NC(H)C(H)N}$), 50.5 (s, OCH_3), 28.7 (s, $C(CH_3)_3$), 24.2 (s, $C(CH_3)_3$), 24.1 (s, $C(CH_3)_3$), −0.4 (s, C_{SiMe2}); **HRMS (ESI+)**: calcd for $[M + H]^+$ = $C_{39}H_{52}N_2O_2Si^+$: 805.3474. Found: 805.3464.

4.4.19 Kinetic Measurements

4.4.19.1 Experimental Details

Reaction 1: methyl propiolate + complex **8**

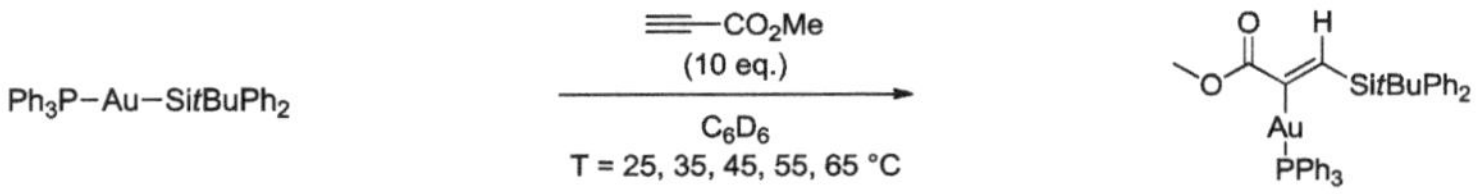

8 **9**

In a glovebox, 200 μL of a standard stock solution (mesitylene (40.3 μL, 0.290 mmol) in C_6D_6 (1959.7 μL)) and 200 μL of a stock solution of methyl propiolate (259.0 μL (2.898 mmol) in C_6D_6 (1741 μL)) were added to a solution of $Ph_3PAuSi^tBuPh_2$ (20 mg, 0.029 mmol) in C_6D_6 (200 μL). The tube was placed immediately inside the NMR machine and the reaction was followed by ^{1}H NMR spectroscopy over time. Spectra were referenced to the aromatic signal of the internal standard (mesitylene, δ = 6.71 ppm). By integration of the methyl signal of the internal standard (δ = 2.15 ppm, [mesitylene] = 0.0483 M) and integration of the signals corresponding to reactants (alkyne: H_{OCH3}, δ = 3.12 ppm; silylgold complex: $H_{C(CH3)3}$, δ = 1.50 ppm) and product (vinylgold complex: $H_{C=CH}$, δ = 8.40 ppm), their concentrations were determined. The kinetic profile of the reaction was established at 25, 35, 45, 55 and 65 °C, respectively.

Reaction 2: methyl propiolate + complex **19**

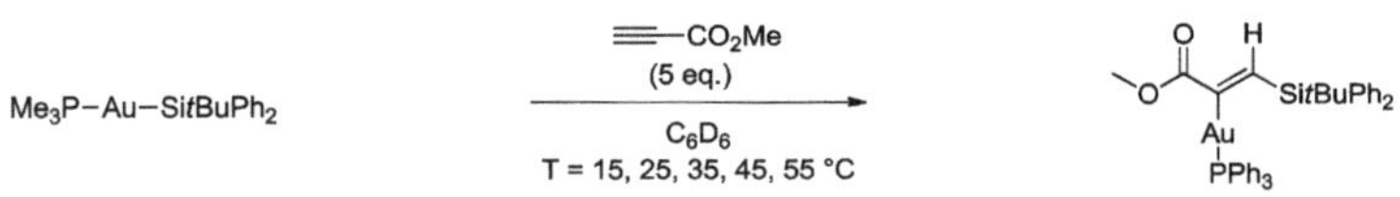

19 **20**

In a glovebox, 200 μL of a standard stock solution (mesitylene (40.3 μL, 0.290 mmol) in C_6D_6 (1959.7 μL)) and 200 μL of a stock solution of methyl propiolate (129.0 μL (1.450 mmol) in C_6D_6 (1871 μL)) were added to a solution of $Ph_3PAuSi^tBuPh_2$ (15 mg, 0.029 mmol) in C_6D_6 (200 μL). The tube was placed immediately inside the NMR machine and the reaction was followed by ^{1}H NMR spectroscopy over time. Spectra were referenced to the aromatic signal of the internal standard (mesitylene, δ = 6.71 ppm). By integration of the methyl signal of the internal standard (δ = 2.15 ppm, [mesitylene] = 0.0483 M) and integration of the signals corresponding to reactants (alkyne: H_{OCH3}, δ = 3.12 ppm; silylgold complex: $H_{P(CH3)3}$, δ = 0.48 ppm) and product (vinylgold complex: $H_{P(CH3)3}$, δ = 0.22 ppm), their concentrations were determined. The kinetic profile of the reaction was established at 15, 25, 35, 45 and 55 °C, respectively.

4.4.19.2 Discussion of the Reaction Order

A zero-order rate law for both reactions (1 and 2) is excluded, since plotting the conversion for the silylgold complex ([P–Au–Si]) or for the vinylgold complex ([Au-vinyl]) versus time does not result in a linear slope (Fig. 4.14).

Plotting ln([P–Au–Si]$_0$/[P–Au–Si]) versus time for reactions 1 and 2 does result in a linear slope, as expected for a reaction which is first-order dependant with regard to the silylgold complex (Fig. 4.15).

Both reactions (1 and 2) obey a second-order rate law, as plotting the natural logarithm for the concentration of methyl propiolate over the concentration of the silylgold complex (ln([alkyne]/[PAuSi])) versus time (t) results in a linear slope, according to Eq. 5.1, describing the behavior of a second-order reaction with unequal reactant concentrations [29].

$$\ln\frac{[\text{alkyne}]}{[\text{P-Au-Si}]} = k_{\text{obs}}([\text{alkyne}]_0 - [\text{P-Au-Si}]_0)t + \ln\frac{[\text{alkyne}]_0}{[\text{P-Au-Si}]_0} \quad (5.1)$$

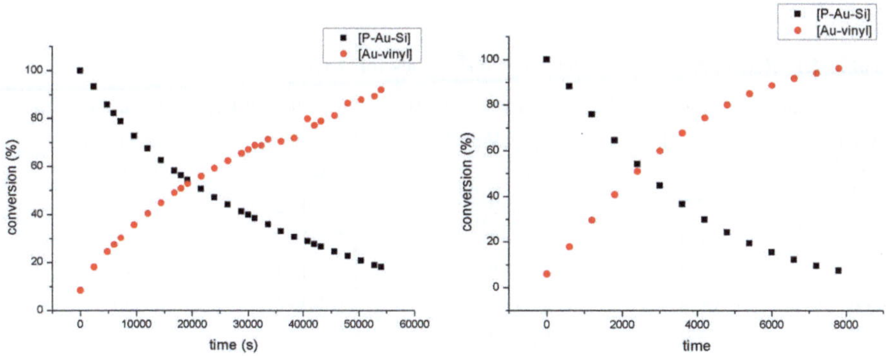

Fig. 4.14 Kinetic profile for reaction 1 (*left*) and 2 (*right*) at 25 °C

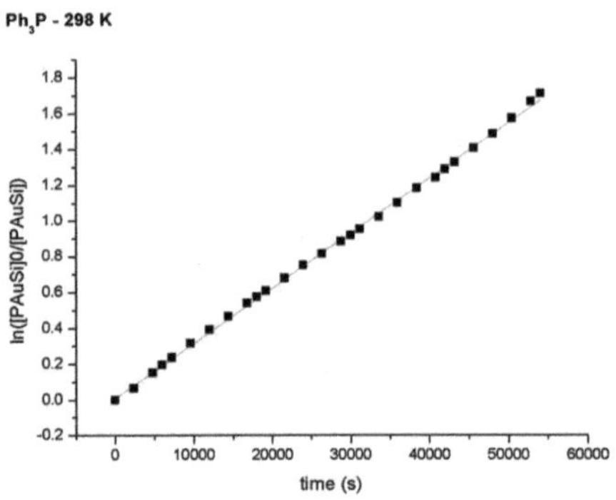

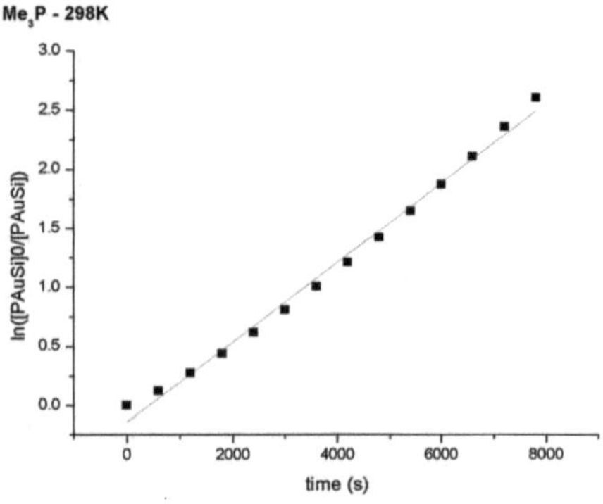

Fig. 4.15 ln([P–Au–Si]$_0$/[P–Au–Si]) over time for reaction 1 (*left*) and reaction 2 (*right*) at 25 °C

Via the slope of the straight line, the rate constants k_{obs} for the respective temperatures were determined (see Table 4.1).

NB: Determination of k_{obs} via k_1 obtained from the pseudo-first order plots (taking the concentration of alkyne into account) resulted in similar values (Figs. 4.16, 4.17, 4.18, 4.19, 4.20, 4.21, 4.22, 4.23, 4.24 and 4.25).

4.4.19.3 Activation Parameters

The activation parameters $\Delta G^{\neq}$, $\Delta H^{\neq}$ and $\Delta S^{\neq}$ for reaction 1 and 2 were obtained by means of the determined rate constants.

(a) ***Determination of $\Delta G^{\neq}$ using the Eyring equation***

Insertion of k_{obs} in the Eyring equation results in $\Delta G^{\neq}$ for the respective temperature (Table 4.1).

$$\Delta \mathrm{G}^{\neq} = RT(\ln\left(\frac{k_B T}{h}\right) - \ln(k_{\mathrm{obs}})) \tag{5.2}$$

Table 4.1 Rate constants and calculated $\Delta G^{\neq}$ values for reactions 1 and 2

Temperature	Reaction 1: k_{obs}	Reaction 1: $\Delta G^{\neq}$	Reaction 2: k_{obs}	Reaction 2: $\Delta G^{\neq}$
[°C]	[s^{-1} M^{-1}]	[kcal/mol]	[s^{-1} M^{-1}]	[kcal/mol]
15	–	–	6.34E-4	21.05
25	6.77E-5	23.13	1.60E-3	21.26
35	1.17E-4	23.59	4.71E-3	21.33
45	2.58E-4	23.88	7.85E-3	21.72
55	3.95E-4	24.37	1.74E-2	21.90
65	1.33E-3	24.32	–	–
Mean value	–	23.9	–	21.5

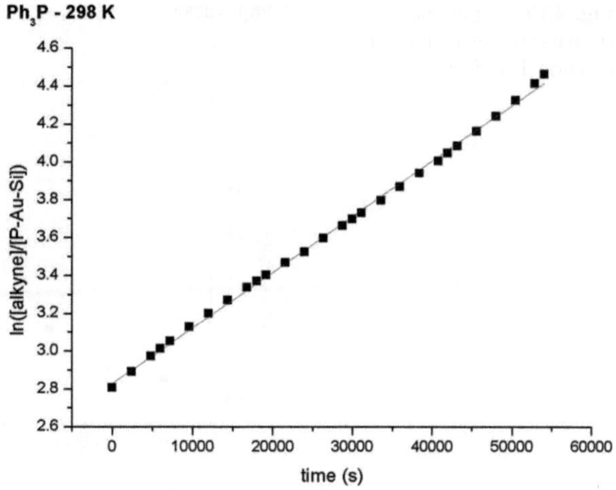

Fig. 4.16 ln([alkyne]/[P–Au–Si]) over time for reaction 1 at 25 °C

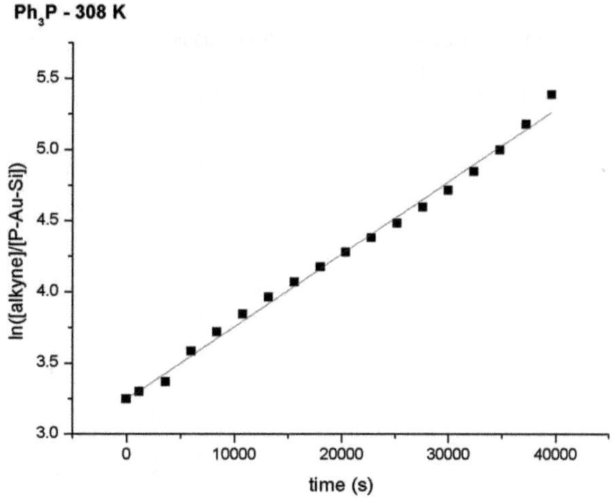

Fig. 4.17 ln([alkyne]/[P–Au–Si]) over time for reaction 1 at 35 °C

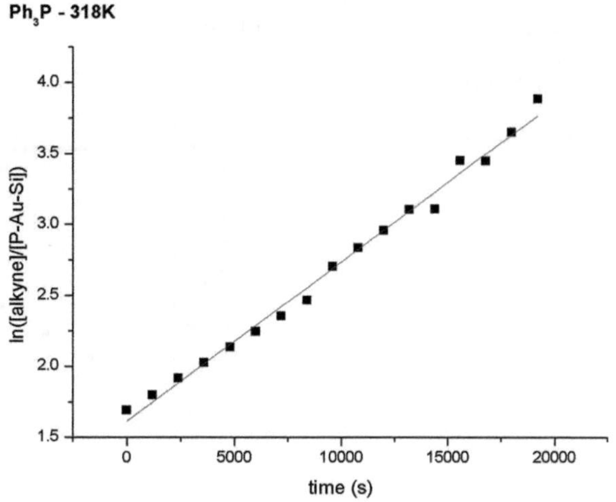

Fig. 4.18 ln([alkyne]/[P–Au–Si]) over time for reaction 1 at 45 °C

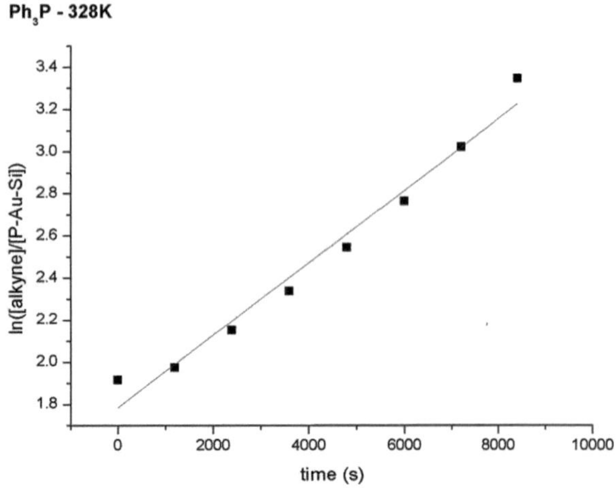

Fig. 4.19 ln([alkyne]/[P–Au–Si]) over time for reaction 1 at 55 °C

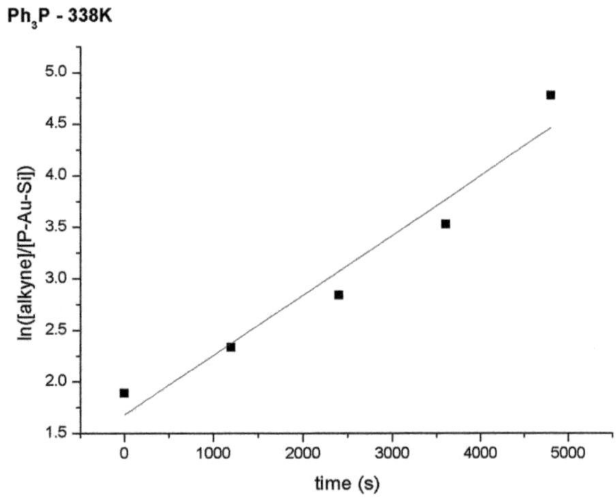

Fig. 4.20 ln([alkyne]/[P–Au–Si]) over time for reaction 1 at 65 °C

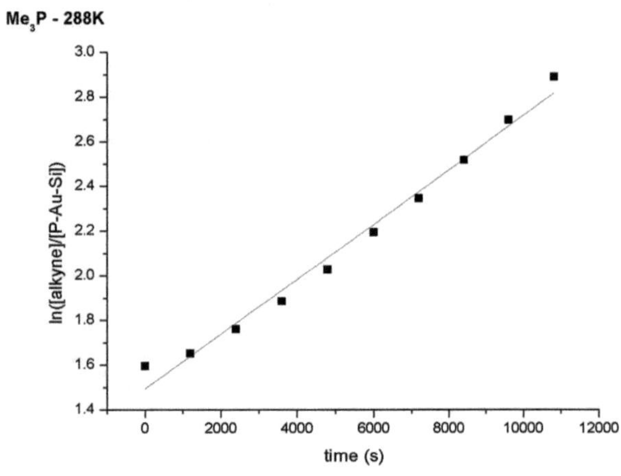

Fig. 4.21 ln([alkyne]/[P–Au–Si]) over time for reaction 2 at 15 °C

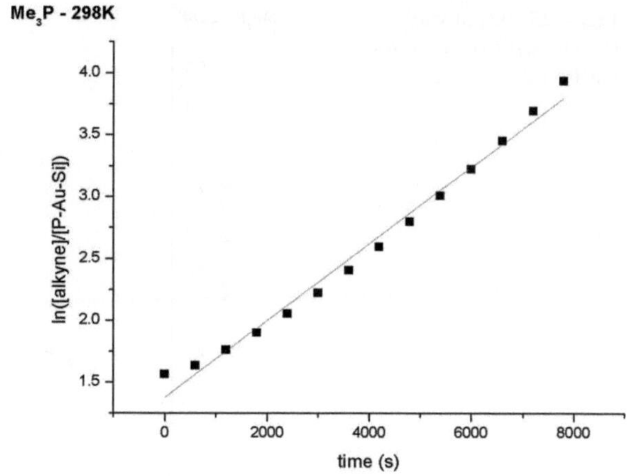

Fig. 4.22 ln([alkyne]/[P–Au–Si]) over time for reaction 2 at 25 °C

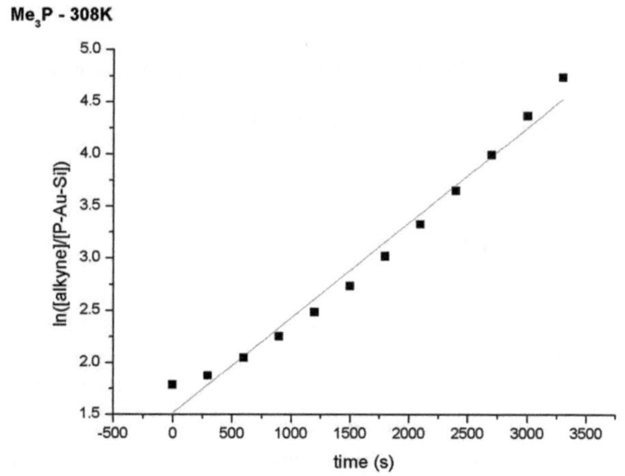

Fig. 4.23 ln([alkyne]/[P–Au–Si]) over time for reaction 2 at 35 °C

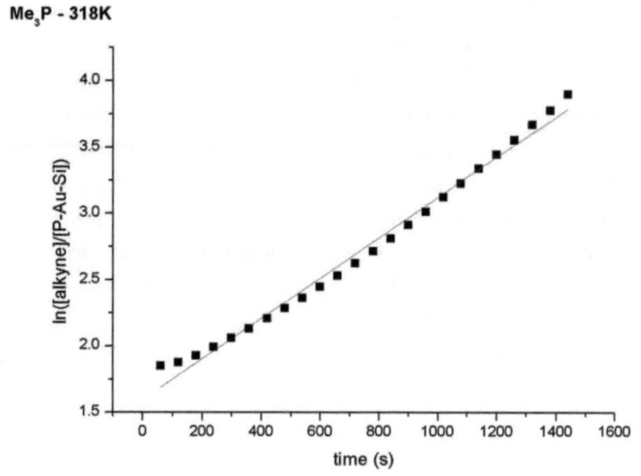

Fig. 4.24 ln([alkyne]/[P–Au–Si]) over time for reaction 2 at 45 °C

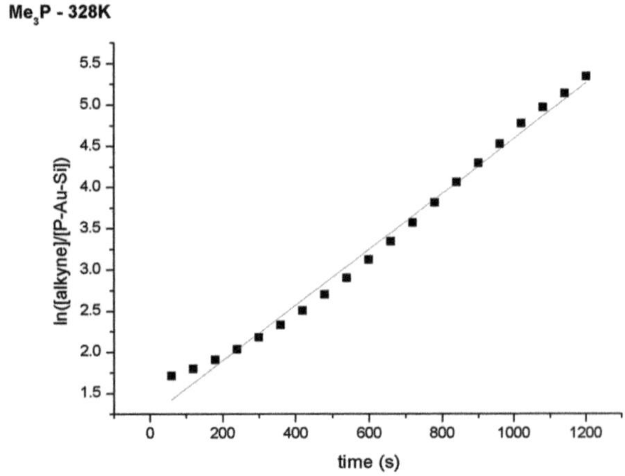

Fig. 4.25 ln([alkyne]/[P–Au–Si]) over time for reaction 2 at 55 °C

(b) ***Determination of*** $\Delta G^{\neq}$, $\Delta H^{\neq}$ ***and*** $\Delta S^{\neq}$ ***by means of an Eyring plot***

See Figs. 4.26, 4.27 and Table 4.2.

4.4.20 Computational Details

Calculations were carried out with the Gaussian 09 program suite to locate minima and transition structures on the potential energy surface of the systems studied [30]. The functionals B97D [22], M06 [31], B3PW91 [32], and PBE0 [33] were tested against the experimental data obtained by kinetic measurements.

The gold atom was described by the relativistic electron core potential (ecp-60-mwb) and the associated basis set [34], augmented by a set of f-orbital polarization functions [35]; and the 6–31G** basis set was employed for all other atoms. All stationary points involved were fully optimized. Frequency calculations were undertaken to confirm the nature of the stationary points (one imaginary frequency for transition states, zero for minima). The connectivity of the transition states and their adjacent minima was confirmed by intrinsic reaction coordinate (IRC) calculations [36, 37]. Zero-point energy (ZPE) corrections were carried out for all computed energies. Gibbs free energies were calculated by using the harmonic approximation and standard textbook procedures.

Natural bond orbital analyses were performed to compute the natural atomic charges and to obtain the electron charge transfer for selected structures computed with the B97D functional [38–41]. Molecular orbitals were drawn with Molekel 5.4 [42].

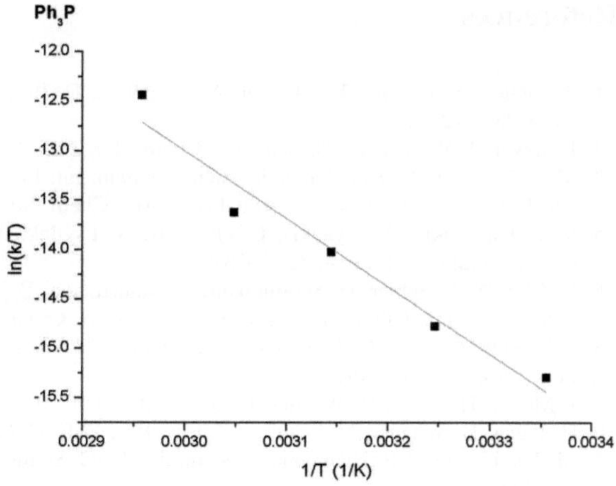

Fig. 4.26 Eyring plot for reaction 1. Temperature range: 25–65 °C. Slope: (−6874.075 ± 804.316) K; intercept: 7.624 ± 2.537

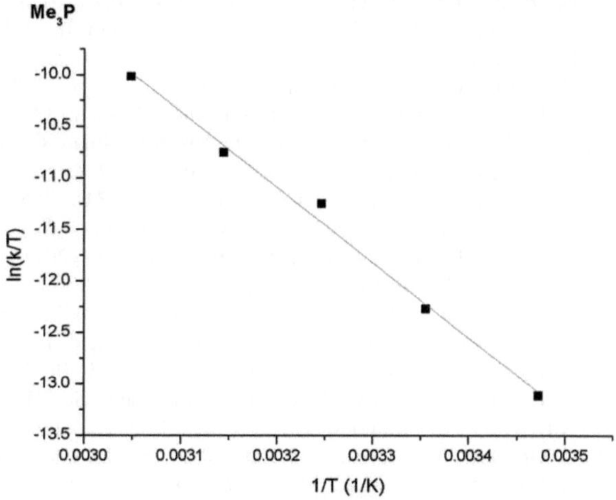

Fig. 4.27 Eyring plot for reaction 2. Temperature range: 15–55 °C. Slope: (−7302.783 ± 355.088) K; intercept: 12.278 ± 1.157

Table 4.2 Activation parameters for reactions 1 and 2 obtained by means of the corresponding Eyring plots

	Reaction 1	Reaction 2
$\Delta H^{\neq}$ [kcal/mol]	13.7 ± 1.6	14.5 ± 0.7
$\Delta S^{\neq}$ [cal/(mol·K)]	−32.0 ± 5.0	21.4 ± 2.3
$\Delta G^{\neq}$ [kcal/mol]	23.9 ± 2.3 (for 318 K) 23.2 (for 298 K)	21.5 ± 1.0 (for 308 K) 20.9 (for 298 K)

References

1. P. Gualco, S. Ladeira, K. Miqueu, A. Amgoune, D. Bourissou, Angew. Chem. Int. Ed. **50**, 8320–8324 (2011)
2. J. Meyer, J. Willnecker, U. Schubert, Chem. Ber. **122**, 223–230 (1989)
3. M. Wilfling, K.W. Klinkhammer, Angew. Chem. Int. Ed. **49**, 3219–3223 (2010)
4. I.N. Lykakis, A. Psyllaki, M. Stratakis, J. Am. Chem. Soc. **133**, 10426–10429 (2011)
5. V. Kotzabasaki, I.N. Lykakis, C. Gryparis, A. Psyllaki, E. Vasilikogiannaki, M. Stratakis, Organometallics **32**, 665–672 (2013)
6. O. Monge, A. Schier, H. Schmidbaur, Z. Naturforsch. B **54**, 26–29 (1999)
7. A.S.K. Hashmi, T.D. Ramamurthi, F. Rominger, J. Organomet. Chem. **694**, 592–597 (2009)
8. J.C. Tebby (ed.), *CRC Handbook of Phosphorus-31 Nuclear Magnetic Resonance Data* (CRC Press, Boca Raton, 1991)
9. J. Meyer, H. Piana, H. Wagner, U. Schubert, Chem. Ber. **123**, 791–793 (1990)
10. H. Piana, H. Wagner, U. Schubert, Chem. Ber. **124**, 63–67 (1991)
11. M. Theil, P. Jutzi, B. Neumann, A. Stammler, H.-G. Stammler, J. Organomet. Chem. **662**, 34–42 (2002)
12. D.S. Weinberger, M. Melaimi, C.E. Moore, A.L. Rheingold, G. Frenking, P. Jerabek, G. Bertrand, Angew. Chem. Int. Ed. **52**, 8964–8967 (2013)
13. E.Y. Tsui, P. Müller, J.P. Sadighi, Angew. Chem. Int. Ed. **47**, 8937–8940 (2008)
14. D.S. Laitar, P. Müller, T.G. Gray, J.P. Sadighi, Organometallics **24**, 4503–4505 (2005)
15. D.A. Ruiz, G. Ung, M. Melaimi, G. Bertrand, Angew. Chem. Int. Ed. **52**, 7590–7592 (2013)
16. G. Ung, G. Bertrand, Angew. Chem. Int. Ed. **52**, 11388–11391 (2013)
17. P. de Frémont, N.M. Scott, E.D. Stevens, S.P. Nolan, Organometallics **24**, 2411–2418 (2005)
18. E. Liepiņš, Y. Goldberg, I. Iovel, E. Lukevics, J. Organomet. Chem. **335**, 301–311 (1987)
19. H.-M. Chen, J.P. Oliver, J. Organomet. Chem. **316**, 255–260 (1986)
20. Y. Shi, S.D. Ramgren, S.A. Blum, Organometallics **28**, 1275–1277 (2009)
21. Y. Nakao, T. Hiyama, Chem. Soc. Rev. **40**, 4893 (2011)
22. S. Grimme, J. Comput. Chem. **27**, 1787–1799 (2006)
23. A.A.C. Braga, G. Ujaque, F. Maseras, Organometallics **25**, 3647–3658 (2006)
24. R.E.M. Brooner, R.A. Widenhoefer, Angew. Chem. Int. Ed. **52**, 11714–11724 (2013)
25. A. Fürstner, P.W. Davies, Angew. Chem. Int. Ed. **46**, 3410–3449 (2007)
26. X.-F. Zhu, C.E. Henry, O. Kwon, J. Am. Chem. Soc. **129**, 6722–6723 (2007)
27. K.D. Hesp, M. Stradiotto, J. Am. Chem. Soc. **132**, 18026–18029 (2010)
28. C. Gryparis, M. Kidonakis, M. Stratakis, Org. Lett. **15**, 6038–6041 (2013)
29. K.A. Connors, *Chemical Kinetics: The Study of Reaction Rates in Solution* (VCH, New York, 1990)
30. M.J. Frisch, G.W. Trucks, H.B. Schlegel, G.E. Scuseria, M.A. Robb, J.R. Cheeseman, G. Scalmani, V. Barone, B. Mennucci, G. A. Petersson et al., *Gaussian 09, Revision A.1.* (Gaussian, Inc., Wallingford, Connecticut, USA, 2009)
31. Y. Zhao, D.G. Truhlar, Theor. Chem. Acc. **120**, 215–241 (2008)
32. A.D. Becke, J. Chem. Phys. **98**, 5648–5652 (1993)
33. J.P. Perdew, K. Burke, M. Ernzerhof, Phys. Rev. Lett. **77**, 3865–3868 (1996)
34. D. Andrae, U. Häußermann, M. Dolg, H. Stoll, H. Preuß, Theor. Chim. Acta **77**, 123–141 (1990)
35. A.W. Ehlers, M. Böhme, S. Dapprich, A. Gobbi, A. Höllwarth, V. Jonas, K.F. Köhler, R. Stegmann, A. Veldkamp, G. Frenking, Chem. Phys. Lett. **208**, 111–114 (1993)
36. K. Fukui, Acc. Chem. Res. **14**, 363–368 (1981)
37. H.P. Hratchian, H.B. Schlegel, in Theory and Application of Computer Chemistry eds. by C.E. Dykstra, G. Frenking, K.S. Kim, G.E. Scuseria. (Elsevier, Amsterdam, 2005), pp. 195–249
38. J.K. Glendening, A.E. Badenhoop, J.E. Carpenter, J.A. Bohmann, C.M. Morales, F. Weinhold, NBO 5.0., Theoretical Chemistry Institute. (University of Wisconsin, Madison, Wisconsin, USA, 2001)

39. A.E. Reed, L.A. Curtiss, F. Weinhold, Chem. Rev. **88**, 899–926 (1988)
40. J.P. Foster, F. Weinhold, J. Am. Chem. Soc. **102**, 7211–7218 (1980)
41. A.E. Reed, F. Weinhold, J. Chem. Phys. **83**, 1736–1740 (1985)
42. U. Varetto, *Molekel 5.4*. (Swiss National Supercomputing Centre, Manno, Switzerland, 2009)

Chapter 5
Gold: Oxidative Addition to Au(I)

In this chapter, the results obtained during the course of this Ph.D. work concerning intermolecular σ-bond activation processes with gold(I) are exposed: (i) the oxidative addition of σ-SiSi bonds at monoligated cationic gold(I) complexes, (ii) the design of diphosphine gold(I) complexes suitable for oxidative addition reactions and (iii) application of the latter in σ-bond activation processes (strained carbocycles and simple aryl halides) will be discussed.

5.1 Intermolecular SiSi Bond Activation at Au(I)

As mentioned above, it was found that disilanes undergo oxidative addition to gold (I) in an intramolecular manner (see Sect. 3.3.3). In that work, the product was stabilized by chelate assistance of two (Scheme 3.13) or only one (Scheme 3.15) phosphine donor atoms pre-orientating the σ-SiSi bond and stabilizing the ensuing oxidative addition product. At the same time, chelate assistance raises the question of the role and importance of the anchoring sites used to promote disilane activation.

We therefore started to investigate if similar transformations would also proceed intermolecularly. Besides elucidating the role of chelate assistance, the study of an intermolecular oxidative addition to gold(I) should pave the way for subsequent transformations (transmetallation, CH activation, reductive elimination) as encountered for example in 2-electron redox catalysis based on group-10 metals. Prominent illustrations are the palladium-catalyzed cross-coupling reactions of aryl halides and p-block element-based carbon-nucleophiles (e.g. arylboronic acids, arylstannanes, arylsilanes, etc.) for which oxidative addition of a carbon-halogen bond to a palladium complex is the entry point in the catalytic cycle, before transmetallation and reductive elimination yields CC-coupled products.

Consequently our efforts concentrated on the identification of gold(I) complexes that are (i) able to undergo oxidative addition of disilanes and (ii) yield gold(III) complexes which are sufficiently stable to study their reactivity.

M. Joost, *Synthesis and Original Reactivity of Copper and Gold Complexes*,
Springer Theses, DOI 10.1007/978-3-319-18690-0_5

We started to explore the reactions of simple monophosphine gold complexes with disilanes.

5.1.1 Experimental Results

As a first control experiment, one equivalent of disilane $(PhMe_2Si)_2$ was added to the phosphine gold complex $(Ph_3P)AuCl$ in toluene (Scheme 5.1).

No reaction occurred over days at room temperature and progressive heating up to 100 °C only led to decomposition of the gold precursor. Neutral (L)AuCl complexes are commonly activated with chloride abstractors, and thus we sought to generate a more electrophilic gold species using $GaCl_3$. Upon addition of $GaCl_3$ (1 eq.), the ^{31}P NMR signal of $(Ph_3P)AuCl$ shifted from $\delta = 33$ ppm to $\delta = 31$ ppm. The resulting adduct is stable for hours at −80 °C, but rapidly decomposes at higher temperatures. Upon addition of $(PhMe_2Si)_2$ to a 1:1 mixture of $(Ph_3P)AuCl$ and $GaCl_3$ in CD_2Cl_2 at −90 °C (Scheme 5.2), the solution immediately turned to light yellow. Analysis of the reaction mixture by ^{31}P NMR spectroscopy at −80 °C indicated the formation of a new species **24a**.

The ^{31}P resonance signal is significantly shifted downfield (δ = 60.9 ppm) compared to the starting material (δ = 31 ppm). Compound **24a** is the major P-containing species (estimated yield by ^{31}P NMR: ~66 %). Another ^{31}P NMR signal is observed at $\delta = 43.7$ ppm, indicating the formation of the cationic complex $[Au(Ph_3P)_2]^+$ as a side-product [1].

Compound **24a** is thermally unstable. Decomposition occurs rapidly above −60 °C and according to ^{29}Si NMR spectroscopy, the main silicon containing decomposition products are the chlorosilane $PhMe_2SiCl$ and the disiloxane $PhMe_2SiOSiMe_2Ph$ [2]. Further spectroscopic characterization was thus performed at −80 °C. The ^{29}Si NMR spectrum of the reaction mixture displays three resonance signals corresponding to the starting disilane (δ = −21.0 ppm), the chlorosilane $PhMe_2SiCl$ (δ = 21.0 ppm) and complex **24a** (δ = 40.7 ppm). The signal for **24a** appears as a doublet (with a J_{SiP}

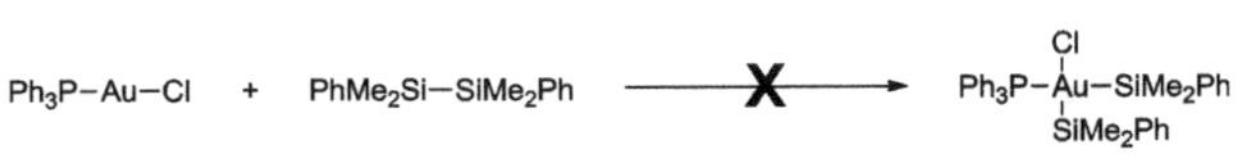

Scheme 5.1 Attempted oxidative addition of a disilane to triphenylphosphine gold(I) chloride

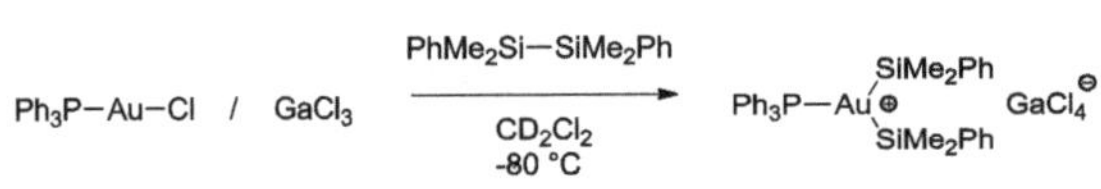

24a (~66 %)

Scheme 5.2 Formation of the phosphine bis(silyl)gold(III) complex **24a** via oxidative addition of $(SiMe_2Ph)_2$

coupling constant of 39.8 Hz) and correlates with the ^{31}P NMR signal at $\delta = 60.9$ ppm, as established by 2D HMBC(^{31}P–^{29}Si) and ^{29}Si{selective ^{31}P} NMR experiments.

Even if the ^{31}P and ^{29}Si NMR data of **24a** indicate the formation of a phosphine silylgold complex, the attribution to either a phosphine silylgold(I) species or a phosphine bis(silyl)gold(III) complex (in a T-shape or symmetric Y-shape form) as shown in Fig. 5.1 is not unequivocal.

Phosphine silylgold(I) complexes are relatively rare (compare Sect. 4.1) [3–5]. The reported ^{31}P and ^{29}Si NMR chemical shifts are in the same region than those observed for **24a**, but the associated ^{29}Si–^{31}P coupling constants are much larger (J_{SiP} *trans* = 165–200 Hz). On the other hand, the only known examples of bis (silyl) Au(III) complexes adopt square-planar geometry and display two very different ^{29}Si–^{31}P coupling constants (J_{SiP} *trans* = 110–150 Hz and J_{SiP} *cis* = 4–10 Hz) [6, 7]. The magnitude of the J_{SiP} coupling constant observed for **24a** is not consistent with a phosphine silyl gold(I) structure, but at this stage, it was not possible to establish unambiguously the presence of two silyl groups per gold center.

We thus envisioned to count the silicon atoms per PPh_3 moiety, and to this end, we recorded a ^{31}P NMR spectrum with a long acquisition time. Apart from the main signal, three sets of satellites, corresponding to ^{29}Si and ^{13}C isotopomers of **24a**, were distinguished (Fig. 5.2).

These isotopomers were unambiguously assigned based on the ^{29}Si and ^{13}C NMR spectra and on the values of the respective J_{SiP} and J_{CP} coupling constants. After deconvolution of the overlapping signals, the area of each satellite was precisely determined and by taking into account the natural abundance of the ^{13}C (1.11 %) and ^{29}Si (4.70 %) isotopes, the number of silyl groups per PPh_3 moiety

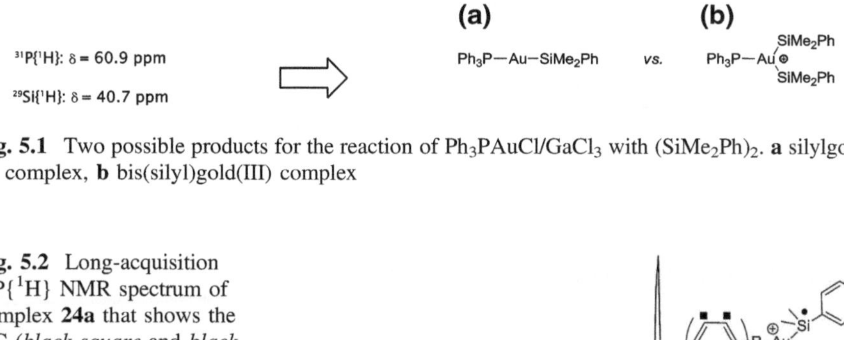

Fig. 5.1 Two possible products for the reaction of $Ph_3PAuCl/GaCl_3$ with $(SiMe_2Ph)_2$. **a** silylgold (I) complex, **b** bis(silyl)gold(III) complex

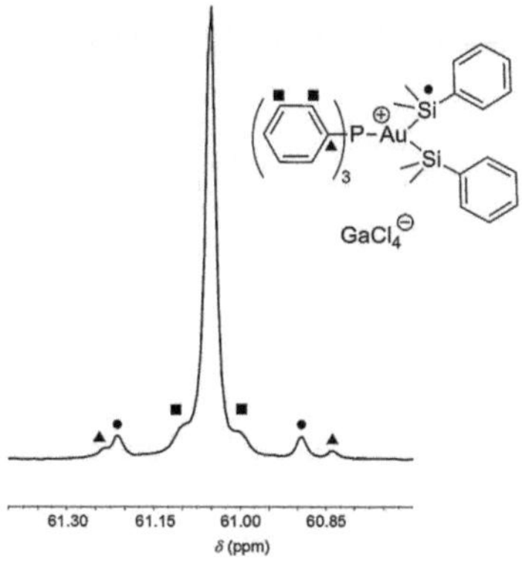

Fig. 5.2 Long-acquisition ^{31}P{^{1}H} NMR spectrum of complex **24a** that shows the ^{13}C (*black square* and *black triangle*) and ^{29}Si (*black circle*) satellites

was determined. The results are consistent with the phosphine bis(silyl)gold formulation of **24a**, for which the presence of a unique ^{29}Si NMR signal and the value of J_{SiP} suggest a Y-shape structure. Compounds such as **24a** are unprecedented, but Markó has recently disclosed an isoelectronic *N*-heterocyclic carbene bis(silyl) Pt complex (Scheme 5.3) [8].

To support further the formation of phosphine bis(silyl)gold(III) species upon oxidative addition of σ-SiSi bonds, we synthesized silylated phosphine **25**. This ligand in which one of the phenyl substituents at phosphorus bears a $SiPhMe_2$ group in *ortho* position, was then reacted with $(Me_2S)AuCl$ to yield the corresponding gold chloride complex **26** (Scheme 5.4).

The spectator silyl group on the phosphine ligand represents an internal ^{29}Si NMR standard for the direct counting of the silyl groups at gold. In complex **26**, the corresponding ^{29}Si NMR signal appears as a doublet ($^3J_{SiP}$ = 8.0 Hz) at δ = −5.5 ppm. Addition of the disilane $(PhMe_2Si)_2$ to a 1:1 mixture of complex **26** and $GaCl_3$ at −80 °C resulted in the instantaneous formation of complex **27** (~60 % yield according to ^{31}P NMR) along with some side products (Scheme 5.5).

Complex **27** is thermally unstable and decomposes within hours at −80 °C. As expected, its ^{29}Si NMR spectrum displays two resonance signals apart from unreacted disilane (Fig. 5.3).

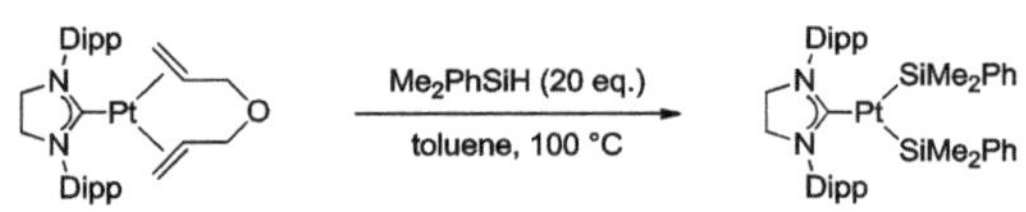

Scheme 5.3 Synthesis of a bis(silyl)platinum(II) complex (isoelectronic to **24a**), reported by Markó et al. Dipp = (2,6-diisopropyl)phenyl

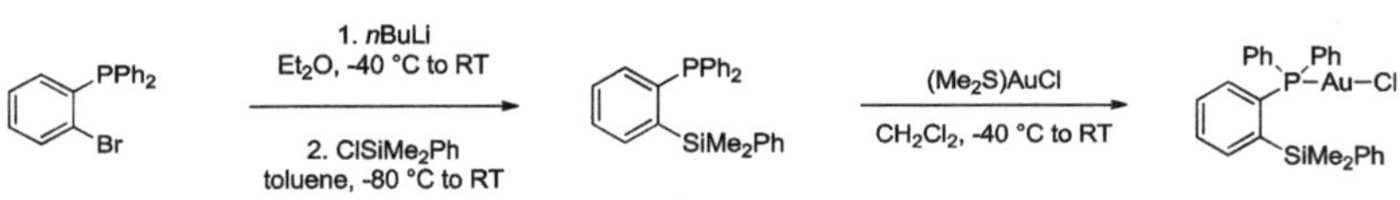

25 (65 %) **26** (83 %)

Scheme 5.4 Synthesis of gold chloride complex **26**

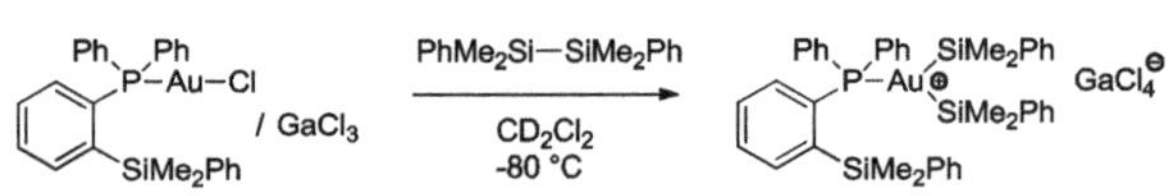

27 (~60 %)

Scheme 5.5 Formation of bis(silyl)gold(III) complex **27** featuring an internal ^{29}Si NMR standard by ortho-silylation of the spectator phosphine ligand

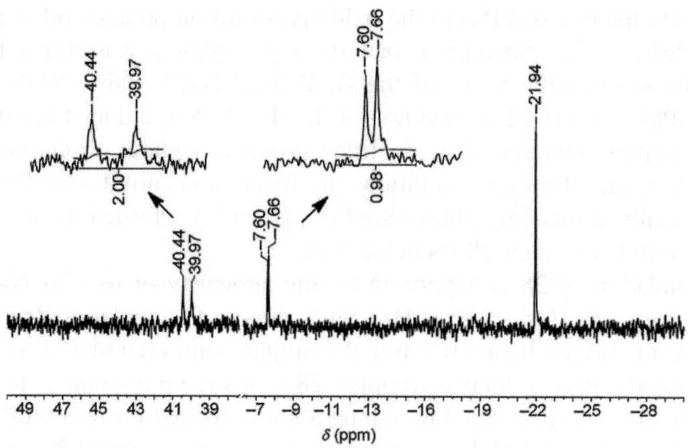

Fig. 5.3 Quantitative $^{29}Si\{^1H\}$ spectrum for bis(silyl)gold(III) complex **27**. The signal at δ = −21.9 ppm corresponds to unreacted $(SiMe_2Ph)_2$

One signal (δ = −7.6 ppm, d, $^3J_{SiP}$ = 4.6 Hz, 1 Si) appears in the region of $ArSiPhMe_2$ derivatives and is analogous to that of the starting complex **26**. The other signal (δ = 40.2 ppm, d, $^2J_{SiP}$ = 38.3 Hz, 2 Si) appears in the region of silylgold species and is similar to that of complex **24a**. The relative integration of these two signals unequivocally confirms the formulation of **27** as a phosphine bis (silyl)gold(III) complex.

The oxidative addition process has been explored with other disilanes (Scheme 5.6). The reaction proceeds similarly with the disilane $(Ph_2MeSi)_2$ and the ensuing complex **24b** (obtained in ~40 % yield according to ^{31}P NMR) displays a very similar ^{29}Si NMR pattern (δ = 34.5 ppm, d, $^2J_{SiP}$ = 39.8 Hz). Interestingly, the non-symmetrical disilane $Ph_2MeSi–SiMe_2Ph$ is also activated at gold. The resulting complex **24c** is highly unstable even at low temperature (~12 % yield according to ^{31}P NMR) but its structure could be established by ^{29}Si NMR at −80 °C. As expected from the dissymmetric nature of the disilane, complex **24c** displays two distinctive ^{29}Si NMR signals in the region of silylgold species: $SiMe_2Ph$ (δ = 42.8 ppm, d, $^2J_{SiP}$ = 27.5 Hz) and $SiMePh_2$ (δ = 33.1 ppm, d, $^2J_{SiP}$ = 49.6 Hz).

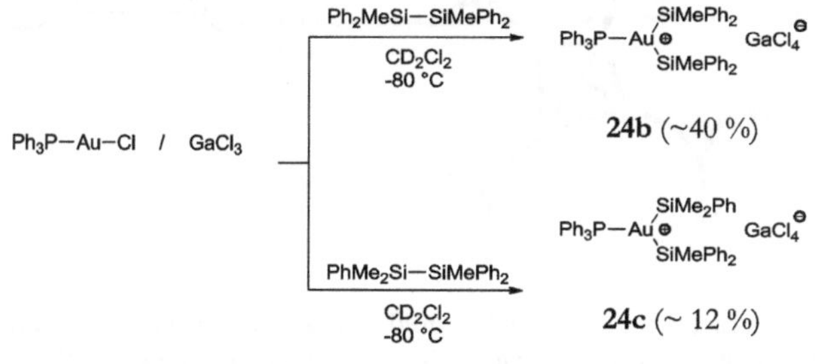

Scheme 5.6 Formation of the phosphine bis(silyl)gold(III) complexes **24b** and **24c** via oxidative addition of $(Ph_2MeSi)_2$ and $PhMe_2SiSiMePh_2$, respectively

To evaluate the role of PPh_3 in the oxidative addition process, other phosphines have been tested. The substitution pattern at phosphorus was found to strongly influence the stability/reactivity of the $(R_3P)AuCl/GaCl_3$ pairs. With phosphites such as $P(OPh)_3$ and $P(OMe)_3$, addition of the $(PhMe_2Si)_2$ induced fast degradation and gave complex mixtures of unidentified products without observation of silyl-gold species. Under the same conditions, no reaction occurred with $P(tBu)_3$ while the less sterically-demanding phosphine $P(Cy)_3$ readily afforded the corresponding gold(III) bis(silyl) complex **28** (Scheme 5.7).

The formulation of **28** is supported by the similarity of its ^{29}Si NMR pattern (δ = 39.5 ppm, d, $^{2}J_{SiP}$ = 35.2 Hz) with those of complexes **24a,b** and **27**. Interestingly, the $P(Cy)_3$ ligand was found to impart somewhat higher stability to the bis(silyl)gold(III) species. Indeed complex **28** is stable up to about −40 °C, but all our attempts to crystallize it at low temperature remained unsuccessful. A minute amount of crystalline material was obtained, and a single crystal X-ray diffraction analysis suggested the formation of a tetranuclear (phosphine)gold cluster compound. Complete refinement of the obtained structure was not possible due to the poor quality of the crystals and only the atom connectivity was determined (Fig. 5.4).

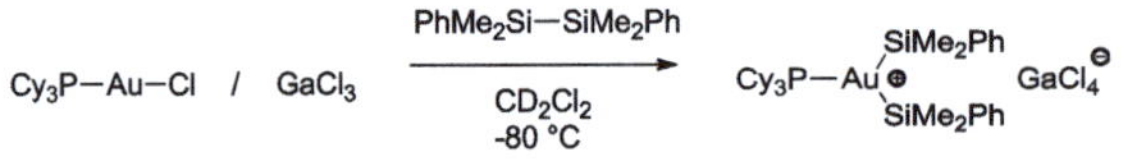

28 (~80 %)

Scheme 5.7 Formation of bis(silyl)gold(III) complex **28** via oxidative addition of $(SiMe_2Ph)_2$

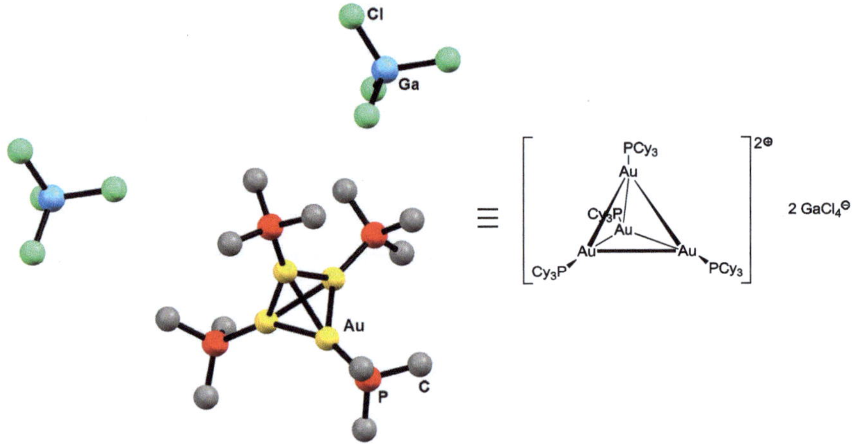

Fig. 5.4 Molecular structure of a tetranuclear gold complex formed upon attempted crystallization of **28** as determined by means of a partial X-ray diffraction analysis. Solvent molecules and hydrogen atoms are omitted for clarity. The cyclohexyl groups on phosphorus are simplified in the ball and stick representation

5.1.2 Computational Analysis

DFT (B3PW91) calculations were carried out in collaboration with the groups of Dr. Karinne Miqueu (Pau) and Prof. Dr. Laurent Maron (Toulouse) to gain more insight into the structure of the bis(silyl)gold(III) complexes and into the mechanism of its formation. Solvent effects (dichloromethane) were included by means of SMD calculations. Reaction of the neutral complex $(Ph_3P)AuCl$ (**I**) was examined first (Fig. 5.5). Activation of the disilane was found to be endothermic (ΔH = 16.0 kcal/mol) and to proceed via a rather high activation barrier ($\Delta H^{\neq}$ = 27.8 kcal/mol).

This is consistent with the fact that no reaction occurred experimentally in the absence of a chloride abstractor. The presence of $GaCl_3$ was thus taken into account computationally (Fig. 5.6).

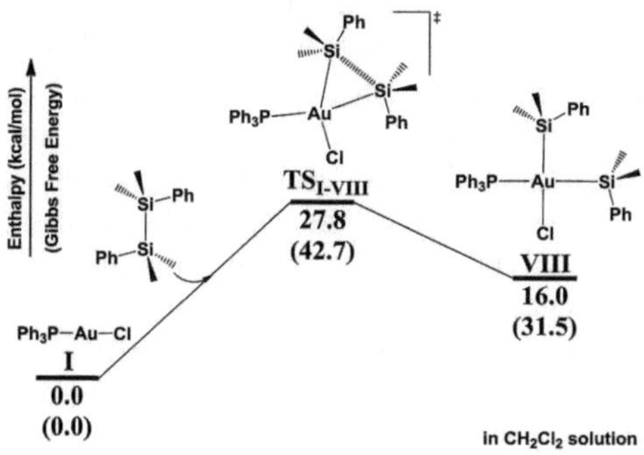

Fig. 5.5 Computed reaction profile for the oxidative addition of $(SiMe_2Ph)_2$ at Ph_3PAuCl (**I**) in absence of $GaCl_3$. Energies in kcal/mol

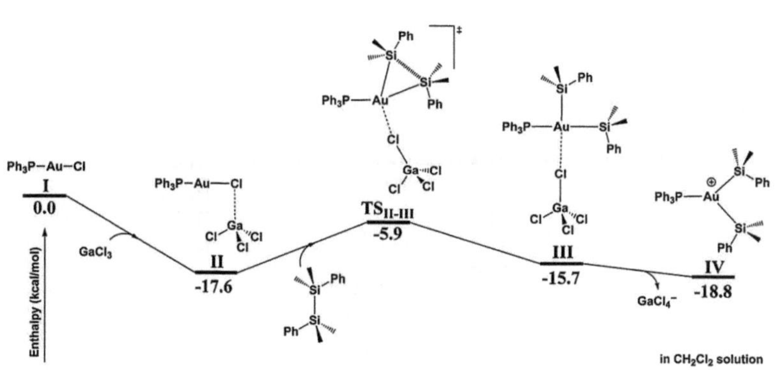

Fig. 5.6 Computed reaction profile for the oxidative addition of $(SiMe_2Ph)_2$ at Ph_3PAuCl (**I**) in presence of $GaCl_3$. Energies in kcal/mol

From **I**, the formation of a close adduct [$(Ph_3P)AuCl\cdots GaCl_3$] (**II**) is thermodynamically favored by 17.6 kcal/mol in ΔH (6.4 kcal/mol in ΔG). The ΔG value will not be discussed further hereafter, as the reaction is accompanied by a decrease in molecularity (three molecules giving a single ion pair), and therefore the computed entropy is not correct. The chlorine atom at gold interacts tightly with the gallium center (Cl–Ga distance: 2.39 Å) and the AuCl bond is noticeably elongated (from 2.30 Å in **I**, to 2.38 Å in **II**). Dissociation of $GaCl_4^-$ from the ion pair **II** is predicted to be endothermic by 19.6 kcal/mol, making the formation of the "naked" cationic gold complex $[Au(PPh_3)]^+$ highly unlikely.

Upon reaction of **II** with the disilane $(PhMe_2Si)_2$, the system reaches the oxidative addition transition state $\mathbf{TS_{II\text{–}III}}$. The corresponding activation barrier is relatively low (11.7 kcal/mol from **II**), in line with a kinetically facile reaction. The $GaCl_4^-$ counter-anion keeps interacting weakly with the gold center in $\mathbf{TS_{II\text{–}III}}$ (Cl–Au distance: 3.20 Å). The σ-SiSi bond is located in the plane defined by the P, Au and Cl atoms, and is substantially elongated (from 2.38 Å in the free disilane, to 2.50 Å in $\mathbf{TS_{II\text{–}III}}$). Following the intrinsic reaction coordinate, $\mathbf{TS_{II\text{–}III}}$ leads to the oxidative addition product **III** in which the σ-SiSi bond is fully broken (3.3 Å). Taking into account the residual contact with the $GaCl_4^-$ counter-anion (Cl–Au distance: 3.09 Å), the gold center is tetracoordinated and adopts distorted square planar geometry. Dissociation of $GaCl_4^-$ from **III** to give the ion pair complex **IV** is exothermic by 3.1 kcal/mol, which makes the whole process of oxidative addition from $(Ph_3P)AuCl/GaCl_3$ slightly favored thermodynamically.

Calculations also suggested reductive elimination to form a SiCl bond as a possible decomposition pathway for the bis(silyl)gold(III) complexes; an activation barrier of 26.1 kcal/mol was predicted for the formation of the chlorosilane and the silylgold(I) complex (Fig. 5.7), in agreement with the experimentally observed chlorosilanes in the reaction mixtures.

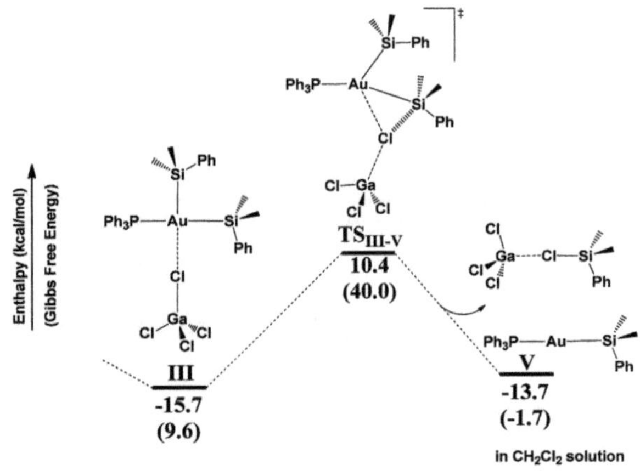

Fig. 5.7 Computed reaction profile for SiCl reductive elimination; a possible decomposition pathway of complex **III**. Energies in kcal/mol

The phosphine bis(silyl)gold(III) complex **IV** adopts a distorted Y-shape geometry (P–Au–Si bond angles of 129.6° and 145.9° on average). The situation differs somewhat from the one encountered for the isoelectronic (NHC) bis(silyl)Pt complex [8, 9]. Careful examination of the Potential Energy Surface (PES) of **IV** revealed the presence of several rotamers of similar structure and energy.

In all of them, the two silyl groups are in different environments. Conversely, a single ^{29}Si NMR signal is observed experimentally (vide supra). The most likely interpretation for the apparent magnetic equivalence of the two silyl groups is a fast isomerization of complex **IV** at the NMR time scale. Indeed, a low energy pathway was identified computationally ($\Delta H^{\neq}$ = 2.0 kcal/mol) (Fig. 5.8).

Finally, the key NMR data of complex **IV** have been evaluated (GIAO calculations) and the computed values of the ^{29}Si (δ = 56.2 ppm) and ^{31}P (δ = 66.6 ppm) chemical shifts, as well as the $^{2}J_{SiP}$ coupling constant (34.5 Hz) compare fairly well with those determined experimentally.

This joint experimental/computational study provides evidence for the oxidative addition of σ-SiSi bonds to gold. Although this reactivity has been observed previously in an intramolecular fashion by application of a chelate assistance strategy [6, 7], it is always questionable to which extent the anchoring sites enforce the observed reactivity. The reactions described above substantiate that oxidative addition can proceed readily at gold as well at the intermolecular level.

However, even though the formation of bis(silyl)gold(III) complexes is thermodynamically favorable and kinetically accessible, their thermal instability limits further exploration. Therefore we aimed at the design of gold(I) complexes allowing for (i) the oxidative addition and (ii) enhanced stability of the ensuing gold (III) complexes. Our hypothesis was that the key to attain at the same time reactivity and stability lies in the choice of the ancillary ligand.

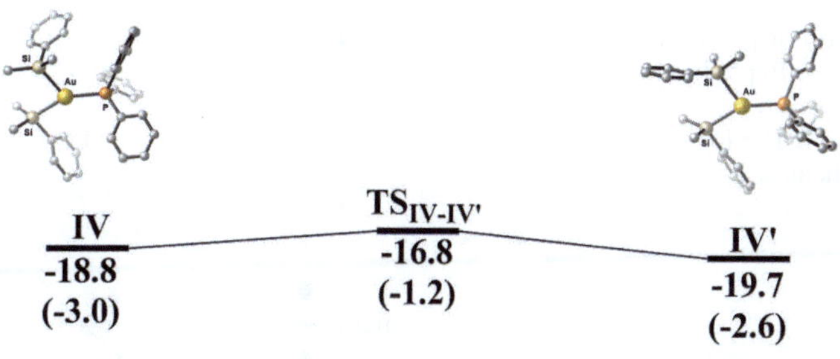

Fig. 5.8 Calculated isomerization process for the bis(silyl)gold(III) complex **IV**. A low energy pathway leads to an isomer **IV′**, stabilized by 0.9 kcal/mol in terms of enthalpy. Energies in kcal/mol

5.2 Design of Gold(I) Complexes Capable of σ-Bond Activation

5.2.1 Design Principles

We reasoned that a major parameter concerning the low thermal stability of the bis (silyl)gold(III) complexes encountered above is their three-coordination, which is to the best of our knowledge unknown for stable, isolated gold(III) complexes. A tetracoordinate, square-planar environment is generally required to impart stability to gold(III) complexes (see Sect. 3.2).

Simple addition of a 2-electron donor ligand (such as a phosphine or nitrile) to the in situ synthesized bis(silyl)gold(III) complexes in order to achieve tetracoordination did not prove to be fruitful and resulted only in complex mixtures of unidentified products.

Starting from dicoordinated gold(I) species to obtain four-coordinated gold(III) complexes after oxidative addition was not possible as well, because dicoordinated gold(I) complexes, either neutral such as $(Ph_3P)AuCl$ (*vide supra*), or cationic such as $[Au(PPh_3)_2]^+X^-$ showed to be completely unreactive towards disilanes. These dicoordinated complexes are linear, by far the most common geometry for gold(I) complexes [10].

It is well known for linear dicoordinated palladium(0) and platinum(0) complexes that significant bending of the L-M-L angle has to occur in order to achieve the transition state of an oxidative addition reaction [11]. This bending is required because in a linear ML_2 complex the orbital interactions with a σ-bond to be cleaved are unfavorable. This situation is illustrated qualitatively in the Walsh diagram depicted below for a PtL_2 complex (Fig. 5.9) [12, 13]. On the left side, in

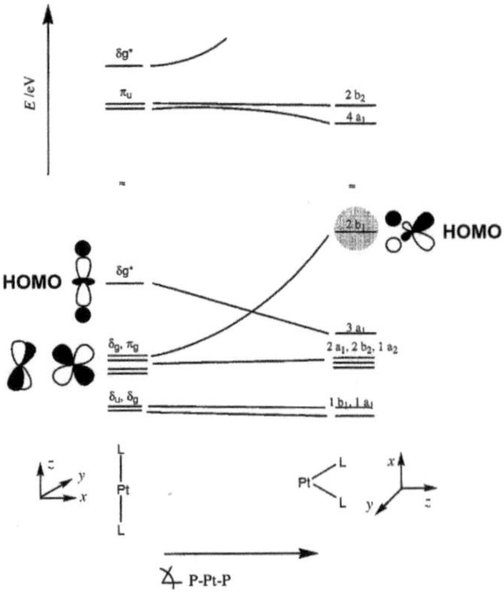

Fig. 5.9 Walsh diagram for PtL_2 complexes. Reproduced from Ref. [13] with permission from The Royal Society of Chemistry

case of the linear ligand environment, the HOMO (δ_g*) does not feature matching symmetry for interaction with the antibonding orbital of a σ-bond. When the L-M-L angle diminishes (going from left to right in the diagram), the δ_g* orbital lowers in energy and instead the $2b_1$ orbital rises in energy, becoming now the HOMO, with suitable symmetry for interaction with the σ* orbital.

As Pt(0) and Au(I) are isoelectronic, the essence of this analysis should be valid as well for dicoordinate gold(I) complexes [14].

We hypothesized that the lack of reactivity for linear, dicoordinated gold(I) complexes towards disilanes might be due to the high deformation energy of the ligand sphere around gold in order to reach the transition state en route to the square-planar geometry in the Au(III) product [10].

In line with the orbital analysis, the use of chelating ligands with L-M-L bite angles close to 90° are known to facilitate the oxidative addition. These bite angle effects are well studied both experimentally and theoretically for palladium and platinum diphosphine complexes [13, 15, 16].

Consequently, so as to attain oxidative addition at gold(I), we decided to start from a prearranged gold(I) complex, in which (i) the spectator ligand already exhibits a geometry well-suited for a square planar gold(III)-like environment and in which (ii) the gold(I) center still features a free coordination site to allow for an interaction with a σ-bond.

A small L-Au-L bite angle imposed by the ligand framework is expected to destabilize the ground state of the gold(I) complex and should therefore significantly decrease the activation barrier for an oxidative addition process, as schematically depicted in Fig. 5.10.

A bent $[L_2Au(I)]^+$ complex could be attained via a gold(I) complex with a trigonal-planar coordination environment, featuring a chelating L_2 ligand, such as a diphosphine or a diamine, exhibiting a small bite angle, i.e. close to 90° as found in square planar gold(III) complexes, and one anionic, one-electron-donating X ligand, by preference a halide. Abstraction of the halide ligand should generate a

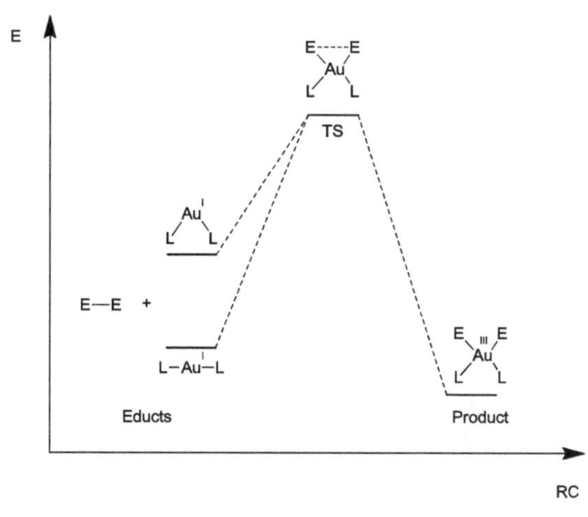

Fig. 5.10 Schematic presentation of the expected decrease of activation energy for an oxidative addition reaction at gold(I) by employing a prearranged bent ligand framework

cationic 14-electron species featuring an open coordination site accessible for interaction with a σ-bond (Scheme 5.8).

Such a ligand environment is highly unusual for gold(I): This is most likely due to the high tendency of gold(I) to form linear complexes (see Sect. 3.2). In fact, many dicoordinating L_2 ligands that are known to form chelating structures with other transition metals, form dinuclear structures in the case of gold(I), so that the linear coordination mode can be maintained. The formation of these dinuclear species is thermodynamically favored by intramolecular aurophilic interactions (Scheme 5.9).

An extensive search in the literature revealed that the choice of reported, suitable gold(I) complexes that fulfill the above-mentioned criteria is rather limited [17].

Among the rare examples that might be well-suited for our project, one complex attracted particularly our attention: the cationic (1,2-bis(diphenylphosphino)-1,2-dicarba-*closo*-dodecaborane-$\kappa^2 P,P'$) (triphenylphosphine) gold(I) complex, published by Laguna et al. exhibits a P-Au-P angle of only 90.2(1)° according to the reported single crystal X-ray diffraction analysis (Fig. 5.11) [18].

The compound was prepared *inter alia* by chloride abstraction from chloro-[1,2-bis(diphenylphosphino)-1,2-dicarba-*closo*-dodecaborane-$\kappa^2 P,P'$]gold(I) with (THT)silver perchlorate and subsequent addition of triphenylphosphine which indicates that halide abstraction is indeed a viable strategy for the generation of the cationic, potentially reactive dicoordinate gold(I) fragment.

Dicarba-*closo*-dodecaboranes (or shortened: carboranes) are icosahedral cluster compounds based on the parent dodecaborate dianion with two of the BH-vertices replaced by CH-units. Among the three isomers of carborane (1,2-, 1,7- or

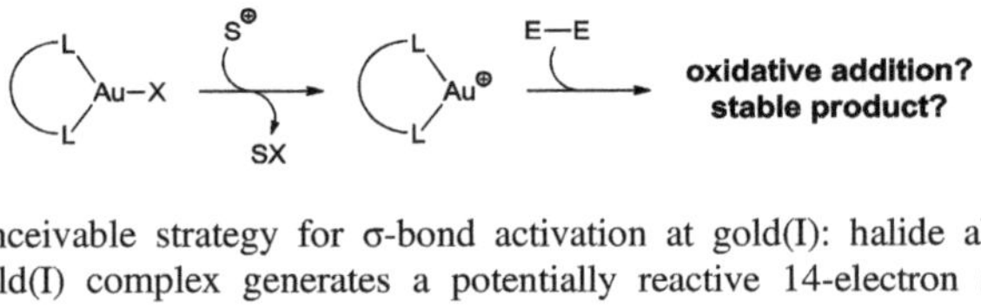

Scheme 5.8 Conceivable strategy for σ-bond activation at gold(I): halide abstraction from a tricoordinated gold(I) complex generates a potentially reactive 14-electron species. *X* halide *S* halide scavenger

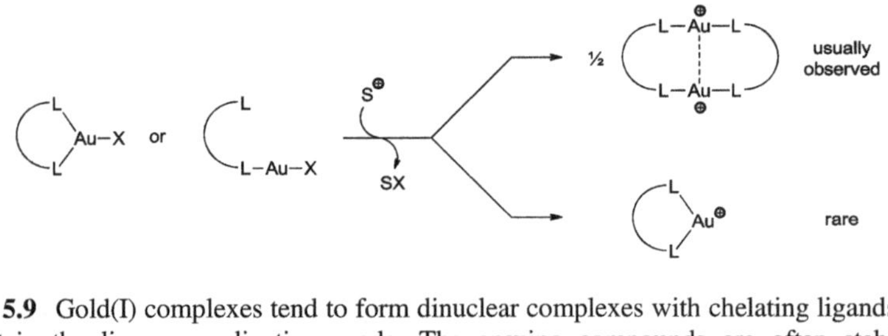

Scheme 5.9 Gold(I) complexes tend to form dinuclear complexes with chelating ligands in order to maintain the linear coordination mode. The ensuing compounds are often stabilized by aurophilic interactions. *X* halide *S* halide scavenger

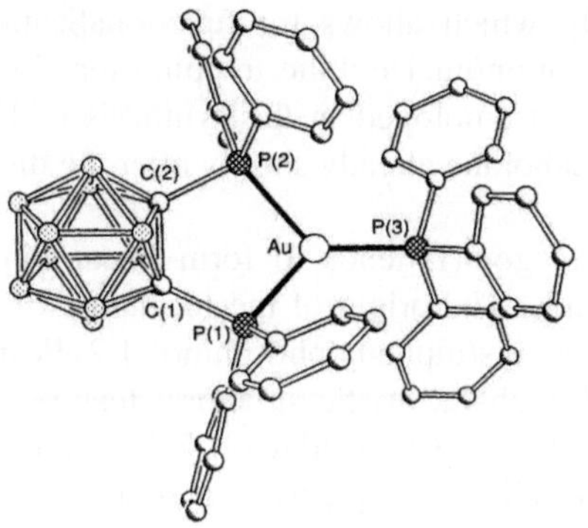

Fig. 5.11 Molecular structure of the (1,2-bis(diphenylphosphino)-1,2-dicarba-closo-dodecaborane-κ^{2}P,P′) (triphenylphosphine) gold(I) cation determined by single crystal X-ray diffraction. The P1-Au-P2 angle is 90.2(1)°. Reproduced from Ref. [18] with permission from The Royal Society of Chemistry

1,12-substitution of BH by CH), the 1,2- or *o*-carborane is the least stable one which transforms at elevated temperatures to the 1,7- and subsequently to the 1,12-isomer (Fig. 5.12) [19].

Carboranes feature a delocalized molecular orbital system involving 3-center 2-electron bonding. Hence, the lines drawn between the boron vertices in the common representation of carborane clusters do not represent 2-electron bonds but illustrate only the atom connectivity to clarify the cluster geometry. The *exo*-icosahedral CH and BH bonds, however, are classical Lewis bonds.

Carboranes are sometimes denoted as "three-dimensional analogues of benzene" and the similarities are indeed pronounced. The size of a carborane corresponds approximately to a rotating benzene molecule. Furthermore, similar to benzene, the highly delocalized cage bonding of carboranes provides notable thermal and (photo)chemical robustness, and leads as well to comparable reactivity [21].

While the BH bonds of the cluster undergo substitution reactions upon attack of electrophiles, the unsaturated and therefore electron-withdrawing character of the cluster (3-center 2-electron bonds!) acidifies the two CH protons (pKa for

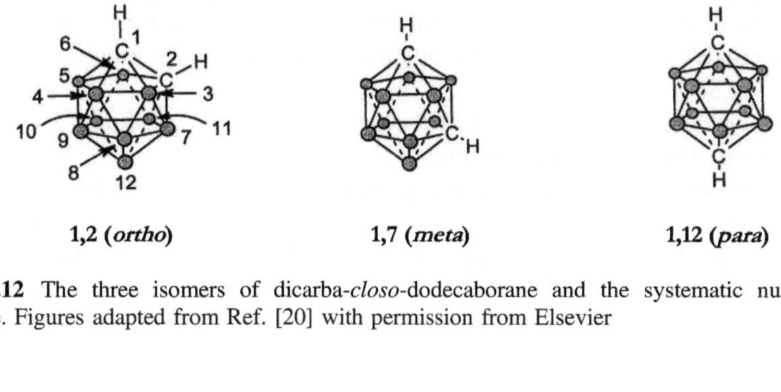

Fig. 5.12 The three isomers of dicarba-*closo*-dodecaborane and the systematic numbering scheme. Figures adapted from Ref. [20] with permission from Elsevier

o-carborane: 22.0) [20], which allows for functionalization of the CH units of carboranes via deprotonation/nucleophilic trapping reaction sequences. This reactivity was recognized and employed in the synthesis of bis(diphenylphosphino)-1,2-dicarba-*closo*-dodecaborane already shortly after the discovery of carboranes in the early 1960s [22].

As mentioned above, gold(I) tends to form linear dinuclear complexes with many diphosphine ligands. The origin of the formation of a monometallic gold(I) complex observed for bis(diphenylphosphino)-1,2-dicarba-*closo*-dodecaborane may be explained by the characteristic structural features of the carborane-based ligand backbone. The very long CC bond (ca. 1.7 Å vs. about 1.4 Å in benzene) and the rigidity of the cluster allows for precise directional control of both phosphorus atoms and puts their electron lone pairs in a geometry suitable for chelation of a metal center. P-M-P bite angles of approximately 90° with *o*-carborane-diphosphine ligands were observed for palladium [23, 24], platinum [25] and coinage metal complexes [7, 26–31].

Although *o*-carborane-diphosphines have been used as ligands for photochemical [32–34] or catalytic [35] applications, their coordination chemistry is by far less developed with regard to comparable, simpler diphosphines such as 1,2-bis(phosphino)ethane or 1,2-bis(phosphino)benzene.

5.2.2 Synthesis and Coordination Chemistry of Diphosphine-o-Carborane Gold(I) Complexes

5.2.2.1 Ligand Synthesis

We considered the *o*-carborane-bis(diphenylphosphine) gold(I) chloride as a most interesting framework for a cationic, dicoordinate gold(I) species potentially active in the activation of σ-bonds. Furthermore, the diphosphine ligand can be readily prepared starting from the parent *o*-carborane by simple deprotonation of the CH-fragments and subsequent electrophilic coupling with a chlorophosphine.

Based on these considerations, we first synthesized 1,2-bis(diphenylphosphino)-1,2-dicarba-*closo*-dodecaborane [22, 36] as well as the more donating ligand 1,2-bis(diisopropylphosphino)-1,2-dicarba-*closo*-dodecaborane [37]. Coordination to a gold(I) precursor complex such as [AuCl(SMe_2)] gave the corresponding [1,2-bis(phosphino)-1,2-dicarba-*closo*-dodecaborane-$\kappa^2 P,P'$]gold(I) chloride complexes **29** [7] and **30** (Scheme 5.10).

The molecular structure of **30** in the solid state was determined by a single crystal X-ray diffraction analysis (Fig. 5.13).

In the solid state, only one of the phosphorus donor arms is involved in the coordination to gold, the second arm being pendant. The coordination environment around the gold atom is quasi-linear with a P1–Au–Cl angle of 175.52(6)°. The distance between the metal center and the coordinated phosphorus atom is with

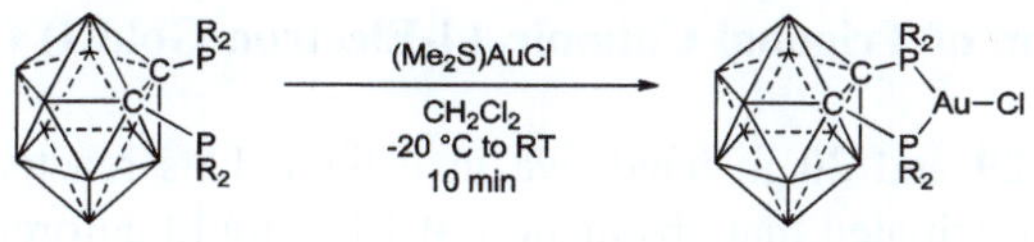

R = Ph: **29** (90 %)

R = *i*Pr: **30** (76 %)

Scheme 5.10 Synthesis of gold(I) chloride **29** and **30**

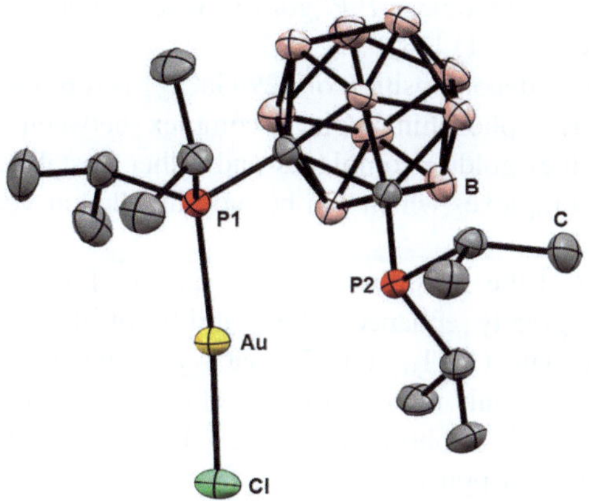

Fig. 5.13 Molecular structure of **30** determined by single crystal X-ray diffraction. Solvent molecules and hydrogen atoms are omitted for clarity. Selected bond lengths (Å) and angles (°): ClAu 2.298(2), AuP1 2.239(1), C2C1 1.763(8), P2Au 3.083(2), C1P1 1.889(6), C2P2 1.878(6), P1AuCl 175.52(6), P1AuP2 79.43(5), P1C1C2 117.3(4), P2C2C1 116.3(3)

2.239(1) Å in the usual range for phosphine gold(I) chloride complexes. For example, the length of the AuP bond (iPr$_3$P)AuCl is 2.239(2) Å [38]. The second phosphorus arm is rather distant from the gold atom [AuP2 = 3.083(2) Å], but well below the sum of the van der Waals radii (sum r_{vdW}(PAu) = 4.05 Å) [39]. The lone pair of the pendant phosphine is orientated in the direction of the gold atom.

By contrast, the solution ^{31}P{^{1}H} NMR spectra of **29** and **30** feature singlet resonances at δ = 35.9 ppm and δ = 68.8 ppm, respectively, indicative either of rapid intramolecular exchange of the phosphines at the metal center on the ^{31}P NMR time scale, or of symmetric coordination of both phosphorus donor arms in solution. For this reason, the trigonal-planar representation in the chemical structures for complexes of this type is given preference over the dicoordinate linear one in this manuscript.

5.2.2.2 Formation of Trigonal Cationic 14-Electron Gold(I) Complexes

With complexes **29** and **30** in hand, we investigated its reactivity after chloride abstraction. We anticipated that abstraction of Cl^- should enforce the coordination of the second phosphine. In order to evaluate the electrophilic character of a putative cationic 14-electron fragment, the diphosphine gold(I) chloride complex **29** was reacted at low temperature with gallium trichloride. Upon addition of $GaCl_3$, the ^{31}P NMR signal for **29** shifts from δ = 35.9 ppm to δ = 58 ppm in agreement with a decrease of electron density at the metal center. {**29**/$GaCl_3$} can be monitored at room temperature, but decomposes slowly in the course of hours into gold metal and the tetracoordinated complex cation bis[1,2-bis(diphenylphosphino)-1,2-dicarba-*closo*-dodecaborane-$\kappa^2 P,P'$]gold(I) as identified by its ^{31}P NMR signal (δ = 42 ppm) (Scheme 5.11) [18].

The relatively slow decomposition of {**29**/$GaCl_3$} positions the thermal stability of this bent cationic diphosphine gold(I) complex between highly stable linear cationic bis(phosphine) gold(I) complexes and rather unstable, 12-electron monophosphine gold(I) complexes which can be only handled at very low temperatures (see Sect. 5.1.1).

The replacement of the substituents on phosphorus for more electron donating isopropyl moieties greatly enhanced the stability of the cationic species after chloride abstraction with $GaCl_3$, and {**30**/$GaCl_3$} is stable for days at room temperature in solution with only minute amounts of decomposition being observed. As for **29**, a downfield shift is observed for the ^{31}P NMR resonance upon chloride abstraction of **30** (δ = 74 ppm).

{**29**/$GaCl_3$} is efficiently stabilized by alkenes, as evidenced by its reaction with styrene (Scheme 5.12). Addition of 5 equivalents of styrene to {**29**/$GaCl_3$} in dichloromethane leads to the tricoordinated gold(I) alkene complex **31** which displays increased stability in solution.

31 was characterized in solution by NMR spectroscopy. In the ^{31}P NMR spectrum, a singlet resonance is observed at δ = 53.6 ppm indicative of symmetric coordination of the diphosphine and fast rotation of the η^2-coordinated styrene in solution.

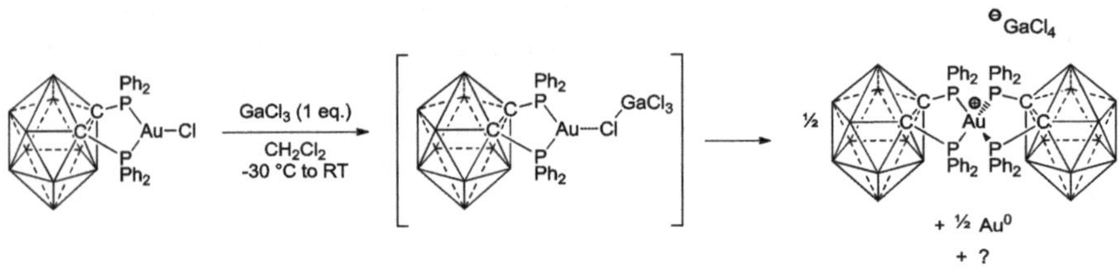

29 {**29** / $GaCl_3$}

Scheme 5.11 Chloride abstraction from gold(I) complex **29** and subsequent decomposition

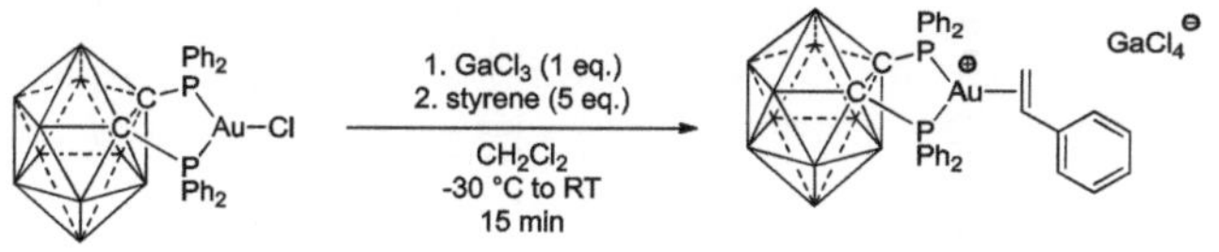

29 **31** (87 %)

Scheme 5.12 Synthesis of gold(I) styrene complex **31**

^{1}H and ^{13}C NMR data for the vinyl group of styrene in complex **31** are summarized in Table 5.1 in comparison with the corresponding values of (i) non-coordinated styrene, (ii) a representative example for a monophosphine η^2-styrene gold(I) complex reported by Widenhoefer [40] and (iii) a bipyridine-based styrene gold(I) complex [41].

As can be seen from Table 5.1, coordination of styrene to gold(I) results for both the monophosphine and the diphosphine gold(I) complex in an upfield shift for the olefinic protons and carbon atoms. While the ^{1}H NMR data do not differ very much, a considerable difference with regard to the chemical shift becomes evident when the ^{13}C values are compared. The shift to higher field with respect to non-coordinated styrene for C1 and C2 is more important for **31** than for the monophosphine gold(I) complex. This behavior may be rationalized by increased backdonation from the gold(I) metal center to the coordinated olefin in case of **31** and thus a more relevant contribution of the auracyclopropane resonance form. For the bipyridine gold(I) styrene complex which was synthesized by Cinellu and coworkers starting from bimetallic gold(III) oxo compounds these upfield shifts are even more pronounced. Indeed, based on a spectroscopic, structural and theoretical analysis it was concluded that the π-backbonding contribution is even higher than that of σ-bonding which is reflected in the large upfield shifts of ^{1}H and ^{13}C NMR resonances of the vinyl group in this compound [41, 42].

Usually, π-backbonding contributes only insignificantly to the coordination of olefins in linear, dicoordinate gold(I) complexes [43]. However, the bonding situation changes for tricoordinate olefin gold(I) complexes (like compound **31**) and the π-backbonding contribution becomes important.

The molecular structure of **31** in the solid state was determined by single crystal X-ray crystallography (Fig. 5.14). The gold atom features a trigonal-planar ligand environment (sum of angles around gold: 360°), i.e. the vinyl moiety of styrene and both phosphorus atoms of the chelating diphosphine ligand are coplanar. The gold atom and the diphosphine span a PAuP angle of 89.13(3)°, the opposite C1AuC2 angle is acute (37.1(1)°). The styrene molecule deviates somewhat from an ideal orthogonal coordination to the PAuP fragment. The angle between the C1C2C$_{ipso}$ plane and the C1C2Au plane is 77°. The C1Au, C2Au and C1C2 bond lengths are 2.170(4), 2.222(4), 1.399(5) Å, respectively, and are comparable to those seen previously in the styrene complexes reported by Cinellu et al. [41, 42].

Table 5.1 Selected ^{1}H and ^{13}C NMR data for free styrene, a monophosphine styrene gold(I) complex and diphosphine styrene complex **31** and a bipyridine styrene gold(I) complex

δ (ppm)		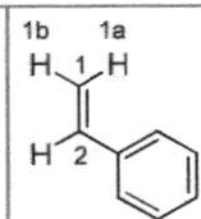	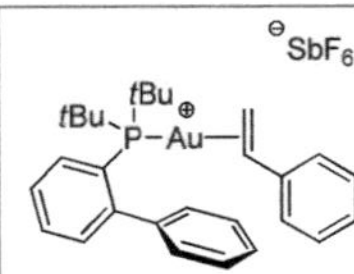	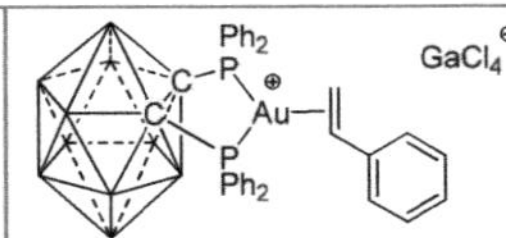 **31**	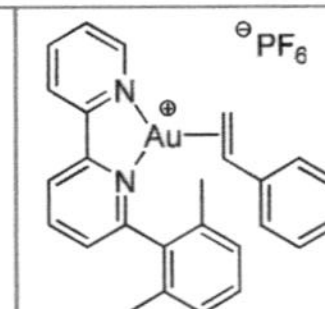
^{1}**H**	1a	5.74	4.84	4.75	2.97
	1b	5.24	4.42	4.56	
	2	6.72	6.84	6.22	5.11
^{13}**C**	1	113.7	91.1	65.9	55.3
	2	136.9	124.5	95.0	81.9

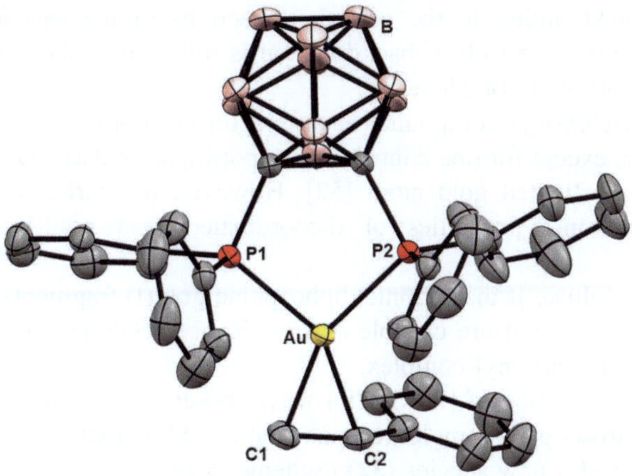

Fig. 5.14 Molecular structure of **31** determined by single crystal X-ray diffraction. The $GaCl_4^-$ counteranion and hydrogen atoms are omitted for clarity. Selected bond lengths (Å) and angles (°): P1Au 2.3821(9), P2Au 2.3788(8), C1Au 2.170(4), C2Au 2.222(4), C1C2 1.399(5), P1AuP 89.13 (3), C2AuC1 37.1(1), AuC1C2 73.4(2), C1C2Au 69.4(2)

The successful synthesis of this tricoordinated diphosphine-alkene gold(I) complex clearly indicates the coordinative unsaturation and electrophilic nature of the 14-electron fragment after chloride abstraction from **29**. According to our experience with monophosphine gold(I) complexes (see Sect. 5.1.1) and σ-SiSi bonds, both are requirements for the intermolecular oxidative addition to gold(I). The spectroscopic characteristics of styrene gold(I) complex **31** indicative of increased π-backbonding prompted us to further investigate the electronic properties the *o*-carborane-diphosphine gold(I) framework.

5.2.2.3 Electronic Properties of Cationic Bent 14-Electron Gold(I) Species

One necessary quality of a metal complex to undergo oxidative addition via σ-bond activation is backdonation of electron density from the metal center into the antibonding σ-orbital of the bond to cleave (see Sect. 2.1.1).

A common method used to quantify the degree of backdonation at a metal center, consists of measuring the CO stretching frequency of a corresponding carbonyl complex by infrared spectroscopy [44]. Carbonyl complexes of gold are exceedingly rare, but more important is the fact that all reported examples are non-classical carbonyl complexes (i.e. the CO stretching frequency experiences a blueshift upon coordination to the metal center, relative to the stretching band of free carbon monoxide at 2143 cm^{-1}) [45–51]. Backdonation in these complexes is non-negligible, but weak. Dias, Frenking and coworkers calculated for a tri(mesityl) phosphine gold(I) carbonyl complex a contribution of 62 and 38 % for the σ-bonding/π-backbonding to the orbital interactions of the (phosphine)Au^{+}–CO bond, respectively. The lack of backdonation is reflected in the poor thermal stability of gold carbonyl complexes.

The studied carbonyl compounds feature a linear coordination environment at the gold center, except for one contribution reporting a gold(I) CO complex with a tetrahedrally coordinated gold atom [52]. However, the carbonyl chemistry and hence the electronic properties of dicoordinated bent gold(I) fragments are unexplored.

In order to evaluate if the cationic diphosphine gold(I) fragments show different electronic properties and are capable of significant backdonation, we synthesized the corresponding carbonyl complex.

To this end, a solution of {**30**/$GaCl_3$} was exposed to CO gas and monitored by ^{31}P NMR spectroscopy. A singlet resonance at δ = 86.1 ppm was indicative of the clean formation of a new species (**32**) (Scheme. 5.13).

Characterization by infrared and NMR spectroscopy confirms the attribution of **32** to a carbonyl complex. Among the analytical data, the ^{13}C NMR signal corresponding to the CO carbon (one might expect a triplet resonance, but likely due to rapid ligand exchange a singlet is observed at δ = 198.7 ppm), as well as the frequency of the CO stretching band (ν(CO): 2115 cm^{-1}) of **32** are particularly noteworthy. The molecular structure of **32** in the solid state was determined by single crystals X-ray analysis (Fig. 5.15).

In the solid state, the gold atom of the complex cation is in a trigonal-planar coordination environment (sum of angles around gold: 360°) with the CO and the diphosphine ligand bound to gold. The carbonyl ligand and the gold atom feature an almost linear arrangement with an AuCO angle of 178.0(3)°. The PAuP bite angle of the diphosphine (94.41(2)°) is bisected by the CO ligand. The AuC and CO distances are 1.990(3) and 1.109(4) Å, respectively.

To put these spectroscopic and structural data into perspective, Table 2.1 lists the key experimental values for the above-mentioned monophosphine gold(I) carbonyl complex, free CO and complex **32** (Table 5.2).

As can be seen from the spectroscopic data, the electronic situation of the carbonyl ligand in complex **32** distinguishes significantly from free CO and [Au(CO)(Mes_3P)]SbF_6. The observed redshift of the IR CO vibration band and the lowfield shift for ^{13}C NMR signal are in line with a decrease of the CO bond strength and π-backbonding from the metal center to the carbonyl ligand.

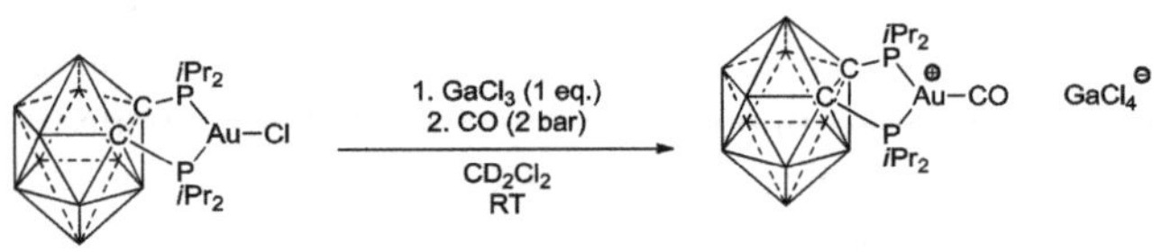

30 **32** (quant.)

Scheme 5.13 Synthesis of gold(I) carbonyl complex **32**

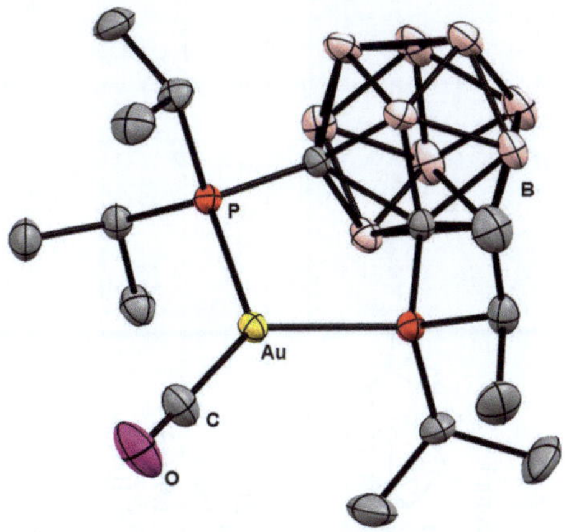

Fig. 5.15 Molecular structure of **32** determined by single crystal X-ray diffraction. The $GaCl_4^-$ counteranion and hydrogen atoms, are omitted for clarity. Selected bond lengths (Å) and angles (°): AuC 1.990(3), CO 1.109(4), P1Au 2.3826(6), P2Au 2.3832(6), AuCO 178.0(3), PAuP 94.41(2)

On the contrary, the structural data do not indicate a significant difference between the two gold(I) carbonyl complexes. However, Krossing et al. pointed out that "*routine X-ray measurements are not appropriate to determine precise CO bond lengths*" [50]. High resolution measurements are necessary to obtain meaningful results. Therefore, the CO bond distances will not be discussed further.

To the best of our knowledge, complex **32** is the first example of a classical carbonyl complex of gold.

Bent diphosphine gold(I) complexes seem to be capable of significant π-backbonding, as suggested by the discussed spectroscopic features of styrene complex **31** and carbonyl complex **32**. Therefore, the cationic, bent diphosphine gold(I) species might be active for oxidative addition of σ-bonds.

5.3 σ-Bond Activation at Cationic Bent 14-Electron Gold(I) Fragments

All of the above mentioned characteristics make **29** and **30** promising candidates for σ-bond activation processes. Consequently, we investigated the oxidative addition of apolar (SiSi, CC) and aryl halide (CX) σ-bonds to these complexes, in order to evaluate their reactivity.

Table 5.2 Selected experimental data comparing a monophosphine gold(I) carbonyl complex, free CO and complex **32**

		$Mes_3P{-}Au^{\oplus}{-}CO$ $^{\ominus}SbF_6$	free CO	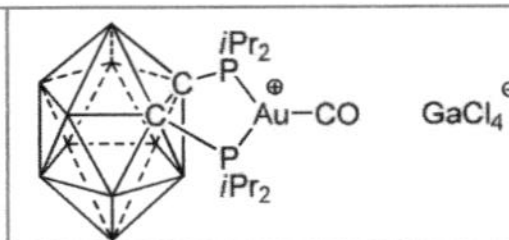
References		[49]	[46, 53]	–
IR: ν(CO) (cm^{-1})		2185	2143	2115
^{13}C NMR: (ppm)		182.6	184	198.7
XRD:	d(AuC) (Å)	2.008(6)	–	1.990(3)
	d(CO) (Å)	1.108(7)	1.128	1.109(4)

5.3.1 Activation of σ–SiSi Bonds

We started to investigate first the activation of relatively weak σ-bonds and therefore studied the reactivity of disilanes towards the cationic, bent diphosphine gold(I) species. 1,1′,2,2′-tetramethyldiphenyldisilane (5 equivalents) was added to a mixture of {**29**/$GaCl_3$} at low temperature. At −30 °C and even at room temperature no reaction takes place, as evidenced by the unchanged ^{31}P NMR spectrum compared to the starting material. However, upon heating the reaction mixture for 60 min at 60 °C (Scheme 5.14), clean formation of a new species **33** is observed exhibiting a ^{31}P resonance signal at δ = 65 ppm. Although the ^{31}P NMR chemical shift is in the range of bis(silyl)gold(III) complexes encountered previously (see Sect. 5.1.1), the formation of the oxidative addition product is not confirmed by the ^{29}Si NMR spectrum of the crude reaction mixture: Instead of the expected doublet of doublets (for a diphosphine bis(silyl)gold(III) complex) in the region of δ = 30–40 ppm as observed previously for silylgold(III) complexes, only unreacted disilane and a new singlet resonance at δ = 20.0 ppm is observed, attributed to $PhMe_2SiCl$ [2].

After isolation of **33** by crystallization, a dark-yellow solid was obtained, which in solution does not display a ^{29}Si NMR signal. In order to elucidate the molecular structure of **33**, single crystals were grown by diffusion of pentane into a concentrated dichloromethane solution of this compound, and a single crystal X-ray diffraction analysis was performed (Fig. 5.16).

As expected from the NMR spectroscopic data, the molecular structure does not contain silicon. Instead of the envisioned bis(silyl)gold(III) complex (Scheme 5.14), the reaction with the disilane gave rise to a cationic, dinuclear gold(I)-gold(I) complex, bridged by a formally anionic $GaCl_2$-moiety. The complex cation is paired with $GaCl_4^-$ as the counteranion. A related GaI_2-motif bridging two gold atoms was observed previously in a neutral, trinuclear gold complex [54].

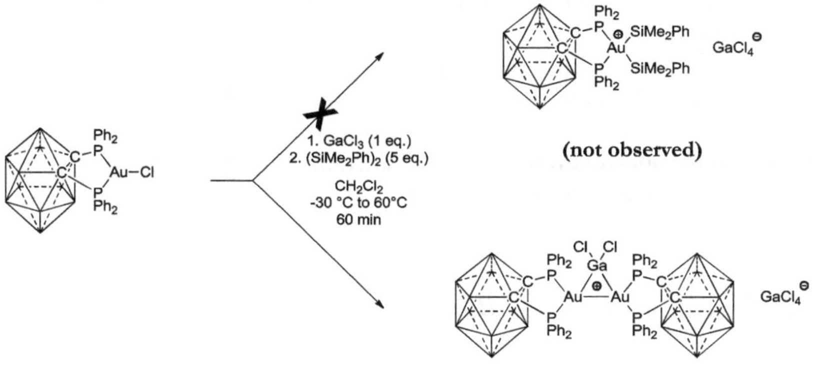

29 **33**

Scheme 5.14 Attempted synthesis of an unsupported bis(silyl) gold(III) complex and formed dinuclear product

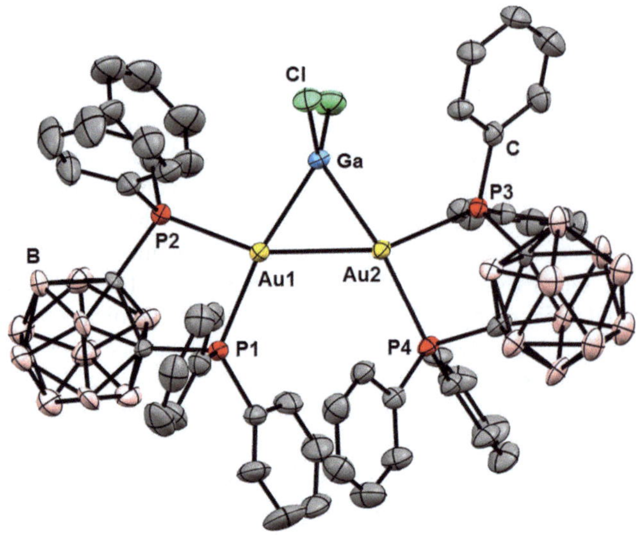

Fig. 5.16 Molecular structure of **33** determined by single crystal X-ray diffraction. The counteranion, solvent molecules and hydrogen atoms, are omitted for clarity. Selected bond lengths (Å) and angles (°): Au1Au2 2.7112(6), Au1Ga 2.475(2), Au2Ga 2.472(1), P1Au1 2.432 (4), P2Au1 2.379(3), P3Au2 2.385(3), P4Au2 2.448(4), P1Au1P2 89.8(1), P3Au2P4 88.3(1), Au1Au2Ga 56.83(4), Au2Au1Ga 56.72(4)

Although the reaction of disilane and the cationic gold(I) fragment did not lead to the envisioned unsupported bis(silyl)gold(III) complex, we hypothesize that the gold(III) species might occur as an intermediate in the formation of **33**. Oxidative addition of the disilane at half an equivalent of the cationic diphosphine gold(I) complex would lead to a transient bis(silyl)gold(III) complex that subsequently reacts with the counteranion and is finally trapped by a second half equivalent of cationic diphosphine gold(I) complex to lead to the dinuclear complex and chlorosilane (Scheme 5.15).

The strong *trans* influence of silyl groups is well known [44] and it is likely that diphosphine bis(silyl)gold(III) complexes cannot be isolated without chelate assistance.

However, the apparent reactivity of **29** after chloride abstraction prompted us to extend our investigation to other σ-bonds.

5.3.2 Activation of σ–CC Bonds

Oxidative addition of CC bonds to transition metals is challenging. From a thermodynamic point of view, σ–CC bonds are usually strong with a mean bond dissociation energy of 83 kcal/mol [55]. Furthermore, oxidative addition is hindered

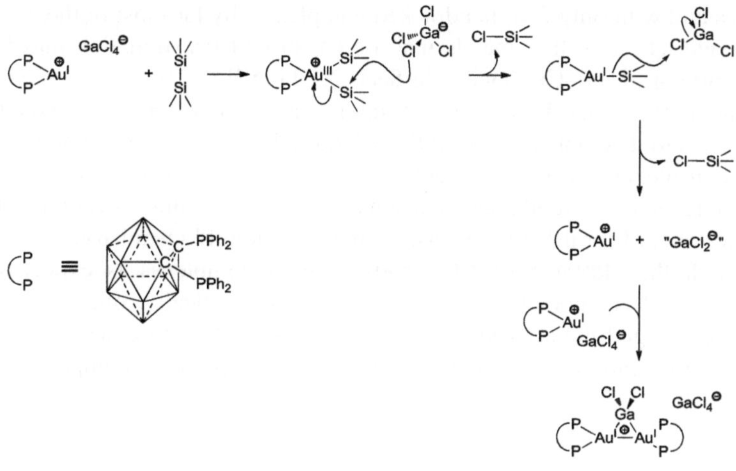

Scheme 5.15 Possible mechanism for the formation of dinuclear complex **33**

by kinetic inertness, due to steric congestion on both ends of the CC bond and to the high directionality of its σ-orbital (Fig. 5.17), hampering the interaction with a metal center [56].

Therefore, CC bond activation by oxidative addition is considered to be one of the most challenging reactions in the field of organometallic chemistry [57].

The mentioned obstacles concerning the cleavage of CC bonds can be overcome by applying different strategies [57–59], one of them being the increase of the energy state of the starting materials. Strained carbocycles possess significantly weakened CC bonds. The release of ring strain in the course of CC bond cleavage can make this process thermodynamically feasible.

Even though transition metal mediated σ–CC bond activation in stoichiometric or catalytic fashion presents an interesting synthetic approach for the transformation of organic molecules [58, 60], it has been achieved merely for a handful of

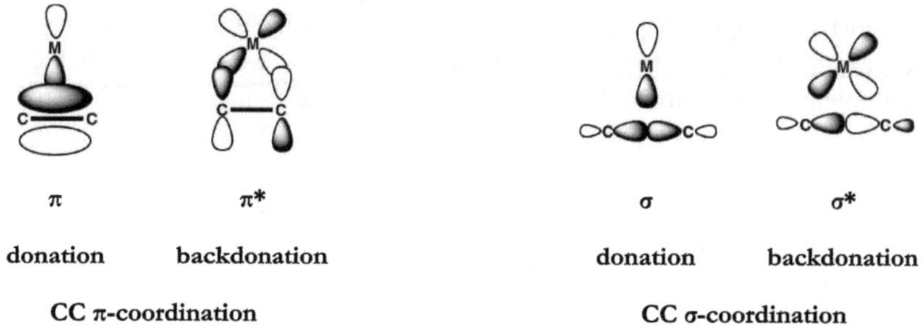

Fig. 5.17 Schematic bonding scheme for the coordination of CC multiple bonds (*left*) and single bonds (*right*) to transition metals, illustrating the unfavorable overlap of the orbitals engaged in the activation process of σ–CC bonds. Figures taken from Ref. [56]

substrates and with only few, handpicked complexes, by far most of them based on electron-rich late transition metals that are capable of backdonation into the anti-bonding orbital of a σ–CC bond (see Sect. 2.1.1) [57].

To the best of our knowledge, oxidative addition of σ–CC bonds to gold complexes is to date not reported in the scientific literature. As CC bond activation at transition metals can be considered as a key example for σ-bond activation of non-polar reagents, a fundamental mechanistic study in this direction with gold complexes should furnish further insight into organogold chemistry.

Although the cleavage of CC bonds at gold complexes is expected to be non-trivial, gold organometallics are often thermodynamically robust [61–65]. We therefore envisaged the activation of σ–CC bonds, as the subsequent formation of two strong gold-carbon bonds may lead to more stable products compared to σ-SiSi bond activation.

5.3.2.1 Oxidative Addition of Biphenylene to Gold(I)

We focused firstly on the activation of biphenylene. The CC bond fusing the aromatic rings is relatively weak (65.4 kcal/mol) [66], due to ring strain and the Hückel antiaromatic character (biphenylene can be viewed as a dibenzo derivative of cyclobutadiene).

The carborane-based diphosphine gold(I) complex **29** was reacted with $GaCl_3$ at low temperature in order to generate an open coordination site and to increase the electrophilicity at the metal center, and biphenylene (5 equivalents) was added in dichloromethane. Slowly heating up the reaction mixture to 120 °C in a pressurizable reaction vessel resulted in a quick color change from pale yellow to green (Scheme 5.16).

The reaction was monitored by ^{31}P NMR spectroscopy: In addition to the signals corresponding to the cationic bis[1,2-bis(diphenylphosphino)-1,2-dicarba-*closo*-dodecaborane-$\kappa^2 P,P'$]gold(I) complex (δ = 42 ppm) in minor quantity, stemming from thermal decomposition of the starting material, {**29**/$GaCl_3$} (δ = 58 ppm), formation of a new compound **34** is observed. **34** features a singlet resonance at δ = 69.9 ppm, in the region of the previously reported bis(diphenylphosphino)bis(silyl)gold(III) complex [7]. Complete conversion is achieved after 3 h. Separation of the product from the formed side-product furnishes a grey-greenish solid that was completely characterized in solution and in the solid state. The structural

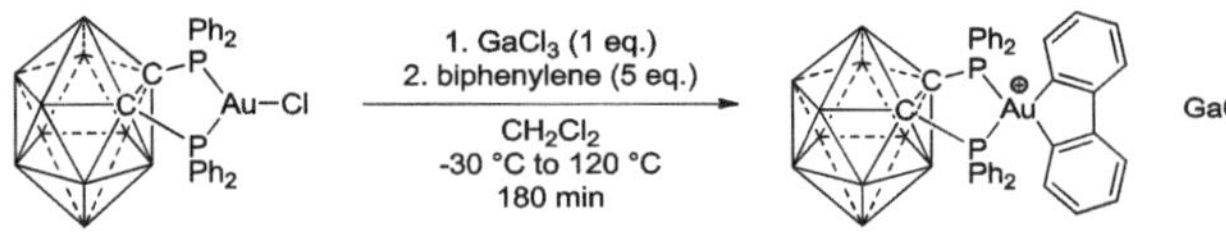

29 **34** (63 %)

Scheme 5.16 Synthesis of dibenzoaurole **34** by oxidative addition of biphenylene at gold(I)

information that can be obtained from ^{1}H NMR spectroscopy is rather limited due to the presence of several overlapping signals in the aromatic region. $^{13}C\{^{1}H\}$ NMR spectroscopy proved to be more meaningful. Noteworthy is the presence of a doublet of doublets at low field (δ = 165.1 ppm). This multiplet turns into a singlet in the corresponding $^{13}C\{^{1}H, ^{31}P\}$ spectrum and features therefore two different C–P coupling constants of 111.2 and 9 Hz. The chemical shift is in the range for an sp^2-hybridized carbon directly bound to gold(III) [67], while the large difference of the two observed coupling constants is typical of a *trans* and *cis* coupling to phosphorus, suggesting that biphenylene underwent oxidative addition at the cationic gold(I) fragment. The proposed structure is further corroborated by the complete ^{13}C NMR spectroscopic data and high resolution mass spectroscopy in the positive mode (ESI+) confirms formation of the cationic bis(diphenylphosphino) dibenzoaurole **34**.

Single crystals of **34** were grown by slow diffusion of pentane into a concentrated solution of this compound at −30 °C, and an X-ray diffraction analysis was performed (Fig. 5.18).

34 crystallizes in the orthorhombic spacegroup *Pnma*. The asymmetric unit contains half the compound lying on a mirror plane. The complex cation and the tetrachlorogallate counteranion are well separated (closest contact: Cl–H, 2.940 (4) Å). The molecular structure of the cationic dibenzoaurole exhibits a square-planar coordination geometry around the gold atom (sum of angles around gold: 360°), as expected for an Au(III) complex. The PAuP angle is 89.44(5)°, while the opposite CAuC angle is slightly acute with 81.1(3)°. The gold(III)-phosphorus distance (2.369(2) Å) and the AuC distance (2.074(7) Å) are comparable to those previously seen (usually in the range of 2.30 Å–2.38 Å and 2.02 Å–2.08 Å, respectively) [68–75].

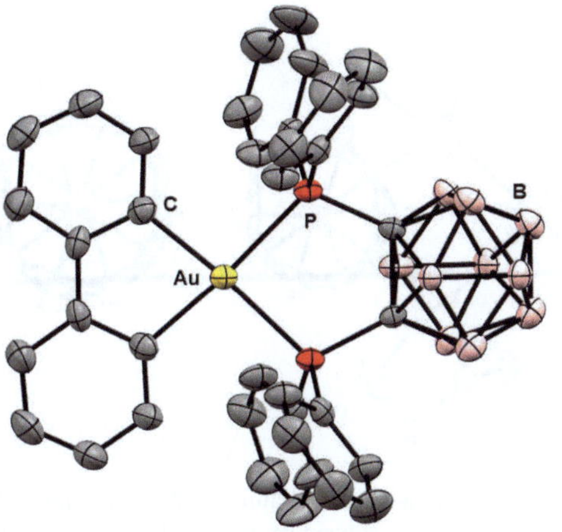

Fig. 5.18 Molecular structure of **34** determined by single crystal X-ray diffraction. The $GaCl_4^-$ counteranion, solvent molecules and hydrogen atoms are omitted for clarity. Selected bond lengths (Å) and angles (°): AuC 2.074(7), AuP 2.369(2), CAuC 81.1(3), PAuP 89.44(5)

By analogy to its phenylated counterpart, the reactivity of gold(I) chloride **30** towards biphenylene was tested (Scheme 5.17): Using $GaCl_3$ as the chloride scavenger with subsequent addition of biphenylene (5 equivalents) in dichloromethane and heating up to 120 °C results in a quick color change from orange to bright yellow, indicative of the formation of a new product **35**, as evidenced as well by ^{31}P NMR monitoring: in the course of the reaction the intensity of a new singlet resonance at δ = 89.5 ppm is steadily increasing, while the signal corresponding to {**30**/$GaCl_3$} (δ = 74 ppm) is decreasing. Unlike as observed for **29**, no side-product is observed during the reaction. Complete conversion is achieved after 5 h. **35** was isolated as a yellow crystalline solid, that displays similar characteristics as observed for compound **34** in its ^{13}C NMR spectrum (e.g. a doublet of doublets at δ = 165.6 ppm with two associated J_{CP} coupling constants of 114.3 and 9.7 Hz). The spectroscopic data are in accord with the dibenzoaurole structure shown in Scheme 5.17 stemming from the oxidative addition of biphenylene at the cationic Au(I) fragment. A single crystal X-ray diffraction analysis (Fig. 5.19) further confirms this attribution to the gold(III) complex.

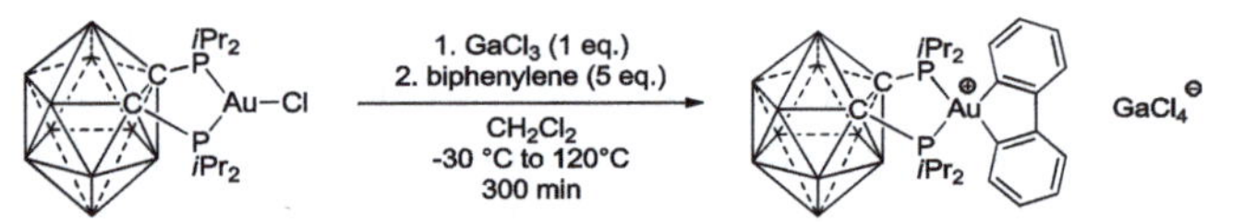

30 **35** (95 %)

Scheme 5.17 Synthesis of dibenzoaurole **35** by oxidative addition of biphenylene at gold(I)

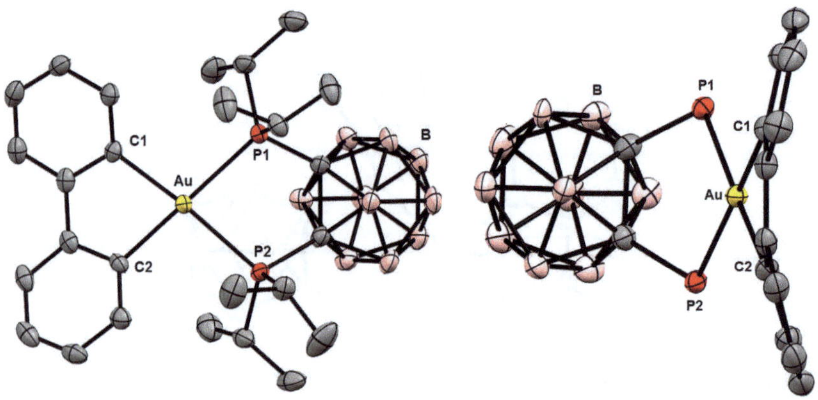

Fig. 5.19 Two views of the molecular structure of **35** determined by single crystal X-ray diffraction. The $GaCl_4^-$ counteranion, solvent molecules, hydrogen atoms and (only for the view on the right) isopropyl substituents are omitted for clarity. Selected bond lengths (Å) and angles (°): AuC1 2.093(8), AuC2 2.086(8), AuP2 2.434(2), AuP1 2.422(2), C1AuC2 79.1(3), P1AuP2 85.25(7)

The molecular structure of dibenzoaurole **35** in the solid state features a square-planar coordination environment for gold (sum of angles around gold: 360°) with symmetric coordination of both the diphosphine and the biphenyl ligand. The bite angle of the diphosphine ligand is acute [PAuP = 85.25(7)°]. The gold-phosphorus and gold-carbon bonds are comparable to those of **34**. In contrast to the molecular structure of **34**, the two aromatic rings of the biphenyl ligand are slightly bent up in a bowl-shape (the planes defined by the two aromatic rings span an angle of ca. 127°), likely due to steric pressure of the opposite isopropyl moieties. The $GaCl_4^-$ counteranion is around 6 Å away from the gold atom.

The oxidative addition of biphenylene to a transition metal is a prime example for a σ-bond activation process of an unpolar bond.

The here employed conditions for the σ–CC bond cleavage in the formation of benzoauroles **34** and **35** may seem rather drastic compared to analogous reactions with other late transition metal complexes (based on group-9 and group-10 metals). However, even though in general lower temperatures are required for the bond cleavage of biphenylene with these complexes (usually between 0 and 80 °C), reaction times of several days are often reported for this reaction, while the oxidative addition at gold(I) described above reaches complete conversion in a few hours [66].

5.3.2.2 Oxidative Addition of Benzocyclobutenone to Gold(I)

Encouraged by these interesting results, we queried whether the CC bond cleavage reaction is limited to biphenylene or if other strained carbocycles might undergo oxidative addition at gold(I) as well. Common targets for transition metal mediated CC bond cleavage reactions are cyclobutenones that are activated by cobalt and rhodium complexes [76–78].

We therefore tempted the activation of benzocyclobutenone at gold(I). To this end, we reacted gold(I) chloride **30** with $GaCl_3$ and added benzocyclobutenone (10 equivalents). Already at room temperature a reaction takes place, as attested by ^{31}P NMR spectroscopy: the starting material (δ = 74 ppm) slowly converts into a new compound (**36**), featuring an AM spin system. Two doublets at δ = 79.7 ppm and δ = 78.3 ppm with a common P–P′ coupling constant of 13.9 Hz point at desymmetrization of the diphosphine environment as would be expected for oxidative addition of the carbocycle to the metal center. Complete conversion is achieved after ca. 24 h at room temperature. In the late stage of this reaction, traces of another product (**37**) are observed, according to ^{31}P NMR data: a second pair of associated doublets at δ = 91.8 ppm and δ = 74.4 ppm (J = 14.4 Hz) is indicative of either further conversion of **36** into **37** or an alternative direct reaction of benzocyclobutenone with {**30**/$GaCl_3$} to yield **37**. In any case, the formation of **37** displays a higher activation barrier compared to **36**, since at room temperature the latter represents by far the major species.

In order to identify the products formed and to elucidate the nature of this process, the reactions conditions were optimized. Both compounds were selectively prepared and isolated (Scheme 5.18).

Heating {**30**/$GaCl_3$} with 5 equivalents of benzocyclobutenone for 1 h at 60 °C furnishes exclusively **36**. This compound was characterized inter alia by multinuclear NMR spectroscopy and single crystal X-ray crystallography. The NMR data are in agreement with the molecular structure in the solid state (Fig. 5.20).

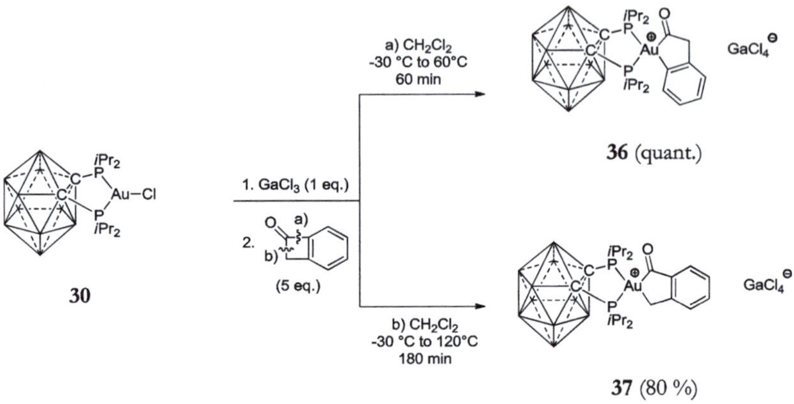

Scheme 5.18 Synthesis of isomeric aurol-2-ones **36** (pathway a) and **37** (pathway b) by oxidative addition of benzocyclobutenone at gold(I)

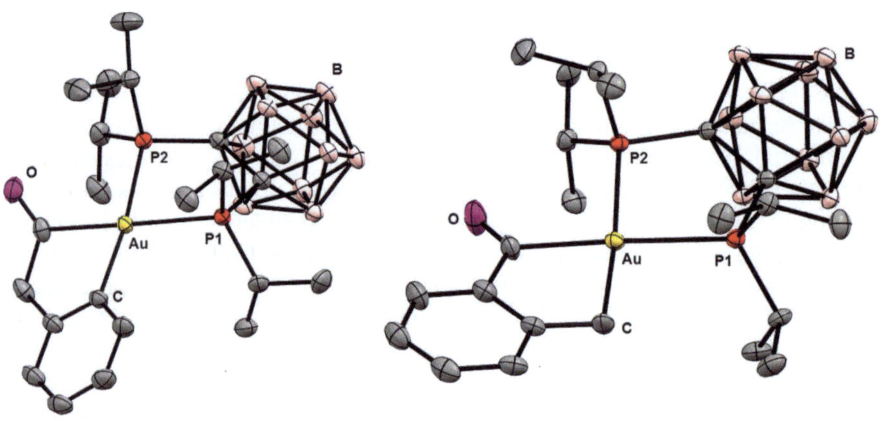

36 **37**

Fig. 5.20 Molecular structures of isomeric complexes **36** (*left*) and **37** (*right*) determined by single crystal X-ray diffraction. $GaCl_4^-$ counteranions and hydrogen atoms are omitted for clarity. Selected bond lengths (Å) and angles (°):**36**: C(O)Au 2.112(3), AuC_{Ar} 2.081(3), AuP1 2.4537(7), AuP2 2.3931(7), CO 1.188(4), CAuC 78.0(1), PAuP 85.61(3) **37**: C(O)Au 2.107(4), CH_2Au 2.105 (4), AuP1 2.436(1), AuP2 2.388(1), CO 1.206(6), CAuC 80.8(2), P1AuP2 89.69(3)

Accordingly, **36** corresponds to a gold(III) complex stemming from the oxidative addition of benzocyclobutenone at the cationized metal center of **30**. The gold atom inserted into the CC bond between the carbonyl carbon and the neighboring C(*sp*^{2}) (bond a) in Scheme 5.18).

37 formed exclusively when heating up the reaction mixture to 120 °C for 3 h under otherwise identical conditions. After isolation, this compound was identified as a regioisomer of **36**, formed by oxidative addition of the CC bond between the methylenic carbon and the carbonyl group (bond b) in Scheme 5.18). Its molecular structure in the solid state was determined by means of single crystal X-ray diffraction (Fig. 5.20).

In both **36** and **37**, the metal center has a square-planar coordination environment, as expected for gold(III) complexes. The formed five-membered auracycles feature an envelope conformation in which the gold center is positioned above the plane spanned by the four surrounding carbon atoms (**36**: 0.76 Å; **37**: 0.47 Å). Gold-carbon and gold-phosphorus bond lengths are in typical ranges (see Sect. 5.3.2.1).

Interestingly, **36** in its pure form is stable in the solid state, but decomposes in solution slowly back to the starting material. Furthermore, heating a sample of **36** in solution leads to conversion to **37** which does not evolve further. These two findings explain now the initial observation when performing the reaction at room temperature: complex **36**, the apparent kinetic product of the reaction of benzocyclobutenone with {**30**/$GaCl_3$}, stands in equilibrium with the starting material. The educts are slightly more stable, but in excess of benzocyclobutenone the gold (III) complex is formed quantitatively via a low energy pathway featuring a barrier that is attained easily even at room temperature. This process is a relatively rare example of a room-temperature σ–CC bond activation [60, 78]. At higher temperature, the thermodynamic product **37** is formed irreversibly.

5.3.3 Activation of Aryl–Iodine Bonds

Transition-metal mediated catalyzed cross-coupling of aryl halides with nucleophiles is nowadays an important tool in organic synthesis, both in academic research and industry [79, 80]. The first step in these catalytic cross-coupling reactions, mostly based on palladium complexes, is the oxidative addition of an aryl halide at a 12- or 14-electron Pd(0) species to give the corresponding Pd(II) complex. This elementary step has been extensively studied, and ligand- and substrate-influence are in general well understood. There is no doubt about the concerted nature of the σ-bond activation of C(*sp*^{2})-halogen bonds, and radical pathways, proposed in the phase of discovery of this reaction, are ruled out today. Platinum complexes show similar reactivity [44].

As mentioned in Chap. 3, there has been some effort to replace group-10 metals by gold in these catalytic reactions, as the isoelectronic nature of Pd(0) and Au(I) might suggest that analogous redox processes are as well possible for gold.

However, solid evidence for homogeneous redox-catalysis by gold(I) complexes involving aryl halides is not available and is sometimes even considered to be impossible (see Sect. 3.3.4).

Recent studies on the intramolecular C(sp^2)–I and C(sp^2)–Br bond activation performed in our laboratory proved that the oxidative addition of aryl halides to gold(I) is in principle possible (see Sect. 3.3.4). The successful activation of σ–CC bonds at a cationic diphosphine gold(I) complex, prompted us to further explore this reactivity and we became curious about the intermolecular oxidative addition of aryl halides at gold(I).

5.3.3.1 Preliminary Experiments

To this end, we attempted the oxidative addition of iodobenzene by adapting the established methodology: reacting gold(I) complex **29** with a chloride scavenger ($GaCl_3$) in order to generate an electrophilic, cationic species, followed by addition of iodobenzene in dichloromethane at low temperature resulted in a rapid color change to bright yellow (Scheme 5.19). Monitoring by ^{31}P NMR spectroscopy revealed that a new phosphorus-containing compound (**38**) formed: two associated doublet resonances at δ = 61.8 ppm and δ = 44.6 ppm (J_{PP} = 29.1 Hz) are reminiscent of the desymmetrization of the ligand framework by oxidative addition of benzocyclobutenone at {**30**/$GaCl_3$} (see Sect. 5.3.2.2). In contrast to the products stemming from the CC-bond cleavage which are thermally robust, **38** decomposes rapidly at room temperature to gold metal, the bis[1,2-bis(diphenylphosphino)-1,2-dicarba-*closo*-dodecaborane-$\kappa^2 P,P'$]gold(I) complex (see Sect. 5.3.2.1) and unidentified products. Even at temperatures as low as −80 °C the complex is not stable and decomposes over days. **38** had to be characterized in situ by means of a low temperature multinuclear NMR analysis. Besides the ^{31}P NMR spectrum, the ^{13}C NMR data are most informative. Noteworthy is a doublet resonance at δ = 144.8 ppm featuring a large C–P coupling constant (J_{CP} = 122.0 Hz), its magnitude being in the range of *trans* carbon-phosphorus couplings (see Sect. 5.3.2).

These spectroscopic data for complex **38** are in agreement with a gold(III) complex stemming from the C(sp^2)–I bond activation as shown in Scheme 5.19.

The analogous reaction with chloro-[1,2-bis(di-*iso*-propylphosphino)-1,2-dicarba-*closo*-dodecaborane-$\kappa^2 P,P'$]gold(I) complex **30** proved to be very fast,

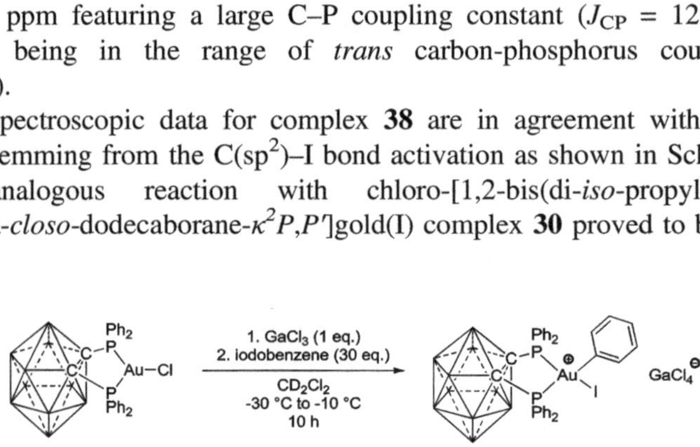

29 **38**

Scheme 5.19 Synthesis of gold(III) complex **38** by oxidative addition of iodobenzene

and a gold(III) complex stemming from oxidative addition of phenyl iodide was not unambiguously detected due to the instability of the formed species and rapid decomposition to unidentified products even at low temperature. Although no strong evidence for the oxidative addition reaction to {**30**/$GaCl_3$} is available, a color change to orange-yellow hints at a possible gold(I)-gold(III) transition.

5.3.3.2 Influence of the Counteranion

As the gold(III) complex stemming from the oxidative addition of phenyl iodide to {**29**/$GaCl_3$} was not stable enough for further characterization, we studied the influence of the counteranion on the thermal stability of this species by variation of the chloride scavenger to generate the reactive gold(I) complex.

We suspected that the origin of the poor stability of the oxidative addition product could be linked to the tetrachlorogallate counteranion as a potential chloride source. Nucleophilic attack of Cl^- or other related side-reactions might lead to decomposition of the formed gold(III) complex.

Efficient chloride abstraction is not limited to the use of gallium trichloride, but can be performed as well by salt elimination with sodium or silver salts containing weakly coordinating anions. Sodium tetrakis((3,5-trifluoromethyl)phenyl)borate, silver trifluoromethanesulfonate or silver bis(trifluoromethane)sulfonimide were efficient in activating the gold(I) chloride complexes for oxidative addition of iodobenzene. We found that the bis(trifluoromethane)sulfonimide anion (triflimidate, NTf_2^-) allows for a good compromise between reactivity and stability of the generated cationic gold(I) complex.

Reaction of $AgNTf_2$ with **29** yields the corresponding gold(I) triflimidate complex which was employed in situ in the reaction with phenyl iodide after separation of silver salts and furnishes the corresponding gold(III) complex **39** which exhibits an increased stability as compared to its tetrachlorogallate equivalent **38** and could be manipulated at room temperature over brief periods (minutes) without complete decomposition. Although it could not be isolated in pure form on a preparative scale, a low temperature NMR analysis unambiguously confirms the attribution to the gold(III) complex shown in Scheme 5.20.

Furthermore, a small number of crystals suitable for an X-ray diffraction analysis were obtained by slow diffusion of pentane into a concentrated solution of **39** at

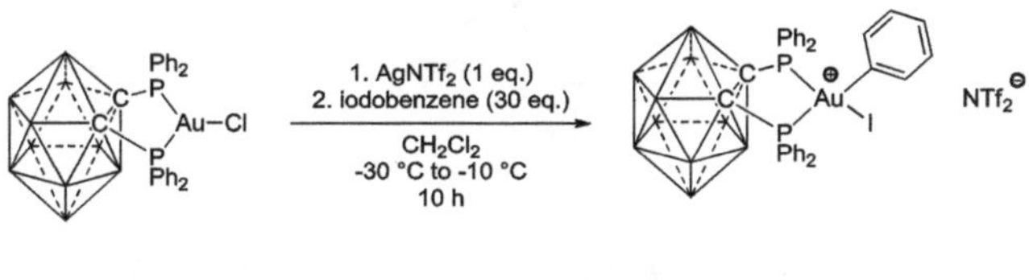

29 **39**

Scheme 5.20 Synthesis of gold(III) complex **39** by oxidative addition of iodobenzene

−60 °C. Despite the sensitivity of these crystals, the solved crystal structure is of reasonable quality and further confirms the formation of the gold(III) complex (Fig. 5.21).

In the complex cation **39**, iodide and phenyl ligands complete with the diphosphine ligand the coordination sphere of the gold atom. The coordination environment of the metal center is square-planar (sum of angles around Au: 360°). The C_{ipso}–I distance is long (3.33(2) Å), indicating complete bond cleavage of the aryl–iodine bond. The plane of the phenyl ring directly bound to gold is almost perpendicular (83.1°) to the plane spanned by the phosphorus atoms, I and C_{ipso}.

5.3.3.3 Influence of the Substitution Pattern at P on the Oxidative Addition Reaction

Aiming at increasing the thermal stability of the oxidative addition product, we investigated the influence of the substitution pattern of the phosphorus donor arms, both in terms of steric bulk and electronic properties, while maintaining the *o*-carborane backbone.

In addition to gold(I) chloride complexes **29** and **30**, featuring phenyl and isopropyl substituents at the P atoms, a third member in this series was synthesized: the 1,2-bis(diaminophosphino)-1,2-dicarba-*closo*-dodecaborane (**40**) was prepared by reacting chloro-diisopropyl-diazaphospholidine with dilithio-1,2-dicarba-*closo*-dodecaborane. This chelating ligand possesses both an important donor potential due to the electron rich substituents on P and significant steric bulk. Diaminophosphines have been successfully employed in the stabilization of

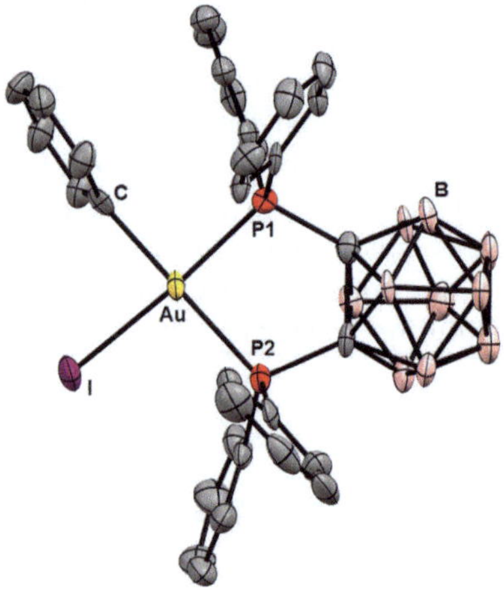

Fig. 5.21 Molecular structure of **39** determined by single crystal X-ray diffraction. The triflimidate counteranion, solvent molecules and hydrogen atoms, are omitted for clarity. Selected bond lengths (Å) and angles (°): P1Au 2.329(4), P2Au 2.395(4), AuC 2.14(2), AuI 2.596(1), P1AuP2 91.7(1), IAuC 88.8(4), P1AuC 87.4(4), P2AuI 92.1(1)

reactive species [81–85]. Coordination of ligand **40** to [(THT)AuCl] yielded the corresponding gold(I) chloride complex **41** (Scheme. 5.21).

Next, **41** was tested for the oxidative addition of iodobenzene after chloride abstraction with $GaCl_3$ (Scheme 5.22).

The resulting gold(III) complex **42** was obtained in 88 % yield. Gratifyingly, **42** proved to exhibit an increased stability compared to its bis(diphenylphosphino) counterpart. Complex **42** can be prepared and handled for short periods of time in solution even at room temperature, even though upon prolonged standing (hours) significant decomposition is observed as well. The increased thermal stability allowed the complete characterization of this compound in solution and in the solid state. In order to elucidate its molecular structure, single crystals of **42** were grown at −60 °C and an X-ray diffraction analysis was performed (Fig. 5.22).

The key metric parameters (AuP, AuC and AuI bond distances, angles around gold) of the molecular structure of the complex cation of **42** are comparable to those of **39**. The square-planar coordination environment of the metal center is slightly distorted. The *ipso* carbon of the phenyl ring bound to gold lies about 0.27 Å above the PAuP plane, the iodide ligand below (0.29 Å). The phenyl ring and the PAuP plane span an angle of 60.5°.

In order to verify if the NTf_2^- anion would have a similar beneficial impact on the stability of the ensuing gold(III) complexes, **41** was reacted with $AgNTf_2$ to yield the corresponding gold(I) triflimidate complex **43** which can be isolated and handled conveniently at room temperature (Scheme 5.23).

43 was fully characterized. A single resonance in its ^{31}P NMR spectrum (δ = 138.3 ppm) shifted lowfield with respect to the chloride precursor indicates at least time-averaged a symmetric coordination environment around gold. The

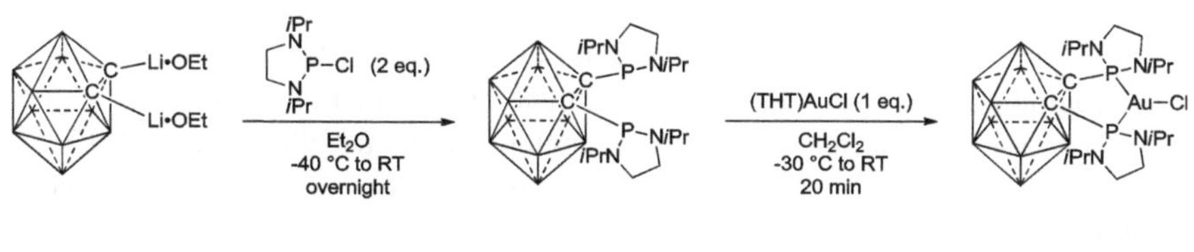

40 (75 %) **41** (88 %)

Scheme 5.21 Synthesis of gold(I) chloride complex **41**

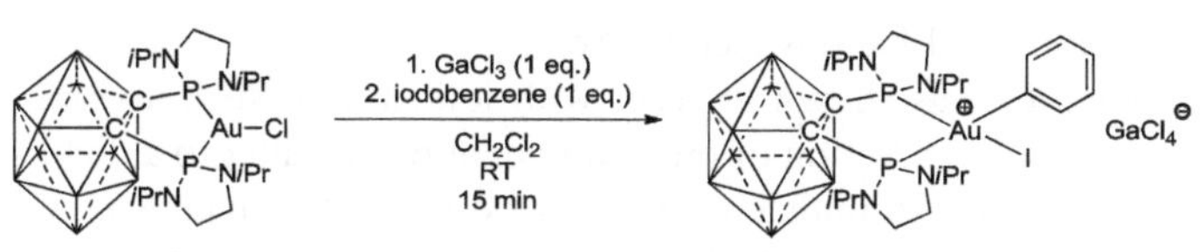

41 **42** (88 %)

Scheme 5.22 Synthesis of gold(III) complex **42** by oxidative addition of iodobenzene

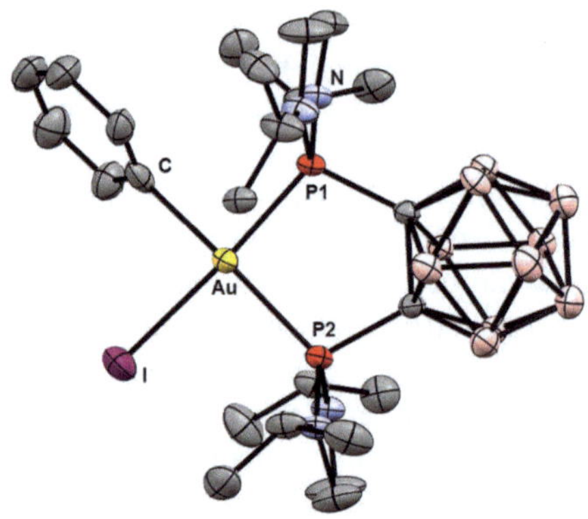

Fig. 5.22 Molecular structure of complex **42** determined by single crystal X-ray diffraction. The $GaCl_4^-$ counteranion, solvent molecules and hydrogen atoms are omitted for clarity. Selected bond lengths (Å) and angles (°): P1Au 2.304(1), P2Au 2.382(1), AuC 2.142(6), AuI 2.6106(4), P1AuP2 93.74(4), IAuC 88.1(1), P1AuC 89.0(1), P2AuI 90.02(3)

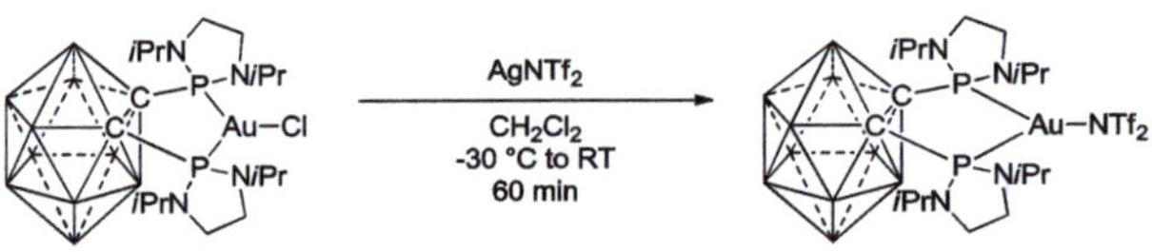

41

43 (90 %)

Scheme 5.23 Synthesis of gold(I) triflimidate complex **43**

molecular structure of **43** in the solid state, as determined by single crystal X-ray crystallography, displays indeed a tricoordinate environment around gold, with the triflimidate ion coordinated to the gold(I) center. The PAuP bite angle is 100.73(2)° which comes close to the 90° for an ideal square-planar coordination in a corresponding gold(III) compound (Fig. 5.23).

The triflimidate ion is tightly bound to the gold atom in the solid state as indicated by a short Au-N distance (AuN = 2.221(1) Å); somewhat longer than the AuN bond in Gagosz's catalyst ([$(Ph_3P)AuNTf_2$]) (AuN = 2.102(3) Å) [86]. As for triphenylphosphine gold(I) triflimidate which is in solution an active catalyst, the counteranion of **43** shows a labile behavior in solution to allow for the oxidative addition of various aryl halides (see below).

For example, **43** was shown to react with iodoarenes, such as iodobenzene or *p*-fluoro-iodobenzene. Addition of one equivalent of the latter to a dichloromethane solution of the gold(I) triflimidate resulted in a quick color change to bright yellow, pointing at the formation of a gold(III) complex by oxidative addition of *p*-F-C_6H_4I. Gratifyingly, complex **44** exhibits reasonable stability and can be handled for short periods of time in solution at room temperature without excessive decomposition.

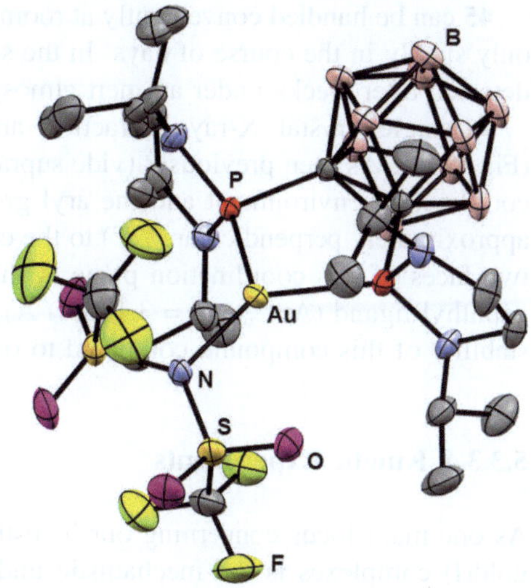

Fig. 5.23 Molecular structure of complex **43** determined by single crystal X-ray diffraction. Solvent molecules and hydrogen atoms are omitted for clarity. Selected bond lengths (Å) and angles (°): P1Au 2.3215(5), P2Au 2.3172(5), AuN 2.221 (2), P1AuP2 100.73(2), P1AuN 129.52(4), P2AuN 129.74(4)

A multinuclear NMR-spectroscopic analysis confirms the formulation as the gold (III) complex depicted in Scheme 5.24.

Further stabilization of the gold(III) aryl iodide products was achieved by increasing the steric bulk around the metal center. Reaction of 1-iodo-8-methyl-naphthalene with **43** furnishes cleanly the deep-red gold(III) complex **45** after oxidative addition of the CI bond (Scheme 5.25).

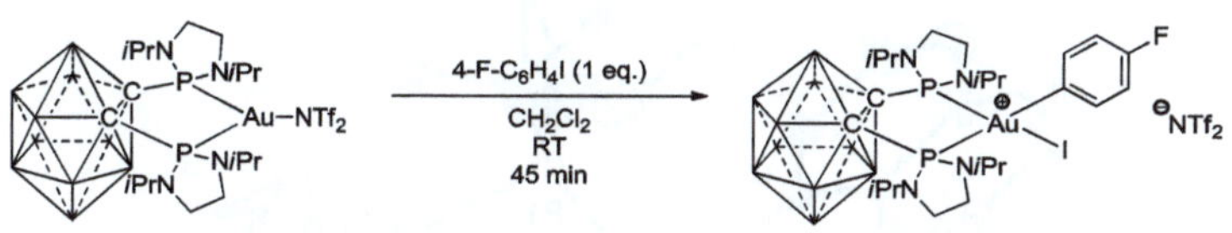

43 **44** (96 %)

Scheme 5.24 Synthesis of gold(III) complex **44** by oxidative addition of *p*-FC_6H_4I to gold(I)

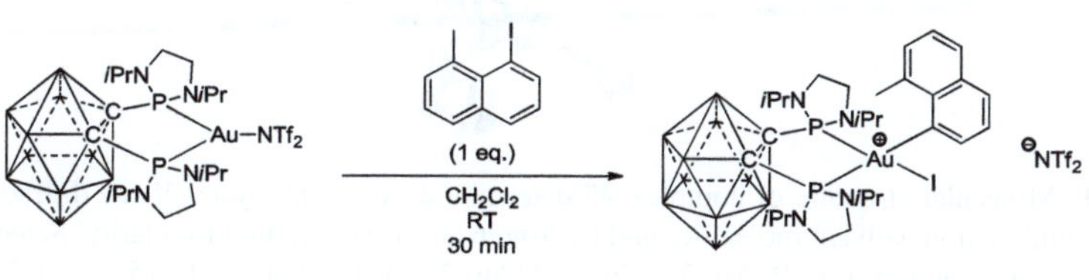

43 **45** (92 %)

Scheme 5.25 Synthesis of gold(III) complex **45** by oxidative addition of 1-iodo-8-methylnaphthalene to gold(I)

45 can be handled conveniently at room temperature and decomposes in solution only slowly in the course of days. In the solid state, no sign of decomposition was detected after weeks under an inert atmosphere.

A single crystal X-ray diffraction analysis of complex **45** was performed (Fig. 5.24). As seen previously (vide supra), the gold atom features a square-planar coordination environment and the aryl group bound to the gold atom is arranged approximately perpendicular (78°) to the coordination plane around Au. One of the two faces of this coordination plane is shielded by the CH_3-group of the methyl-naphthyl ligand (AuC_{methyl} = 3.153(7) Å) which might contribute to the increased stability of this compound compared to other examples in this series.

5.3.3.4 Kinetic Experiments

As one main focus concerning our investigation of oxidative addition reactions at gold(I) complexes is the mechanistic understanding of these transformations, we performed kinetic experiments in order to gain insight into the mechanism operating.

We investigated the impact of the donor capability of the phosphine ligand on the rate of the carbon-iodine bond activation. Phenyl- and bis(amino)-substituted phosphine complexes **29** and **41** were compared with regard to the oxidative addition of iodobenzene after chloride abstraction with $GaCl_3$.

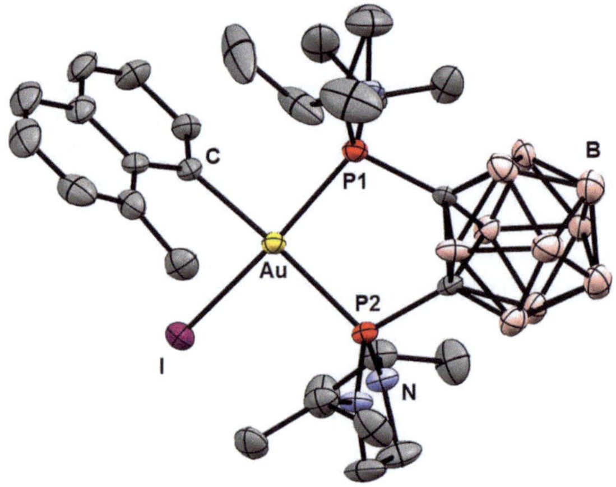

Fig. 5.24 Molecular structure of complex **45** determined by single crystal X-ray diffraction. The $GaCl_4^-$ counteranion, solvent molecules and hydrogen atoms are omitted for clarity. Selected bond lengths (Å) and angles (°): P1Au 2.327(2), P2Au 2.381(1), AuC 2.092(5), AuI 2.6218(5), AuC_{methyl} 3.153(7), P1AuP2 93.67(5), IAuC 86.4(2), P1AuC 89.8(2), P2AuI 90.46(4)

The reactions were monitored by ^{31}P NMR spectroscopy. At −10 °C and in presence of 5 equivalents of phenyl iodide, complexes **29** and **41** transform cleanly into the corresponding gold(III) complexes **38** and **42**, respectively (Schemes 5.22 and 5.26).

Both reactions were monitored at −10 °C until complete conversion was reached. Plotting the natural logarithm of the concentration of the gold(I) complex over time resulted for both reactions in a straight line (Fig. 5.25), indicating a first-order dependence for the gold(I) complex under these conditions.

This rough kinetic estimation indicates that the diaminophosphine-ligated gold (I) complex reacts about one order of magnitude faster than the corresponding complex featuring the diphenylphosphine substitution pattern (k_{obs}(**29**) = 5.62×10^{-5} s^{-1} vs. k_{obs}(**41**) = 2.43×10^{-4} s^{-1}).

This behavior can be rationalized by increased backdonation of electron density from the metal center to the antibonding molecular orbital of the σ-CI bond in case of the electron-rich bis(diaminophosphine)-ligated complex **41**, thus facilitating the bond cleavage.

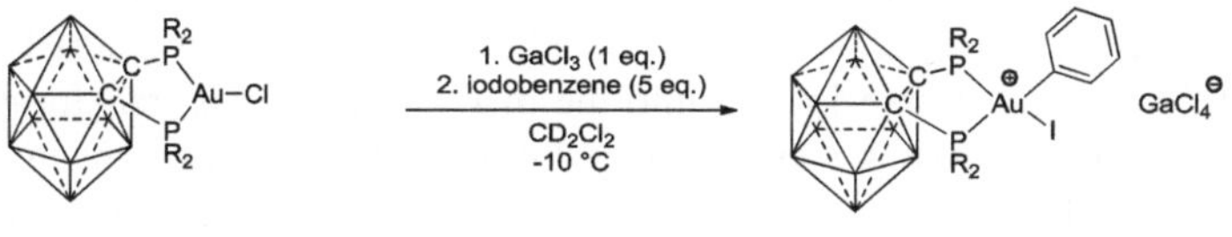

29 (R = Ph)

38 (R = Ph)

41 (R = *i*PrN(CH_2CH_2)N*i*Pr)

42 (R = *i*PrN(CH_2CH_2)N*i*Pr)

Scheme 5.26 Semi-quantitative comparison of reaction rates for the oxidative addition of iodobenzene at gold(I) complexes depending on the substitution pattern on P. R=Ph, *i*PrN (CH_2CH_2)N*i*Pr. For R=*i*Pr only decomposition products have been observed

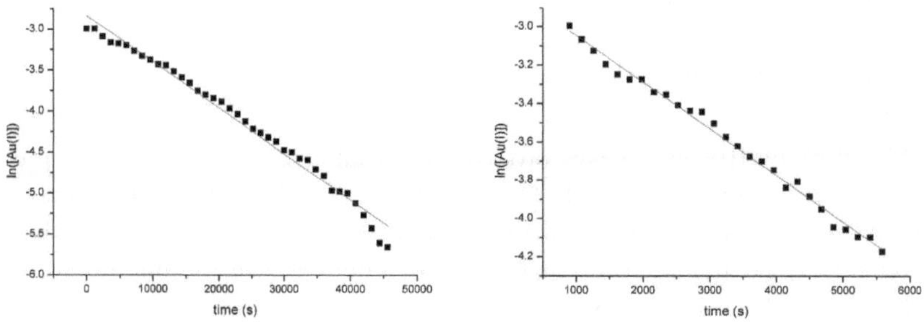

29 + $GaCl_3$ + PhI

41 + $GaCl_3$ + PhI

Fig. 5.25 Kinetic data for the oxidative addition of iodobenzene at gold(I) complexes. Natural logarithmic plots of the concentration of gold(I) complexes over time. *Left* **29**. *Right* **41**

5.4 Conclusion

In conclusion, the here presented results provide unambiguous evidence for the oxidative addition of σ–SiSi, σ–CC and σ–C(sp^2)–I bonds to gold and prove that intermolecular oxidative additions proceeding via concerted σ-bond activation pathways are not excluded for gold(I) complexes.

Disilanes react readily at low temperature with phosphine gold(I) chloride complexes in the presence of $GaCl_3$ to give unusual cationic bis(silyl)gold(III) complexes. A joint computational analysis revealed a Y-shape structure for these thermally unstable species. The concerted cleavage of the σ–SiSi bond was found to proceed via a low energy pathway in agreement with the experimentally observed spontaneous oxidative addition of the disilane to gold(I).

In contrast to linear, dicoordinated gold(I) complexes, bis(phosphine) gold(I) complexes, featuring a bidentate, *o*-carborane-based ligand with a small bite-angle, were shown to exhibit substantial backdonation, as exemplified by the synthesis of a classical gold(I) carbonyl complex.

Cationic, dicoordinate gold(I) complexes ligated by such an (*o*-carboranyl) diphosphine ligand were shown to undergo facile oxidative addition of strained σ–CC bonds and σ–C(sp^2)–I bonds to yield the corresponding gold(III) complexes.

In order to gain deeper insight into the mechanism of this transformation, more detailed kinetic studies as well as computational investigations are in progress.

Even though the averseness of gold(I) complexes to undergo oxidative addition via concerted σ-bond activation pathways is not yet fully understood, it is clear now that an analysis based solely on the high redox potential of gold is too simplistic [87, 88]. As shown in this chapter, the ligand sphere strongly impacts the reactivity of gold(I) complexes. Careful evaluation of ligand influence may help to gain deeper insight into the factors which govern the gold(I)-gold(III) transition.

5.5 Experimental Part

5.5.1 General Remarks

The bis(silyl)gold(III) complexes proved too unstable to be isolated and were thus characterized in situ at −80 °C. In the ^1H and ^{13}C NMR spectra, not all the signals could be unequivocally attributed due to multiple overlapping with unidentified byproducts. Even though the reagents were employed in equimolar quantities, unreacted disilane was usually detected.

5.5.2 Au(III) Bis(Silyl) Complex 24a

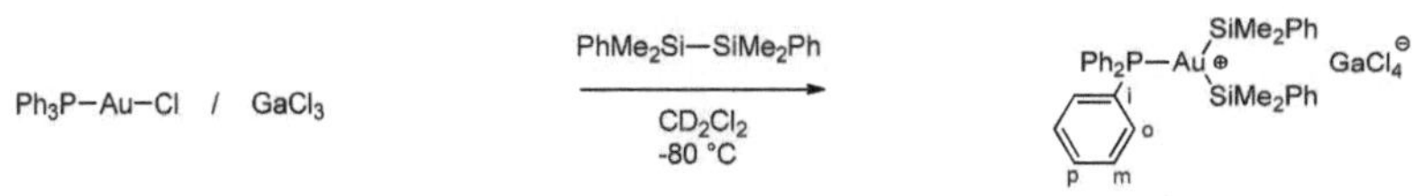

24a

In a glovebox, Ph_3PAuCl (30 mg, 0.06 mmol) were transferred into a routine Norell NMR tube and dissolved in dichloromethane-d_2 (0.3 mL). The tube was closed with a rubber septum assured with a piece of parafilm. $(PhMe_2Si)_2$ (16.5 mg, 0.06 mmol) and $GaCl_3$ (10.7 mg, 0.06 mmol) were transferred into small glass vials and solubilized in dichloromethane-d_2 (0.3 mL), respectively. The prepared solutions were loaded into plastic syringes equipped with stainless steel needles. The syringes were closed by blocking the needles with a septum. Outside of the glovebox, the NMR tube was put under positive Argon pressure and cooled down to −90 °C (acetone/N_2 coldbath). At this temperature, the solution of $GaCl_3$ was added. The tube was gently shaken avoiding the warm-up of the reaction mixture. Subsequently, the solution of the disilane was added at −90 °C, which was accompanied by a quick color change of the mixture from colorless to yellow-brown. The tube was kept at −90 °C and immediately introduced into an NMR machine for analysis. **24a** was formed with about 66 % yield along with unidentified byproducts and could be observed by multinuclear NMR spectroscopy at −80 °C. In addition to the ^{31}P and ^{29}Si NMR data, only the ^{1}H signals for the $SiCH_3$ groups and the ^{13}C signals for the $SiCH_3$ groups as well as for the C_i and $C_{m/o}$ of the phenyl groups bound to phosphorus could be unambiguously identified.

^{1}H NMR (400 MHz, 193 K, CD_2Cl_2): δ 1.02 (d, $^{4}J_{HP} = 1.1$ Hz, $SiCH_3$); **^{31}P{^{1}H} NMR** (162 MHz, 193 K, CD_2Cl_2): δ 60.9 (s, ^{13}C satellite signals: d, $^{1}J_{PCi} = 48.1$ Hz; d, $J_{PCo/m} = 11.5$ Hz; ^{29}Si satellite signals: d, $^{2}J_{PSi} = 38.7$ Hz); **^{29}Si {^{1}H} NMR** (79 MHz, 193 K, CD_2Cl_2): δ 40.7 (d, $^{2}J_{SiP} = 38.7$ Hz); **^{13}C{^{1}H} NMR** (100 MHz, 193 K, CD_2Cl_2): δ 129.7 (d, $J_{CP} = 11.5$ Hz, C_o or C_m), 127.1 (d, $^{1}J_{CP} = 48.1$ Hz, C_i), 1.4 (d, $^{3}J_{CP} = 4.1$ Hz, $SiCH_3$).

5.5.3 Au(III) Bis(Silyl) Complex 24b

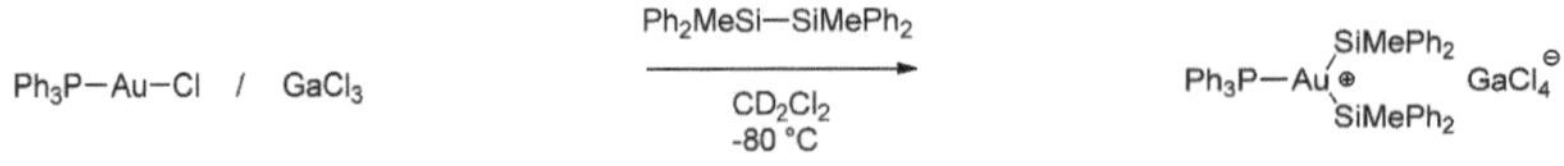

24b

In a glovebox, Ph_3PAuCl (30 mg, 0.06 mmol) was transferred into an NMR tube and dissolved in dichloromethane-d_2 (0.3 mL). The tube was closed with a rubber septum assured with a piece of parafilm. Dimethyltetraphenyldisilane $(Ph_2MeSi)_2$ (16.5 mg, 0.06 mmol) and $GaCl_3$ (10.7 mg, 0.06 mmol) were transferred into small glass vials and solubilized in dichloromethane-d_2 (0.3 mL), respectively. The prepared solutions were loaded into plastic syringes equipped with stainless steel needles. The syringes were closed by blocking the needles with a septum. Outside of the glovebox, the NMR tube was put under positive argon pressure and cooled down to −90 °C (acetone/N_2 coldbath). At this temperature, the solution of $GaCl_3$ was added. The tube was gently shaken avoiding the warm-up of the reaction mixture. Subsequently, the solution of the disilane was added at −90 °C, which was accompanied by a quick color change of the mixture from colorless to yellow-brown. The tube was kept at −90 °C and immediately introduced into an NMR machine for analysis. According to ^{31}P NMR, **24b** was formed in about 24 % yield along with unidentified byproducts. In addition to the ^{31}P and ^{29}Si NMR data, only the ^{1}H and ^{13}C signals for the $SiCH_3$ groups could be unambiguously identified. **^{1}H NMR** (400 MHz, 193 K, CD_2Cl_2): δ 0.97 (sl, 6 H, $SiCH_3$); **^{31}P{^{1}H} NMR** (162 MHz, 193 K, CD_2Cl_2): δ 62.0 (s); **^{29}Si{^{1}H} NMR** (79 MHz, 193 K, CD_2Cl_2): δ 34.5 (d, $^{2}J_{SiP} = 39.8$ Hz); **^{13}C{^{1}H} NMR** (100 MHz, 193 K, CD_2Cl_2): δ 2.6 (d, $^{2}J_{CP} = 4.0$ Hz, $SiCH_3$).

5.5.4 Au(III) Bis(Silyl) Complex 24c

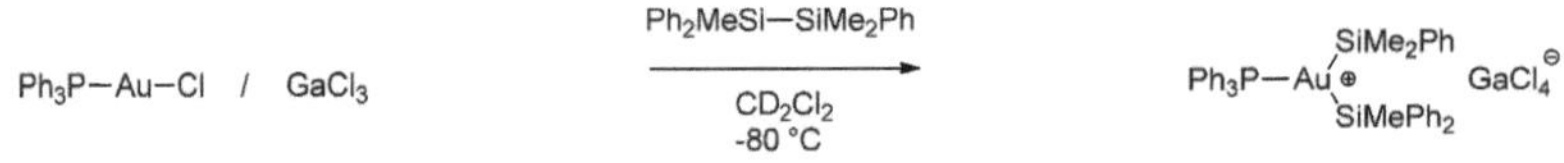

24c

In a glovebox, Ph_3PAuCl (30 mg, 0.06 mmol) was transferred into an NMR tube and dissolved in dichloromethane-d_2 (0.3 mL). The tube was closed with a rubber septum assured with a piece of parafilm. $Ph_2MeSiSiMe_2Ph$ (95 μL of a stock solution in dichloromethane-d_2 (c = 214 mg/mL), 0.06 mmol) and $GaCl_3$ (10.7 mg, 0.06 mmol) were transferred into small glass vials and solubilized in dichloromethane-d_2 (0.3 mL), respectively. The further experimental procedure was analogous to that for **24a**. According to ^{31}P NMR, **24c** was formed in about 12 % yield along with unidentified byproducts. In addition to the ^{31}P and ^{29}Si NMR data, only the ^{1}H signals for the $SiCH_3$ groups could be unambiguously identified. **24c** proved to be too unstable even at −80 °C to record ^{13}C and quantitative ^{29}Si{^{1}H} NMR data. **^{1}H NMR** (500 MHz, 193 K, CD_2Cl_2): δ 1.37 (sl, 3H, $SiCH_3$), 0.72 (sl, 6H, $SiCH_3$); **^{31}P{^{1}H} NMR** (203 MHz, 193 K, CD_2Cl_2): δ 61.7 (s); **^{29}Si{^{1}H} NMR** (99 MHz, 193 K, CD_2Cl_2): δ 42.8 (d, $^{2}J_{SiP}$ = 27.5 Hz), 33.1 (d, $^{2}J_{SiP}$ = 49.6 Hz).

5.5.5 *(o-(Dimethylphenylsilyl)Phenyl)Diphenylphosphine (25)*

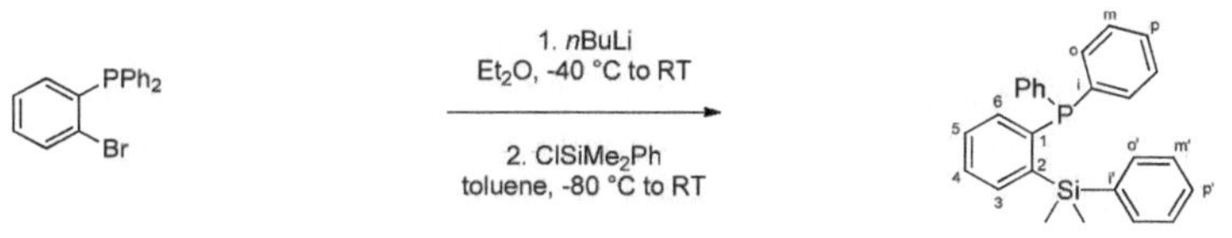

25

A solution of *n*-butyllithium (1.6 M in hexane, 2.2 mL, 3.52 mmol) was added at −40 °C to a solution of 1-bromo-2-diphenylphosphinobenzene (1 g, 2.93 mmol) in diethyl ether (6 mL). The solution was allowed to warm to room temperature for 1 h and then cooled down to −20 °C. The supernatant was filtered off and the residue dried under vacuum for 30 min to give an off-white solid that was subsequently solubilized in cold toluene (6 mL). The solution was then cooled down to −80 °C and chloro(dimethyl)phenylsilane (0.6 mL, 3.52 mmol) was added dropwise. The solution was allowed to warm up to room temperature overnight. The yellowish reaction mixture was filtered over Celite, volatiles were removed in vacuo and the crude yellowish residue purified by column chromatography over silica (eluent: pentane, R_f: 0.38). The obtained colorless oil was triturated with pentane (2 × 3 mL) to give the phosphinosilane as a white solid. Yield: 755 mg (65 %). **Mp**: 63 °C; **^{1}H NMR** (300 MHz, $CDCl_3$): δ 7.60–6.97 (m, 19H, H_{ar}), 0.63 (d, $^{5}J_{HP}$ = 1.6 Hz, 6H, $SiCH_3$); **^{31}P{^{1}H} NMR** (121 MHz, $CDCl_3$): δ −10.9 (s); **^{29}Si{^{1}H} NMR** (60 MHz, C_6D_6): δ −7.7 (d, $^{3}J_{SiP}$ = 11.2 Hz); **^{13}C{^{1}H} NMR** (121 MHz, $CDCl_3$): δ 146.3 (d, $^{1}J_{CP}$ = 47.5 Hz, C_1), 143.7 (d, $^{2}J_{CP}$ = 12.1 Hz, C_2), 139.7 (d, $^{4}J_{CP}$ = 2.8 Hz, $C_{i'}$), 138.1 (d, $^{1}J_{CP}$ = 12.1 Hz, C_i), 135.9 (d, $^{2}J_{CP}$ = 15.9 Hz, C_6), 135.6 (d, $^{3}J_{CP}$ = 1.2 Hz, C_3 or C_5), 134.4 (d, $^{3}J_{CP}$ = 1.8 Hz, C_3 or C_5), 133.2 (d, J_{CP} = 18.6 Hz, C_o or C_m), 129.5 (s, C_4), 128.6 (s, C_{ar}), 128.4 (s, C_{ar}), 128.2

(d, J_{CP} = 6.3 Hz, C_o or C_m), 128.0 (s, C_p), 127.5 (s, C_{ar}), 0.2 (d, $^4J_{CP}$ = 10.6 Hz, $SiCH_3$); **HRMS (ESI+)**: calcd for $[M+H]^+$ = $C_{26}H_{26}PSi^+$: 397.1541. Found: 397.1523; **Elt. Anal.**: calcd for $C_{26}H_{25}PSi$: C, 78.75; H, 6.35. Found: C, 78.75; H, 6.19.

5.5.6 Coordination of 25 to Gold(I) (26)

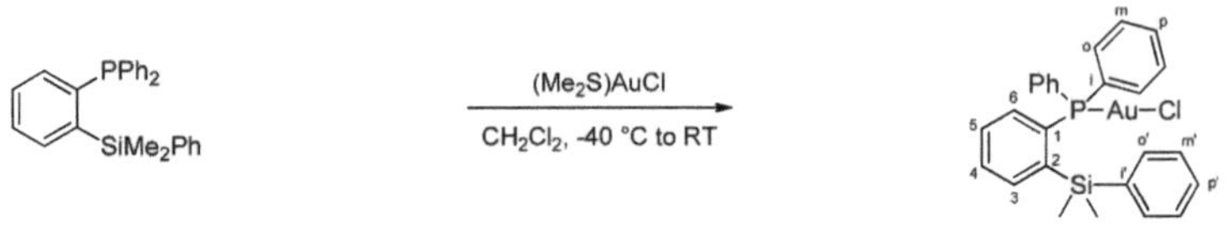

25 **26**

A solution of the phosphinosilane ligand (**2**) (301 mg, 0.76 mmol) in dichloromethane (3 mL) was added at −40 °C to a solution of chloro(dimethylsulfide)gold (I) (224 mg, 0.76 mmol) in dichloromethane (2 mL). The reaction mixture was quickly warmed up to room temperature and stirred for further 20 min. The solution was concentrated by evaporation to a volume of about 1.5 mL. Addition of 5 mL of pentane leads to the precipitation of a colorless solid that was separated by filtration, washed with pentane (2 × 1 mL) and dried under vacuum. Complex **26** is obtained as a white powder. Yield: 398 mg (83 %). **Mp**: 181 °C (decomposition); **^{1}H NMR** (300 MHz, $CDCl_3$): δ 7.87–7.77 (m, 1H, H_{ar}), 7.58 – 7.14 (m, 17H, H_{ar}), 7.00–6.87 (m, 1H, H_{ar}), 0.67 (s, 6H, $SiCH_3$); **^{31}P{^{1}H} NMR** (162 MHz, $CDCl_3$): δ 45.4 (s); **^{29}Si{^{1}H} NMR** (60 MHz, $CDCl_3$): δ –5.5 (t, $^3J_{SiP}$ = 8.0 Hz); **^{13}C{^{1}H} NMR** (75 MHz, $CDCl_3$): δ 144.4 (d, $^1J_{CP}$ = 24.3 Hz, C_1), 138.9 (d, $^2J_{CP}$ = 15.7 Hz, C_6), 138.4 (s, C_2 or $C_{i'}$), 134.9 (d, $^3J_{CP}$ = 9.1 Hz, C_3 or C_5), 134.6 (s, $C_{o'}$), 134.5 (d, J_{CP} = 5.8 Hz, C_o or C_m), 134.2 (s, C_2 or $C_{i'}$), 131.8 (d, $^3J_{CP}$ = 2.6 Hz, C_3 or C_5), 130.6 (s, C_4), 129.6 (d, $^1J_{CP}$ = 24.6 Hz, C_i), 129.5 (s, $C_{p'}$), 129.4 (s, $C_{m'}$), 129.3 (d, J_{CP} = 12.0 Hz, C_o or C_m), 128.0 (s, C_p), 2.4 (d, $^4J_{CP}$ = 1.1 Hz, $SiCH_3$); **HRMS (ESI+)**: calcd for $[M-Cl]^+$ = $C_{26}H_{25}PSiAu^+$: 593.1129. Found: 593.1132.

5.5.7 Au(III) Bis(Silyl) Complex 27

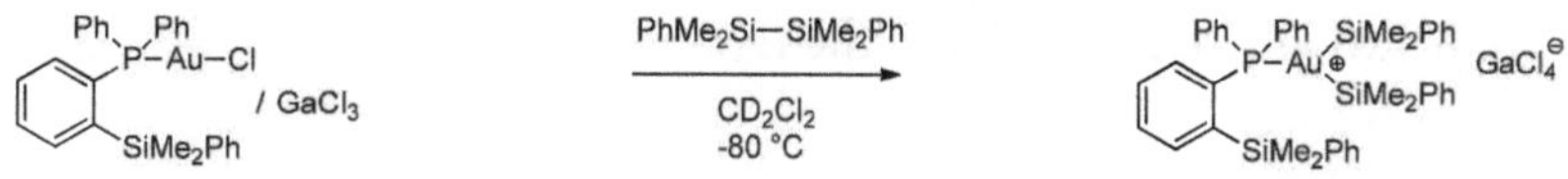

27

In a glovebox, Au(I) complex **3** (30 mg, 0.05 mmol) was transferred into an NMR tube and dissolved in dichloromethane-d_2 (0.3 mL). The tube was closed with a rubber septum assured with a piece of parafilm. $(PhMe_2Si)_2$ (12.9 mg, 0.05 mmol) and $GaCl_3$ (8.4 mg, 0.05 mmol) were transferred into small glass vials and solubilized in dichloromethane-d_2 (0.3 mL), respectively. The further experimental procedure was analogous to that for **24a**. According to ^{31}P NMR, complex **27** was formed in about 59 % yield along with unidentified byproducts. In addition to the ^{31}P and ^{29}Si NMR data, only the ^{1}H and ^{13}C signals for the $SiCH_3$ groups could be unambiguously identified. The complex decomposes completely at −80 °C in about 8 h (mostly to $\{Au[P(o\text{-}C_6H_4SiMe_2Ph)Ph_2]_2\}GaCl_4$ and elemental gold). **^{1}H NMR** (400 MHz, 193 K, CD_2Cl_2): δ 1.05 (sl, $Au(Si(CH_3)_2Ph)_2$, 12H), 0.47 (s, Ar–Si$(CH_3)_2Ph$, 6H); **^{31}P{^{1}H} NMR** (162 MHz, 193 K, CD_2Cl_2): δ 62.4 (s); **^{29}Si{^{1}H} NMR** (79 MHz, 193 K, CD_2Cl_2): δ 40.2 (d, $^{2}J_{SiP}$ = 38.3 Hz, 2Si), −7.6 (d, $^{3}J_{SiP}$ = 4.6 Hz, 1 Si); **^{13}C{^{1}H} NMR** (100 MHz, 193 K, CD_2Cl_2): δ 1.9 (d, $^{3}J_{CP}$ = 4.1 Hz, $Au(Si(CH_3)_2Ph)_2$), 1.8 (s, Ar–$Si(CH_3)_2Ph$).

5.5.8 Au(III) Bis(Silyl) Complex 28

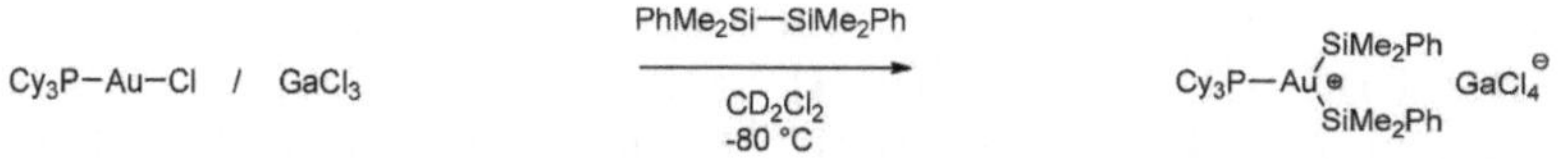

28

In a glovebox, Cy_3PAuCl (30 mg, 0.059 mmol) was transferred into an NMR tube and dissolved in dichloromethane-d_2 (0.3 mL). The tube was closed with a rubber septum assured with a piece of parafilm. $(Ph_2MeSi)_2$ (16.5 mg, 0.06 mmol) and $GaCl_3$ (10.7 mg, 0.06 mmol) were transferred into small glass vials and solubilized in dichloromethane-d_2 (0.3 mL), respectively. The further experimental procedure was analogous to that for **24a**. According to ^{31}P NMR, **28** was formed in about 80 % yield along with unidentified byproducts. In addition to the ^{31}P and ^{29}Si

NMR data, only the ^{1}H and ^{13}C signals for the $SiCH_3$ groups could be unambiguously identified. **^{1}H NMR** (400 MHz, 193 K, CD_2Cl_2): δ 1.02 (sl, 12H, $SiCH_3$); **^{31}P{^{1}H} NMR** (162 MHz, 193 K, CD_2Cl_2): δ 75.7 (s); **^{29}Si{^{1}H} NMR** (79 MHz, 193 K, CD_2Cl_2): δ 39.5 (d, $^2J_{SiP}$ = 35.2 Hz); **^{13}C{^{1}H} NMR** (100 MHz, 193 K, CD_2Cl_2): δ 1.81 (d, $^2J_{CP}$ = 3.81 Hz, $SiCH_3$).

5.5.9 Gold(I) Chloride Complex 30

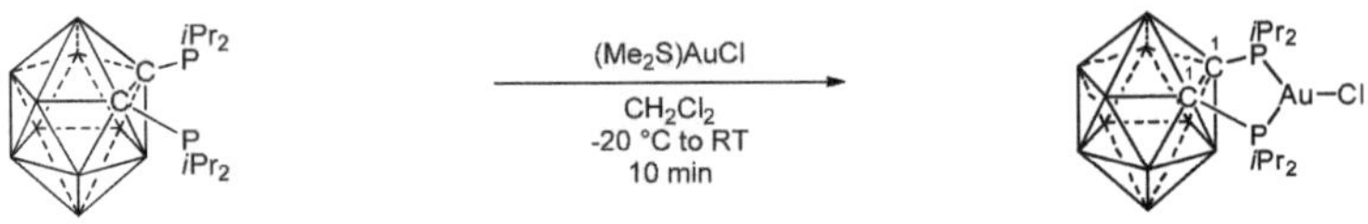

30

A solution of 1,2-(iPr$_2$)$_2$-1,2-$C_2B_{10}H_{10}$ (300 mg, 0.8 mmol, 1 eq.) in dichloromethane (5 mL) was added under stirring to a dispersion of (Me_2S)AuCl (234.7 mg, 0.8 mmol, 1 eq.) in dichloromethane (5 mL) at −20 °C. The reaction mixture was allowed to warm up to room temperature over 10 min. The brown solution was concentrated under vacuum to a volume of about 2 mL and layered with pentane (10 mL). Storing for 2 days at −60 °C furnished a brown crystalline precipitate which was separated by cannula filtration and dried under vacuum to give **30** (240 mg) as a brown solid. The supernatant was concentrated, layered with pentane and stored for 2 days at −60 °C to yield after filtration and drying under vacuum a second crop of **30** (130 mg). Yield (combined fractions): 370 mg (76 %). Crystals suitable for X-ray diffraction were obtained at 25 °C by layering a concentrated solution of **30** with hexamethyldisiloxane.

Mp: 183–185 °C (decomposition); **^{1}H NMR** (300 MHz, $CDCl_3$): δ 2.49 (m, 4H, $CH(CH_3)_2$), 1.3 (m, 24H, $CH(CH_3)_2$), 1.0–3.6 (vbr, 10H, H_{BH}); **^{31}P{^{1}H} NMR** (121 MHz, $CDCl_3$): δ 68.8 (s); **^{13}C{^{1}H}** NMR (75 MHz, $CDCl_3$): δ 98.8 (s, C_1), 29.4 (*pseudo* t, J_{CP} = 4.3 Hz, $\mathit{C}H(CH_3)_2$), 22.8 (*pseudo* t, J_{CP} = 8.1 Hz, $CH(\mathit{C}H_3)_2$), 19.7 (*pseudo* t, J_{CP} = 3.8 Hz, $CH(CH_3)_2$); **HRMS (ESI+)**: calcd for $[M\text{-}Cl]^+$ = $C_{14}H_{38}B_{10}P_2Au^+$: 573.3124. Found: 573.3129. **Elt. Anal.**: calcd for $C_{14}H_{38}AuB_{10}ClP_2$: C, 27.61; H, 6.29. Found: C, 27.79; H, 5.83.

5.5.10 *Gold(I) Styrene Complex 31*

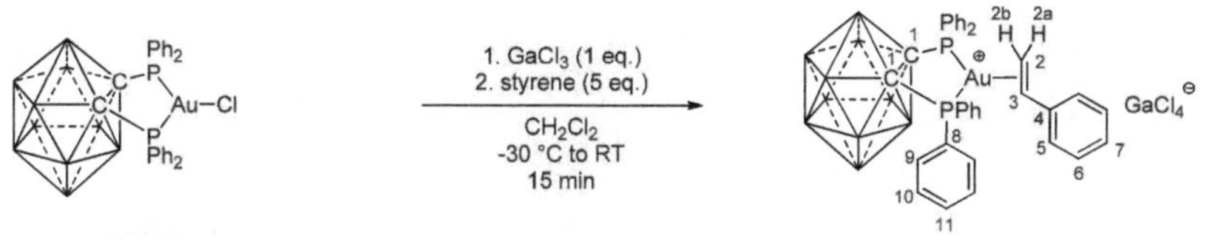

29 **31 (87 %)**

A solution of $GaCl_3$ (9.5 mg, 0.05 mmol) in dichloromethane (2 mL) was added to a solution of [AuCl(1,2-$(PPh_2)_2$-1,2-$C_2B_{10}H_{10}$)] (40 mg, 0.05 mmol, 1 eq.) in dichloromethane (2 mL) at −30 °C and stirred at this temperature for 3 min. Styrene (31 µL, 0.27 mmol, 5 eq.) in dichloromethane (2 mL) was added at −30 °C and the mixture was allowed to warm up to room temperature (10 min). Volatiles were removed in vacuo. The off-white residue was solubilized in dichloromethane (~2 mL). The resulting solution was filtered by cannula filtration, layered with pentane and stored at −20 °C to yield **31** a colorless crystalline solid that was separated by filtration and dried under vacuum. Yield: 44 mg (87 %). Crystals suitable for X-ray diffraction were obtained at −20 °C by layering a saturated solution of **31** with pentane. **Mp**: 155–157 °C (decomposition); **^{1}H NMR** (300 MHz, CD_2Cl_2): δ 7.90–7.15 (m, 25H, H_{ar}), 6.22 (dd, $^3J_{HH2b}$ = 9.1 Hz, $^3J_{HH2a}$ = 14.5 Hz, 1H, H_3), 4.75 (d, $^3J_{HH}$ = 14.5 Hz, 1H, H_{2a}), 4.56 (d, $^3J_{HH}$ = 9.1 Hz, 1H, H_{2b}), 3.26–0.98 (vbr, 10H, H_{BH}); **^{31}P{^{1}H} NMR** (121 MHz, CD_2Cl_2): δ 53.6 (s); **^{13}C{^{1}H}** NMR (75 MHz, CD_2Cl_2): δ 135.40 (s, C_4), 135.34–134.40 (m, $C_{CH\text{-}Ar}$), 134.69 (s, C_5 or C_6), 134.23 (s, C_7 or C_{11}), 130.65–129.9 (m, $C_{CH\text{-}Ar}$), 126.6 (s, C_5 or C_6), 124.96 (dd, $^1J_{CP}$ = 56.7 Hz, $^3J_{CP}$ = 23.9 Hz, C_8), 95.03 (t br, $^2J_{CP}$ = 5.4 Hz, C_3), 83.5 (*pseudo* t, $^1J_{CP}$ = 5.2 Hz, C_1), 65.94 (t br, $^2J_{CP}$ = 5.8 Hz, C_2), due to multiple overlapping not all the signals for the aromatic methine-carbon signals could be unequivocally assigned; **HRMS (ESI+)**: the cation of **31** could not be observed by mass spectrometry. Instead, the bis(diphosphine) complex ([Au(1,2-$(PPh_2)_2$-1,2-$C_2B_{10}H_{10})_2]^+$ = $C_{52}H_{60}B_{20}P_4Au^+$: 1222.5375) was detected as the predominant decomposition species; **Elt. Anal.**: calcd for $C_{33}H_{38}AuB_{10}Cl_4GaP_2$: C, 39.83; H, 3.74. Found: C, 39.23; H, 3.38.

5.5.11 Gold(I) Carbonyl Complex 32

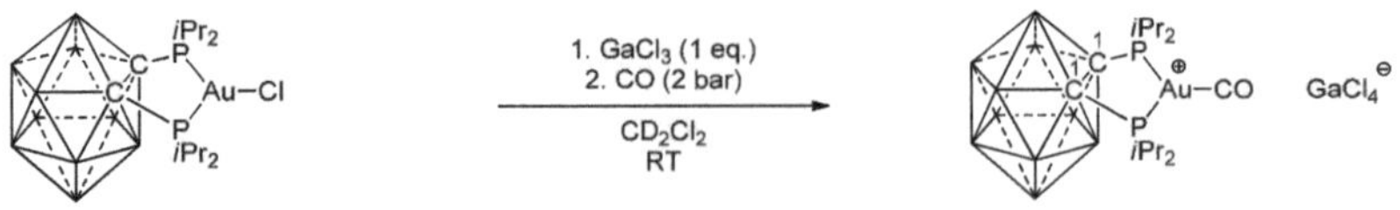

30

32 (quant.)

In a glovebox, a solution of $GaCl_3$ (8.7 mg, 0.05 mmol, 1 eq.) in dichloromethane-d_2 (0.3 mL) was added to a solution of [AuCl(1,2-(P*i*Pr$_2$)$_2$-1,2-$C_2B_{10}H_{10}$)] (30 mg, 0.05 mmol, 1 eq.) in dichloromethane-d_2 (0.3 mL) at room temperature and stirred for 3 min. The orange solution was transferred into a Wilmad quick pressure valve NMR tube. Outside of the glovebox the sample was frozen. The NMR tube was put under vacuum, then closed and the solution thawed, before charging with CO gas (2 bar). The orange solution quickly turned to pale-yellow. NMR analysis was performed after three freeze-pump-thaw cycles and indicated quantitative conversion. Crystals suitable for single crystal X-ray diffraction analysis were grown at room temperature by layering the dichloromethane solution in the NMR tube with pentane. **Mp**: 80–82 °C (color change, probably loss of CO), 141–143 °C (dec.); **^{1}H NMR** (500 MHz, CD_2Cl_2): δ 3.40–1.70 (vbr, 10H, H_{BH}), 2.72 (dsept, $^3J_{HH}$ = 7.2 Hz, 4H, $CH(CH_3)_2$), 1.43 (d, $^3J_{HH}$ = 7.2 Hz, 12H, CH (CH_3)$_2$), 1.30 d, $^3J_{HH}$ = 7.2 Hz, 12H, CH(CH_3)$_2$); **^{31}P{^{1}H} NMR** (203 MHz, CD_2Cl_2): δ 86.12 (s); **^{11}B{^{1}H}** NMR (161 MHz, CD_2Cl_2): δ 0.6, −2.1, −9.6; **^{13}C {^{1}H}** NMR (126 MHz, CD_2Cl_2): δ 198.7 (s, C_{CO}); 92.1 (d, $^1J_{CP}$ = 1.3 Hz, C_1), 29.4 (t, $^1J_{CP}$ = 7.9 Hz, $CH(CH_3)_2$), 23.2 (t, $^2J_{CP}$ = 5.6 Hz, CH(CH_3)$_2$), 19.9 (t, $^2J_{CP}$ = 2.7 Hz, CH(CH_3)$_2$), in order to identify the chemical shift of the CO carbon, the ^{13}C labeled compound has been prepared as described above, using ^{13}CO. For this isotopomer, the signal for the carbon atoms in the carborane backbone could not be detected; **IR (KBr pellet)**: ν(CO): 2115 cm^{-1} (s); **Elt. Anal.**: calcd for $C_{15}H_{38}OAuB_{10}Cl_4GaP_2$: C, 22.16; H, 4.71. Found: C, 21.96; H, 4.27.

5.5.12 Dinuclear Gold(I) Complex 33

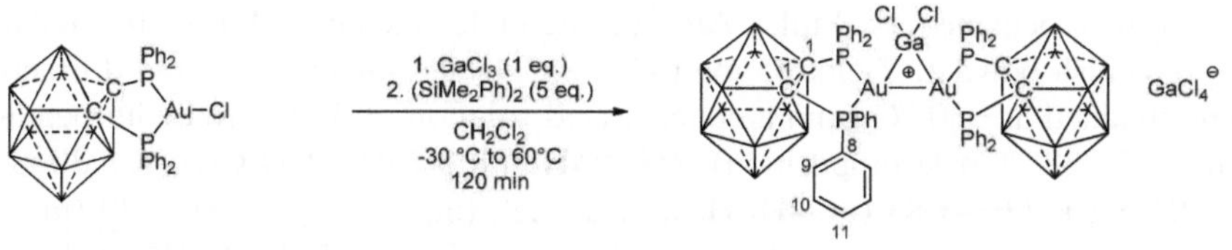

29 33

A solution of $GaCl_3$ (9.45 mg, 0.054 mmol) in dichloromethane (1 mL) was added to a solution of [AuCl(1,2-$(PPh_2)_2$-1,2-$C_2B_{10}H_{10}$)] (40.0 mg, 0.054 mmol, 1 eq.) in dichloromethane (1 mL) at −30 °C, stirred at this temperature for 3 min and then added to a solution of 1,1,2,2-tetramethyldiphenyldisilane (14.5 mg, 0.54 mmol, 1 eq.) in dichloromethane (1 mL) in a rotaflo Schlenk at −30 °C. The reaction mixture is heated in an oilbath at 60 °C for 1 h. The reaction mixture turned quickly dark yellow. After cooling down to room temperature, volatiles were removed in vacuo. The residue was washed with pentane (2 × 2 mL) and dried under vacuum. Crystals suitable for X-ray diffraction were obtained at −60 °C from a concentrated solution of **33** layered with pentane. **$^{31}P\{^1H\}$ NMR** (162 MHz, $CDCl_3$): δ 66.1 (s); **HRMS (ESI+)**: calcd for $[M]^+ = C_{52}H_{60}B_{20}P_4Cl_2GaAu_2^+$:1559.3617. Found: 1559.3633.

5.5.13 Dibenzoaurole 34

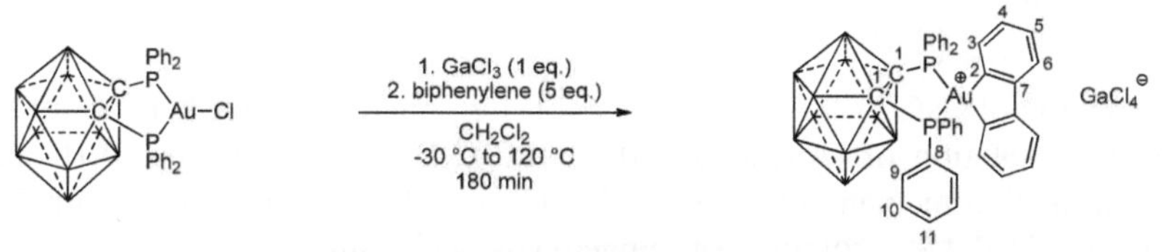

29 34

A solution of $GaCl_3$ (11.8 mg, 0.067 mmol) in dichloromethane (1 mL) was added to a solution of [AuCl(1,2-$(PPh_2)_2$-1,2-$C_2B_{10}H_{10}$)] (50 mg, 0.067 mmol, 1 eq.) in dichloromethane (1 mL) at −30 °C, stirred at this temperature for 3 min and then added to a solution of biphenylene (31 mg, 0.34 mmol, 5 eq.) in dichloromethane (2 mL) in a rotaflo Schlenk at −30 °C. While warming up to room temperature, the solution is concentrated to a volume of ~1 mL. The reaction mixture is heated in an oilbath at 120 °C for 3 h. The initially yellow solution turns green after about 5 min. After cooling down to room temperature, the dark-green dispersion is resolubilized in dichloromethane (3 mL) and filtered via cannula to give a green solution. Volatiles were removed in vacuo and the remaining solid washed with pentane (3 × 2 mL). After drying under vacuum, **34** was obtained as a gray-greenish powder. Yield: 45 mg (63 %). Crystals suitable for X-ray diffraction were obtained at −30 °C from a concentrated solution of **34** layered with pentane. **Mp**: 237–238 °C (decomposition); **^{1}H NMR** (400 MHz, CD_2Cl_2): δ 8.30–8.13 (m, 8H, H_{ar}), 7.93–7.83 (m, 4H, H_{ar}), 7.81–7.68 (m, 8H, H_{ar}), 7.61–7.49 (m, 2H, H_{Ar}), 7.24–7.14 (m, 2H, H_{Ar}), 1.2–3.2 (vbr, 10H, H_{BH}); **^{31}P{^{1}H} NMR** (162 MHz, CD_2Cl_2): δ 69.9 (s); **^{13}C{^{1}H}** NMR (100 MHz, CD_2Cl_2): δ 165.1 (dd, $^2J_{CP\text{-trans}}$ = 111.2 Hz, $^2J_{CP\text{-cis}}$ = 9.0 Hz, C_2), 154.0 (t, $^3J_{CP}$ = 1.3 Hz, C_7), 136.1 (t, J_{CP} = 7.8 Hz, C_3 or C_4 or C_6), 135.9 (m, C_9 or C_{10}), 135.7 (t, $^4J_{CP}$ = 1.2 Hz C_{11}), 130.4 (m, C_9 or C_{10}), 129.6 (s, C_5), 128.3 (*pseudo* t, J_{CP} = 7.1 Hz, C_3 or C_4 or C_6), 123.0 (*pseudo* t, J_{CP} = 2.6 Hz, C_3 or C_4 or C_6), 121.6 (dd, $^1J_{CP}$ = 56.7 Hz, $^3J_{CP}$ = 3.2 Hz, C_8), 83.5 (*pseudo* t, $^1J_{CP}$ = 12.2 Hz, C_1); **HRMS (ESI+)**: calcd for $[M]^+$ = $C_{38}H_{38}B_{10}P_2Au^+$: 862.3109. Found: 862.3127. **Elt. Anal.**: calcd for $C_{38}H_{38}AuB_{10}Cl_4GaP_2$: C, 42.52; H, 3.57. Found: C, 43.33; H, 3.41.

5.5.14 Dibenzoaurole 35

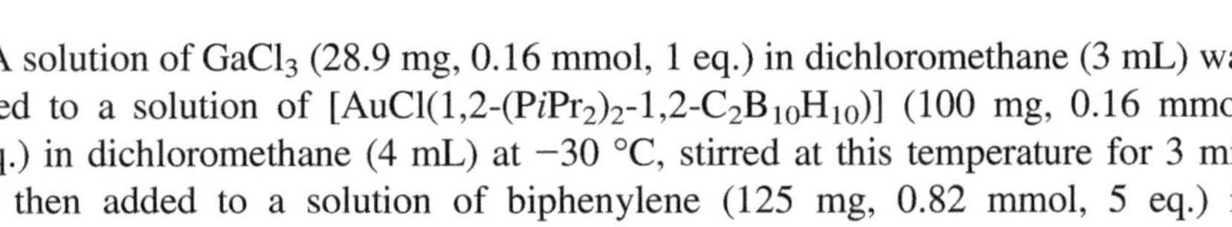

30 **35**

A solution of $GaCl_3$ (28.9 mg, 0.16 mmol, 1 eq.) in dichloromethane (3 mL) was added to a solution of [AuCl(1,2-(P*i*Pr$_2$)$_2$-1,2-$C_2B_{10}H_{10}$)] (100 mg, 0.16 mmol, 1 eq.) in dichloromethane (4 mL) at −30 °C, stirred at this temperature for 3 min and then added to a solution of biphenylene (125 mg, 0.82 mmol, 5 eq.) in dichloromethane (4 mL) in a rotaflo Schlenk at −30 °C. While warming up to room temperature, the solution is concentrated to a volume of ~4 mL. The reaction mixture is heated in an oilbath at 120 °C for 5 h. The initially orange solution turned gradually yellow. After cooling down to room temperature, volatiles were removed in vacuo. The residue was washed with diethyl ether (5 × 2 mL) and dried under vacuum to give **35** as a yellow solid. Yield: 142 mg (95 %). Crystals suitable for X-ray diffraction were obtained at 25 °C from a concentrated solution of **35** by vapor diffusion with pentane. **Mp**: The compound decomposes gradually without melting starting from ~180 °C; **^{1}H NMR** (300 MHz, CD_2Cl_2): δ 7.59–7.49

(m, 4H, H_3 & H_6), 7.37 (tt, J_{HH} = 7.5 Hz, J_{HP} = 1.0 Hz, 2H, H_5), 7.13 (tm, J_{HH} = 7.7. Hz, 2H, H_4), 3.28 (sept, $^3J_{HH}$ = 7.2 Hz, 4H, C*H*(CH_3)$_2$), 1.75 (dd, $^3J_{HH}$ = 7.2 Hz, $^3J_{HP}$ = 18.7 Hz, 12H, CH(C*H*$_3$)$_2$), 1.63 (dd, $^3J_{HH}$ = 7.2 Hz, $^3J_{HP}$ = 17.8 Hz, 12H, CH(C*H*$_3$)$_2$), 1.0–3.6 (vbr, 10H, H_{BH}); **^{31}P{^{1}H} NMR** (162 MHz, CD_2Cl_2): δ 89.5 (s); **^{13}C{^{1}H}** NMR (75 MHz, CD_2Cl_2): δ 165.6 (dd, $^2J_{CP\text{-}trans}$ = 114.3 Hz, $^2J_{CP\text{-}cis}$ = 9.7 Hz, C_2), 153.7 (t, $^3J_{CP}$ = 1.3 Hz, C_7), 134.8 (t, $^4J_{CP}$ = 5.7 Hz, C_6), 129.6 (s, C_5), 128.0 (t, $^4J_{CP}$ = 6.0 Hz, C_4), 123.5 (t, $^3J_{CP}$ = 2.6 Hz, C_3), 81.7 (d, $^1J_{CP}$ = 1.3 Hz,C_1), 31.4 (t, $^1J_{CP}$ = 6.8 Hz, *C*H(CH_3)$_2$), 21.1 (s, CH(*C*H$_3$)$_2$), 20.5 (s, CH(*C*H$_3$)$_2$); **HRMS (ESI+)**: calcd for $[M]^+$ = $C_{26}H_{46}B_{10}P_2Au^+$: 726.3727. Found: 726.3709. **Elt. Anal.**: calcd for $C_{26}H_{46}AuB_{10}Cl_4GaP_2$: C, 33.52; H, 4.95. Found: C, 33.53; H, 4.81.

5.5.15 Benzoaurolone 36

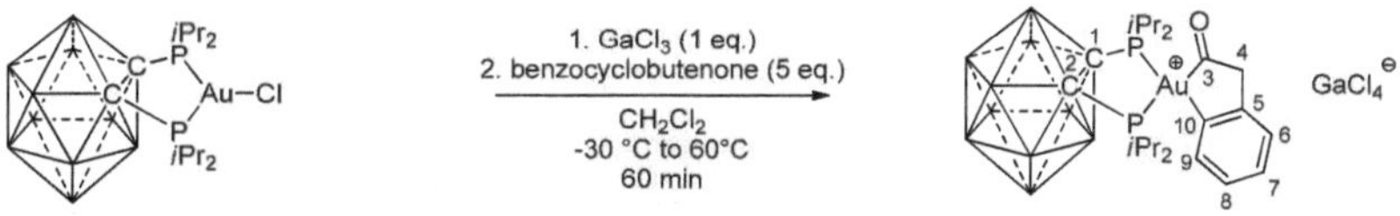

30 **36**

A solution of $GaCl_3$ (14.5 mg, 0.08 mmol) in dichloromethane (3 mL) was added to a solution of [AuCl(1,2-(*i*Pr$_2$)$_2$-1,2-$C_2B_{10}H_{10}$)] (50 mg, 0.08 mmol, 1 eq.) in dichloromethane (3 mL) at −30 °C, stirred at this temperature for 3 min and then added to a solution of benzocyclobutenone (40 μL, 0.41 mmol, 5 eq.) in dichloromethane (3 mL) in a rotaflo Schlenk at −30 °C. While warming up to room temperature, the solution was concentrated to a volume of ~2 mL and heated in an oilbath at 60 °C for 60 min. After cooling down to room temperature, volatiles were removed in vacuo. The residue was washed with diethyl ether (3 × 2 mL) to give **36** as a white solid. Yield: 72 mg (quant.). Crystals suitable for X-ray diffraction were obtained at 25 °C by vapor diffusion of pentane into a concentrated dichloromethane solution of **36**. The isolated compound decomposes in solution slowly back to the starting material. In order to obtain the ^{13}C NMR data, a 10-fold excess of benzocyclobutenone was added, keeping the equilibrium on the product side. **Mp**: 165–167 °C (decomposition); **^{1}H NMR** (300 MHz, CD_2Cl_2): δ 7.66–7.30 (m, 4H, H_{ar}), 4.55 (d, $^4J_{HP}$ = 4.2 Hz, 1H, H_{4a}), 3.98 (s, 1H, H_{4b}), 3.05 (m, $^3J_{HH}$ = 6.8 Hz, 2H, C*H*(CH_3)$_2$), 2.72 (septd, $^3J_{HH}$ = 6.9 Hz, $^2J_{HP}$ = 3.3 Hz, 2H, C*H*(CH_3)$_2$), 1.63 (m, $^3J_{HH}$ = 6.9 Hz, $^3J_{HP}$ = 1.6 Hz, 6H, CH(C*H*$_3$)$_2$), 1.57 (m, $^3J_{HH}$ = 6.9 Hz, $^3J_{HP}$ = 0.9 Hz, 6H, CH(C*H*$_3$)$_2$), 1.43 (m, $^3J_{HH}$ = 6.8 Hz, 6H, CH (C*H*$_3$)$_2$), 1.34 (m, $^3J_{HH}$ = 6.8 Hz, 6H, CH(C*H*$_3$)$_2$), 1.8–3.7 (vbr, 10H, H_{BH}); **^{31}P {^{1}H} NMR** (121 MHz, CD_2Cl_2): δ 79.7 (d, $^2J_{PP}$ = 13.9 Hz, 1P), 78.3

(d, $^2J_{PP}$ = 13.9 Hz, 1P); **^{11}B{^{1}H} NMR** (96 MHz, CD_2Cl_2): δ 2.0, −3.7, −9.5; **^{13}C {^{1}H} NMR** (75 MHz, CD_2Cl_2): δ 205.3 (dd, $^2J_{CP\text{-trans}}$ = 89.2 Hz, $^2J_{CP\text{-cis}}$ = 5.7 Hz, C_3), 156.2 (dd, $^2J_{CP\text{-trans}}$ = 93.0 Hz, $^2J_{CP\text{-cis}}$ = 11.2 Hz, C_{10}), 138.9 (d, $^3J_{CP}$ = 1.0 Hz C_5), 136.1 (dd, $^3J_{CP\text{-trans}}$ = 6.3 Hz, $^3J_{CP\text{-cis}}$ = 4.5 Hz, C_9), 130.7 (d, J_{CP} = 1.4 Hz, C_6 or C_7), 128.5 (d, $^4J_{CP}$ = 8.8 Hz, C_8), 124.1 (d, J_{CP} = 7.5 Hz, C_6 or C_7), 82.6 (dd, $^1J_{CP}$ = 13.7 Hz, $^2J_{CP}$ = 7.2 Hz, C_1 or C_2). 82.6 (dd, $^1J_{CP}$ = 8.6 Hz, $^2J_{CP}$ = 4.8 Hz, C_1 or C_2), 71.8 (dd, $^3J_{CP\text{-trans}}$ = 30.7 Hz, $^3J_{CP\text{-cis}}$ = 0.9 Hz, C_4), 30.9 (dd, $^1J_{CP}$ = 9.0 Hz, $^3J_{CP}$ = 0.8 Hz, *C*H(CH_3)$_2$), 30.5 (dd, $^1J_{CP}$ = 14.9 Hz, $^3J_{CP}$ = 1.0 Hz, *C*H(CH_3)$_2$), 20.9 (d, $^2J_{CP}$ = 1.7 Hz, CH(*C*H$_3$)$_2$), 30.5 (dd, $^1J_{CP}$ = 14.9 Hz, $^3J_{CP}$ = 1.0 Hz, *C*H (CH_3)$_2$), 20.6 (d, $^2J_{CP}$ = 2.1 Hz, CH(*C*H$_3$)$_2$), 20.3 (d, $^2J_{CP}$ = 1.0 Hz, CH(*C*H$_3$)$_2$), 20.0 (d, $^2J_{CP}$ = 2.1 Hz, CH(*C*H$_3$)$_2$); **HRMS (ESI+)**: calcd for $[M]^+$ = $C_{22}H_{44}B_{10}P_2Au^+$: 691.3547. Found: 691.3549; **Elt. Anal.**: calcd for $C_{22}H_{44}AuB_{10}Cl_4GaP_2$: C, 29.26; H, 4.91. Found: C, 29.21; H, 4.64.

5.5.16 Benzoaurolone 37

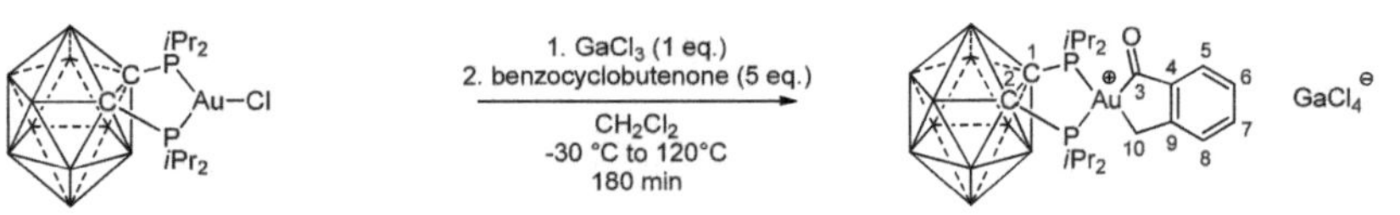

30 **37**

A solution of $GaCl_3$ (14.5 mg, 0.08 mmol) in dichloromethane (3 mL) was added to a solution of [AuCl(1,2-(*i*Pr$_2$)$_2$-1,2-$C_2B_{10}H_{10}$)] (50 mg, 0.08 mmol, 1 eq.) in dichloromethane (3 mL) at −30 °C, stirred at this temperature for 3 min and then added to a solution of benzocyclobutenone (40 µL, 0.41 mmol, 5 eq.) in dichloromethane (3 mL) in a rotaflo Schlenk at −30 °C. While warming up to room temperature, the solution is concentrated to a volume of ~2 mL. The reaction mixture is heated in an oilbath at 120 °C for 180 min. After cooling down to room temperature, volatiles were removed in vacuo and the remaining brown residue is washed with ether (3 × 2 mL) and dried under vacuum to give **37** as an orange-brown solid. Yield: 58 mg (80 %). Crystals suitable for X-ray diffraction were obtained at 25 °C from a concentrated solution of **37** layered with pentane. **Mp**: 184–185 °C (decomposition); **^{1}H NMR** (300 MHz, CD_2Cl_2): δ 7.69–7.50 (m, 2H, H_{ar}), 7.38–7.13 (m, 2H, H_{ar}), 4.27 (*pseudo* t, J_{HP} = 5.2 Hz, 1H, H_{10a}), 3.93 (s, 1H, H_{10b}), 3.16 (m, $^3J_{HH}$ = 6.8 Hz, 2H, C*H*(CH_3)$_2$), 3.01 (m, $^3J_{HH}$ = 7.1 Hz, 2H, C*H*(CH_3)$_2$), 1.61 (m, $^3J_{HH}$ = 7.1 Hz, 6H, CH(C*H*$_3$)$_2$), 1.56 (m, $^3J_{HH}$ = 7.1 Hz, 6H, CH(C*H*$_3$)$_2$), 1.52 (m, $^3J_{HH}$ = 6.8 Hz, 6H, CH(C*H*$_3$)$_2$), 1.51 (m, $^3J_{HH}$ = 6.8 Hz, 6H, CH(C*H*$_3$)$_2$), 0.8–2.8 (vbr, 10H, H_{BH}); **^{31}P{^{1}H} NMR** (121 MHz, CD_2Cl_2): δ 91.8 (d, $^2J_{PP}$ = 14.4 Hz, 1P), 74.4 (d, $^2J_{PP}$ = 14.4 Hz, 1P); **^{11}B{^{1}H} NMR** (96 MHz,

CD_2Cl_2): δ 1.2, −3.2, −9.9; **$^{13}C\{^1H\}$ NMR** (75 MHz, CD_2Cl_2): δ 209.4 (dd, $^2J_{CP\text{-trans}}$ = 101.4 Hz, $^2J_{CP\text{-cis}}$ = 6.8 Hz, C_3), 153.2 (dd, $^3J_{CP\text{-trans}}$ = 4.6 Hz, $^3J_{CP\text{-cis}}$ = 2.4 Hz, C_9), 146.1 (dd, $^3J_{CP\text{-trans}}$ = 32.7 Hz, $^3J_{CP\text{-cis}}$ = 1.8 Hz, C_4), 136.3 (s, $C_{ar\text{-}H}$), 128.7 (d, J_{CP} = 11.3 Hz, $C_{ar\text{-}H}$), 128.0 (d, J_{CP} = 2.0 Hz, $C_{ar\text{-}H}$), 126.1 (d, J_{CP} = 1.8 Hz, $C_{ar\text{-}H}$), 84.8 (*pseudo* dd, J_{CP} = 9.5 Hz, 7.2 Hz, C_1 or C_2), 83.5 (*pseudo* t, J_{CP} = 9.7 Hz, C_1 or C_2), 49.9 (dd, $^3J_{CP\text{-trans}}$ = 69.9 Hz, $^3J_{CP\text{-cis}}$ = 9.1 Hz, C_{10}), 30.8 (dd, $^1J_{CP}$ = 10.5 Hz, $^3J_{CP}$ = 1.2 Hz, *C*H$(CH_3)_2$), 28.8 (d, $^1J_{CP}$ = 14.8 Hz, *C*H $(CH_3)_2$), 21.8 (d, $^2J_{CP}$ = 4.9 Hz, CH(*C*$H_3)_2$), 20.1 (s, *C*H$(CH_3)_2$), 19.8 (s, CH (*C*$H_3)_2$), 19.3 (d, $^2J_{CP}$ = 1.4 Hz, CH(*C*$H_3)_2$); **HRMS (ESI+)**: calcd for $[M]^+$ = $C_{22}H_{44}B_{10}P_2Au^+$: 691.3547. Found: 691.3560. **Elt. Anal.**: calcd for $C_{22}H_{44}AuB_{10}Cl_4GaP_2$: C, 29.26; H, 4.91. Found: C, 29.09; H, 4.41.

5.5.17 Gold(III) Phenyl Complex 38

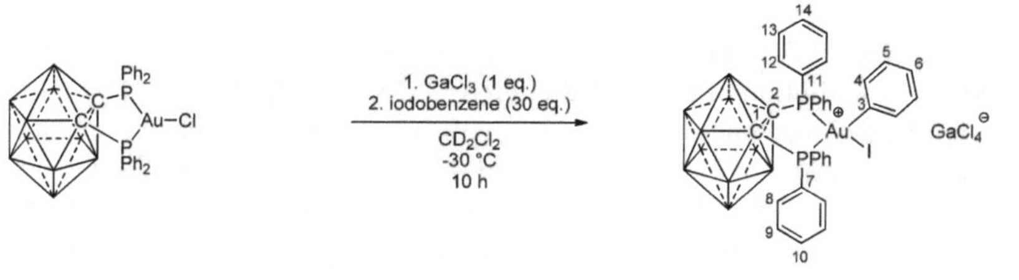

38

A solution of $GaCl_3$ (9.5 mg, 0.054 mmol) in dichloromethane-d_2 (0.4 mL) was added to a solution of $[AuCl(1,2\text{-}(PPh_2)_2\text{-}1,2\text{-}C_2B_{10}H_{10})]$ (40 mg, 0.054 mmol, 1 eq.) in dichloromethane-d_2 (0.6 mL) at −30 °C and stirred at this temperature for 3 min. Iodobenzene (0.2 mL, 30 eq.) was added. Within minutes, the initially colorless solution turned bright yellow. The reaction mixture was stirred for 10 h at −10 °C, then cooled down to −30 °C. An aliquot of the reaction mixture is analyzed by NMR spectroscopy at −30 °C. Despite several attempts, isolation of **38** by crystallization failed, due to the thermal instability of the compound over time even at −80 °C. **^1H NMR** (400 MHz, CD_2Cl_2, 243 K): δ 8.26–8.17 (m, 5H, H_{ar}), 8.03–7.97 (m, 1H, H_{ar}), 7.76–7.56 (m, 12H, H_{ar}), 7.03–6.81 (m, 7H, H_{Ar}), 1.6–3.4 (vbr, 10H, H_{BH}); **$^{31}P\{^1H\}$ NMR** (121 MHz, CD_2Cl_2, 243 K): δ 61.8 (d, $^2J_{PP}$ = 29.1 Hz, 1P), 44.6 (d, $^2J_{PP}$ = 29.1 Hz, 1P); **$^{13}C\{^1H\}$ NMR** (100 MHz, CD_2Cl_2, 243 K): δ 144.8 (d, $^2J_{CP\text{-trans}}$ = 122.0 Hz, C_3), 137.9 (s, C_6), 136.3 (d, J_{CP} = 3.4 Hz, C_{10} or C_{14}), 136.1 (d, J_{CP} = 12.3 Hz, C_8 or C_9 or C_{12} or C_{13}), 135.6 (d, $^4J_{CP}$ = 2.6 Hz, C_{10} or C_{14}), 135.2 (d, J_{CP} = 11.8 Hz, C_8 or C_9 or C_{12} or C_{13}), 132.6 (*pseudo* t, $^3J_{CP}$ = 2.6 Hz, C_4), 131.0 (d, J_{CP} = 10.3 Hz, C_8 or C_9 or C_{12} or C_{13}), 130.4 (d, J_{CP} = 12.9 Hz, C_8 or C_9 or C_{12} or C_{13}), 126.8 (s, C_5), 121.8 (d, $^1J_{CP}$ = 53.9 Hz, C_7 or C_{11}), 117.5 (d, $^1J_{CP}$ = 65.5 Hz, C_7

or C_{11}), 84.8 (*pseudo* dd, J_{CP} = 14.7 Hz, 16.1 Hz, C_1 or C_2), 82.6 (*pseudo* t, J_{CP} = 7.2 Hz, C_1 or C_2).

5.5.18 Gold(III) Phenyl Complex 39

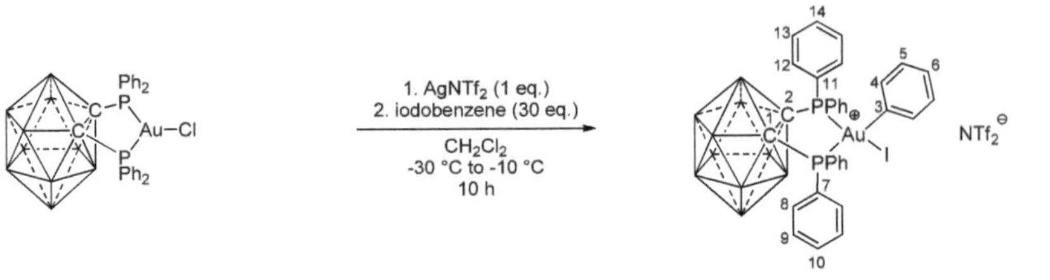

39

A dispersion of $AgNTf_2$ (20.8 mg, 0.054 mmol) in dichloromethane-d_2 (0.5 mL) was added to a solution of $[AuCl(1,2\text{-}(PPh_2)_2\text{-}1,2\text{-}C_2B_{10}H_{10})]$ (40 mg, 0.054 mmol, 1 eq.) in dichloromethane-d_2 (0.5 mL) at −30 °C and stirred at this temperature for 15 min under protection of light. The white dispersion was filtered through a pad of Celite at −30 °C. The filter frit was rinsed with dichloromethane-d_2 (0.5 mL). Iodobenzene (0.2 mL, 30 eq.) was added to the pale yellow solution. The reaction mixture turned quickly bright yellow and was allowed to warm up to −10 °C. At this temperature the mixture was stirred for 10 h. Due to the thermal instability of **39**, all attempts to isolate it in pure form failed. The product was characterized by NMR spectroscopic analysis of an aliquot of the reaction mixture at −30 °C. Crystals suitable for X-ray diffraction were obtained at −60 °C from a concentrated solution of **39** in dichloromethane layered with pentane. These crystals were extremely sensitive to air and moisture in addition to their thermal instability and had to be selected at −30 °C under a microscope by means of a low temperature mounting device [89]. **^{1}H NMR** (400 MHz, CD_2Cl_2, 243 K): δ 8.22–8.16 (m, 4H, H_{ar}), 7.95–7.81 (m, 8H, H_{ar}), 7.73–7.59 (m, 8H, H_{ar}), 6.93–6.77. (m, 5H, H_{Ar}), 1.6–3.2 (vbr, 10H, H_{BH}); **^{31}P{^{1}H} NMR** (121 MHz, CD_2Cl_2, 243 K): δ 61.7 (d, $^2J_{PP}$ = 28.8 Hz, 1P), 44.4 (d, $^2J_{PP}$ = 28.8 Hz, 1P); **^{13}C{^{1}H} NMR** (100 MHz, CD_2Cl_2, 243 K): δ 144.7 (d, $^2J_{CP\text{-}trans}$ = 121.9 Hz, C_3), 137.9 (s, C_6), 136.2 (d, J_{CP} = 3.3 Hz, C_{10} or C_{14}), 136.0 (d, J_{CP} = 12.0 Hz, C_8 or C_9 or C_{12} or C_{13}), 135.6 (d, $^4J_{CP}$ = 2.7 Hz, C_{10} or C_{14}), 135.2 (d, J_{CP} = 11.4 Hz, C_8 or C_9 or C_{12} or C_{13}), 132.6 (*pseudo* t, $^3J_{CP}$ = 2.6 Hz, C_4), 131.0 (d, J_{CP} = 10.6 Hz, C_8 or C_9 or C_{12} or C_{13}), 130.4 (d, J_{CP} = 12.8 Hz, C_8 or C_9 or C_{12} or C_{13}), 126.8 (s, C_5), 121.8 (d, $^1J_{CP}$ = 53.7 Hz, C_7 or C_{11}), 117.5 (d, $^1J_{CP}$ = 66.3 Hz, C_7 or C_{11}), 84.7 (*pseudo* t, J_{CP} = 14.8 Hz, C_1 or C_2), 82.6 (*pseudo* t, J_{CP} = 7.6 Hz, C_1 or C_2); **HRMS (ESI+)**: calcd for $[M]^+$ = $C_{32}H_{35}B_{10}P_2IAu^+$: 914.1915. Found: 914.1945.

5.5.19 Diphosphino-Carborane Ligand 40

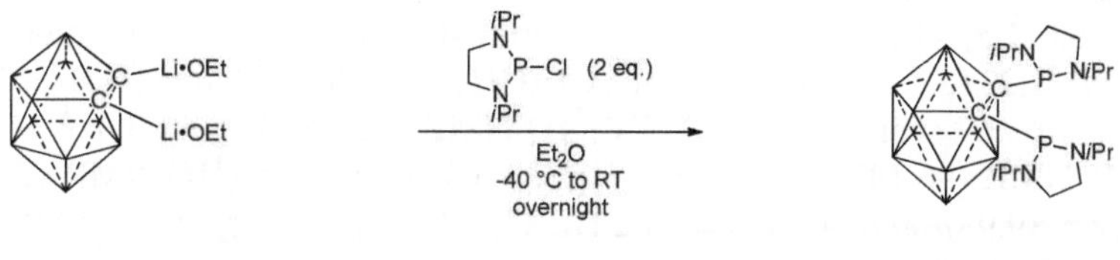

40 (75 %)

A solution of 2-chloro-1,3-diisopropyl-1,3,2-diazaphospholidine (140.5 mg, 0.67 mmol, 1 eq.) in diethyl ether (2 mL) was added to a solution of 1,2-dilithio-1,2-dicarba-*closo*-dodecaborane (102.5 mg, 0.33 mmol, 1 eq.) in diethyl ether (2 mL) at −40 °C via cannula transfer. The cannula was rinsed with additional diethyl ether (2 mL). The reaction mixture was stirred overnight and allowed to warm up to room temperature. Volatiles were removed in vacuo. The residue was redissolved in toluene and the resulting dispersion was filtered over Celite. Volatiles were removed under vacuum. The resulting off-white solid was recrystallized from toluene/pentane to give **40** as colorless crystals. Yield: 120 mg (75 %). **Mp**: 127–129 ° C; **^{1}H NMR** (300 MHz, $CDCl_3$): δ 3.60 (m, 4H, C*H*$(CH_3)_2$), 3.32 (mbr, 4H, N$(CH_2)_2$N), 3.06 (mbr, 4H, N$(CH_2)_2$N), 1.20 (d, $^3J_{HH}$ = 6.6 Hz, 12H, CH$(CH_3)_2$), 1.10 (d, $^3J_{HH}$ = 6.5 Hz, 12H, CH$(CH_3)_2$), 1.4–3.4 (vbr, 10H, H_{BH}); **^{31}P{^{1}H} NMR** (121 MHz, $CDCl_3$): δ 114.2 (s); **^{11}B{^{1}H}** NMR (96 MHz, $CDCl_3$): δ −2.2, −7.7, −10.1, −12.5; **^{13}C{^{1}H}** NMR (100 MHz, $CDCl_3$): δ 89.9 (*pseudo* t, J_{CP} = 59.4 Hz, C_1), 50.0 (*pseudo* t, J_{CP} = 17.0 Hz, CH$(CH_3)_2$), 45.2 (*pseudo* t, J_{CP} = 3.5 Hz, N$(CH_2)_2$N), 22.5 (*pseudo* t, J_{CP} = 4.6 Hz, CH$(CH_3)_2$), 21.5 (*pseudo* t, J_{CP} = 2.1 Hz, CH$(CH_3)_2$); **HRMS (ESI+)**: calcd for $[M]^+$ = $C_{18}H_{47}B_{10}N_4P_2^+$: 489.4288. Found: 489.4278.

5.5.20 Gold(I) Chloride Complex 41

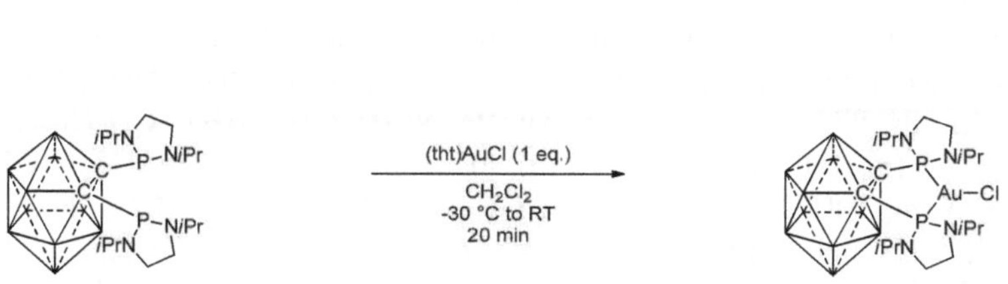

40 **41 (88 %)**

A solution of **40** (100 mg, 0.20 mmol) in dichloromethane (4 mL) was added to a solution of (THT)AuCl (65.6 mg, 0.20 mmol, 1 eq.) in dichloromethane (4 mL) at −20 °C, stirred at this temperature for 5 min. After warming up to room temperature, volatiles were removed in vacuo. The yellow residue was recrystallized from dichloromethane/toluene giving **41** as a fine crystalline off-white solid. Yield: 127 mg (88 %). **Mp**: 112–113 °C; **^{1}H NMR** (300 MHz, $CDCl_3$): δ 3.91 (sept, $^{3}J_{HH}$ = 6.6 Hz, 4H, $CH(CH_3)_2$), 3.26 (mbr, 4H, $N(CH_2)_2N$), 3.19 (mbr, 4H, N $(CH_2)_2N$), 1.23 (d, $^{3}J_{HH}$ = 6.6 Hz, 12H, $CH(CH_3)_2$), 1.19 (d, $^{3}J_{HH}$ = 6.6 Hz, 12H, CH $(CH_3)_2$), 1.6–3.7 (vbr, 10H, H_{BH}); **^{31}P{^{1}H} NMR** (121 MHz, $CDCl_3$): δ 116.6 (s); **^{11}B{^{1}H}** NMR (96 MHz, $CDCl_3$): δ −1.5, −6.8, −10.4; **^{13}C{^{1}H}** NMR (75 MHz, $CDCl_3$): δ 89.9 (*pseudo* t, $^{1}J_{CP}$ = 28.2 Hz, C_1), 49.9 (*pseudo* t, $^{2}J_{CP}$ = 12.8 Hz, *C*H $(CH_3)_2$), 42.7 (s, $N(CH_2)_2N$), 21.6 (*pseudo* t, $^{3}J_{CP}$ = 3.7 Hz, $CH(CH_3)_2$), 20.7 (s, CH $(CH_3)_2$); **HRMS (ESI+)**: calcd for $[M]^+$ = $C_{18}H_{46}B_{10}N_4P_2Au^+$: 685.3875. Found: 685.3885.

5.5.21 Gold(III) Phenyl Complex 42

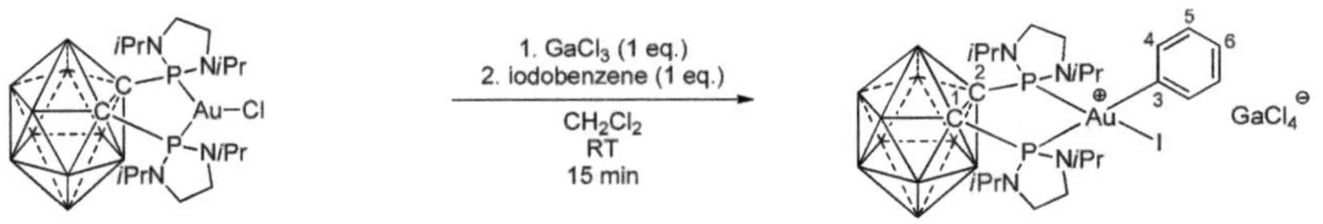

41 **42 (88 %)**

A solution of $GaCl_3$ (12.2 mg, 0.069 mmol) in dichloromethane (1 mL) was added to a solution of **41** (50 mg, 0.069 mmol, 1 eq.) in dichloromethane (2 mL) at room temperature and stirred for 1 min. Iodobenzene (7.7 µL, 0.69 mmol, 1 eq.) was added. The reaction mixture turns immediately bright yellow and is stirred for 15 min. Volatiles were removed in vacuo. The residue is washed with pentane (3 × 2 mL) and dried under vacuum to give **42** as a yellow powder. Yield: 67 mg (88 %). The compound decomposes at room temperature in solution (hours) as well as in the solid state (days). Crystals suitable for X-ray diffraction were obtained at −30 °C from a concentrated solution of **42** in dichloromethane layered with pentane. **Mp**: 84–87 °C (decomposition); **^{1}H NMR** (400 MHz, CD_2Cl_2, 253 K): δ 7.76–7.71 (m, 1H, H_6), 7.42–7.13 (m, 4H, H_4 & H_5), 4.27 (m, 1H, $N(CH_2)_2N$), 3.85–3.23 (m, 10H, $N(CH_2)_2N$ & $CH(CH_3)_2$), 2.14 (m, 1H, $N(CH_2)_2N$), 1.57 (d, $^{3}J_{HH}$ = 6.5 Hz, 6H, $CH(CH_3)_2$), 1.43 (d, $^{3}J_{HH}$ = 6.5 Hz, 6H, $CH(CH_3)_2$), 1.42 (d, $^{3}J_{HH}$ = 6.4 Hz, 6H, $CH(CH_3)_2$), 0.86 (d, $^{3}J_{HH}$ = 6.3 Hz, 6H, $CH(CH_3)_2$), 1.8–4.2 (vbr, 10H, H_{BH}); **^{31}P{^{1}H} NMR** (161 MHz, CD_2Cl_2, 253 K): δ 122.2 (d, $^{2}J_{PP}$ = 21.2 Hz, 1P), 82.3 (d, $^{2}J_{PP}$ = 21.2 Hz, 1P); **^{13}C{^{1}H} NMR** (100 MHz, CD_2Cl_2, 253 K): δ 143.9 (d, $^{2}J_{CP\text{-}trans}$ = 157.7 Hz, C_3), 134.8 (s, C_5), 136.1 (d, $^{3}J_{CP\text{-}trans}$ = 12.8 Hz, C_4), 128.2 (s, C_6), 88.5 (dd, J_{CP} = 31.3 Hz, 12.9 Hz, C_1 or C_2), 84.7 (d, J_{CP} = 19.1 Hz, C_1 or C_2), 50.2 (d, $^{2}J_{CP}$ = 9.0 Hz, $CH(CH_3)_2$), 49.7 (d,

$^2J_{CP}$ = 12.5 Hz, $CH(CH_3)_2$), 43.2 (s, $N(CH_2)_2N$), 42.4 (s, $N(CH_2)_2N$), 22.0 (d, $^3J_{CP}$ = 2.5 Hz, $CH(CH_3)_2$), 21.8 (d, $^3J_{CP}$ = 4.5 Hz, $CH(CH_3)_2$), 21.7 (d, $^3J_{CP}$ = 4.8 Hz, $CH(CH_3)_2$), 20.9 (d, $^3J_{CP}$ = 4.4 Hz, $CH(CH_3)_2$); **HRMS (ESI+)**: calcd for $[M]^+ = C_{24}H_{51}B_{10}P_2N_4IAu^+$: 890.3285. Found: 890.3292; **Elt. Anal.**: calcd for $C_{24}H_{51}B_{10}P_2N_4IAu$: C, 26.18; H, 4.67; N, 5.09. Found: C, 26.23; H, 4.38, N, 4.94.

5.5.22 *Gold(I) Triflimidate Complex 43*

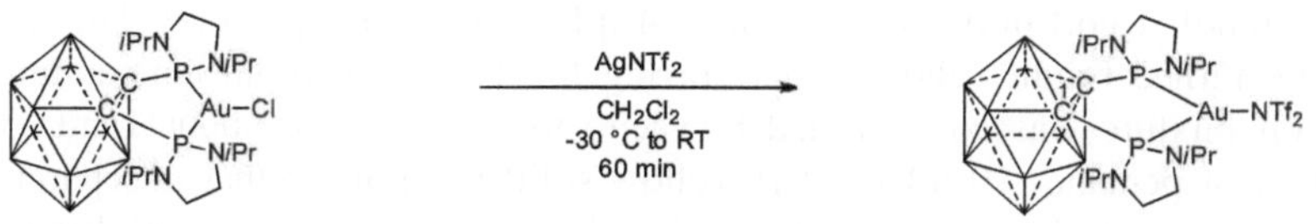

41 **43 (90 %)**

A solution of **41** (500 mg, 0.69 mmol) in dichloromethane (10 mL) was added to a dispersion of $AgNTf_2$ (269.1 mg, 0.69 mmol, 1 eq.) in dichloromethane (5 mL) at −20 °C and stirred for 60 min under protection from light while slowly warming up to room temperature. The reaction mixture was filtered over a pad of Celite to give a clear yellow filtrate. The filter frit was rinsed with dichloromethane (3 × 5 mL). Volatiles were removed in vacuo. The residue is washed with pentane (3 × 3 mL) and dried under vacuum to give **43** as a yellow powder that was of sufficient quality to be used without further purification. Yield: 600 mg (90 %). Crystals suitable for X-ray diffraction were obtained at −30 °C from a concentrated solution of **43** in dichloromethane layered with pentane. **Mp**: 119–121 °C (decomposition); **^{1}H NMR** (300 MHz, $CDCl_3$): δ 3.90 (sept, $^3J_{HH}$ = 6.5 Hz, 4H, $CH(CH_3)_2$), 3.39 (mbr, 4H, $N(CH_2)_2N$), 3.23 (mbr, 4H, $N(CH_2)_2N$), 1.22 (d, $^3J_{HH}$ = 6.5 Hz, 12H, CH $(CH_3)_2$), 1.14 (d, $^3J_{HH}$ = 6.5 Hz, 12H, $CH(CH_3)_2$), 1.3–4.0 (vbr, 10H, H_{BH}); **^{31}P {^{1}H} NMR** (121 MHz, $CDCl_3$): δ 138.3 (s); **^{11}B{^{1}H}** NMR (96 MHz, $CDCl_3$): δ −2.3, −4.1, −10.3, −15.3; **^{13}C{^{1}H} NMR** (75 MHz, $CDCl_3$) δ 119.8 (q, $^1J_{CF}$ = 321.8 Hz, CF_3), 95.3 (*pseudo* t, $^1J_{CP}$ = 24.7 Hz, C_1), 49.7 (*pseudo* t, $^2J_{CP}$ = 11.1 Hz, $CH(CH_3)_2$), 43.5 (m, $N(CH_2)_2N$), 42.6 (m, $N(CH_2)_2N$), 21.5 (sbr, $CH(CH_3)_2$), 20.9 (s, $CH(CH_3)_2$), 20.3 (s, $CH(CH_3)_2$), 20.0 (s, $CH(CH_3)_2$); **HRMS (ESI+)**: calcd for $[M]^+ = C_{18}H_{46}B_{10}N_4P_2Au^+$: 685.3875. Found: 685.3892.

5.5.23 Gold(III) 4-Fluoro-Phenyl Complex 44

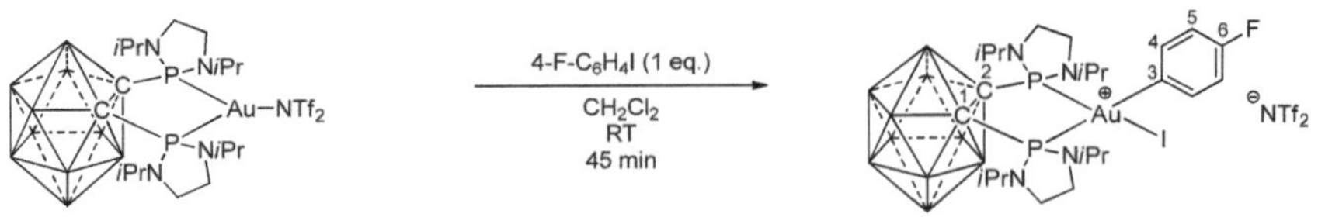

43 44 (96 %)

4-fluoroiodobenzene (7.7 mL, 0.067 mmol) was added to a solution of **43** (65 mg, 0.067 mmol, 1 eq.) in dichloromethane (4 mL) at room temperature. The reaction mixture turned bright yellow. After stirring for 45 min at room temperature, the reaction mixture was concentrated in vacuo to a volume of about 1 mL. Upon addition of pentane (10 mL) a dark-yellow solid precipitates that is separated by cannula filtration. The residue is washed with pentane (3 × 2 mL) and dried under vacuum to give **44** as a yellow powder. Yield: 76 mg (96 %). The compound decomposes in solution (at room temperature: hours, at −30 °C: days) as well as in the solid state (at room temperature: days). **^{1}H NMR** (400 MHz, CD_2Cl_2, 253 K): δ 7.35 (m, $^{5}J_{HP}$ = 2.8 Hz, 2H, H_5), 7.15 (m, $^{4}J_{HP}$ = 3.6 Hz 2H, H_4), 3.77–3.28 (m, 12H, $N(CH_2)_2N$ & $CH(CH_3)_2$), 1.55 (d, $^{3}J_{HH}$ = 6.3 Hz, 6H, $CH(CH_3)_2$), 1.42 (d, $^{3}J_{HH}$ = 6.8 Hz, 6H, $CH(CH_3)_2$), 1.41 (d, $^{3}J_{HH}$ = 6.5 Hz, 6H, $CH(CH_3)_2$), 0.89 (d, $^{3}J_{HH}$ = 6.3 Hz, 6H, $CH(CH_3)_2$), 1.6–3.2 (vbr, 10H, H_{BH}); **^{31}P{^{1}H} NMR** (161 MHz, CD_2Cl_2, 253 K): δ 122.2 (d, $^{2}J_{PP}$ = 30.4 Hz, 1P), 81.7 (d, $^{2}J_{PP}$ = 30.4 Hz, 1P); **^{13}C{^{1}H} NMR** (100 MHz, CD_2Cl_2, 253 K): δ 162.5 (dd, $^{1}J_{CF}$ = 246.5 Hz, $^{5}J_{CP\text{-trans}}$ = 3.0 Hz, C_6), 138.5 (dbr, $^{2}J_{CP\text{-trans}}$ = 162.0 Hz, $^{4}J_{CF}$ not resolved, C_3), 136.0 (m, $^{3}J_{CF}$ = 6.9 Hz, $^{3}J_{CP\text{-trans}}$ not resolved, C_4), 119.8 (q, $^{1}J_{CF}$ = 321.3 Hz, N ($SO_2CF_3)_2$), 117.4 (dd, $^{2}J_{CF}$ = 20.3 Hz, $^{4}J_{CP\text{-trans}}$ = 14.3, C_5), 88.1 (*pseudo* dd, J_{CP} = 27.0 Hz, 13.1 Hz, C_1 or C_2), 84.3 (*pseudo* d, J_{CP} = 20.1 Hz, C_1 or C_2), 50.1 (d, $^{2}J_{CP}$ = 8.8 Hz, $CH(CH_3)_2$), 49.7 (d, $^{2}J_{CP}$ = 12.0 Hz, $CH(CH_3)_2$), 43.0 (d, $^{2}J_{CP}$ = 3.3 Hz, $N(CH_2)_2N$), 42.4 (s, $N(CH_2)_2N$), 22.0 (d, $^{3}J_{CP}$ = 3.2 Hz, CH $(CH_3)_2$), 21.7 (d, $^{3}J_{CP}$ = 4.1 Hz, $CH(CH_3)_2$), 21.5 (d, $^{3}J_{CP}$ = 4.7 Hz, $CH(CH_3)_2$), 20.9 (d, $^{3}J_{CP}$ = 3.7 Hz, $CH(CH_3)_2$); **HRMS (ESI+)**: calcd for $[M]^+$ = $C_{24}H_{50}B_{10}P_2N_4$ $FIAu^+$: 908.3190. Found: 890.3217.

5.5.24 1-Iodo-8-Methylnaphthalene

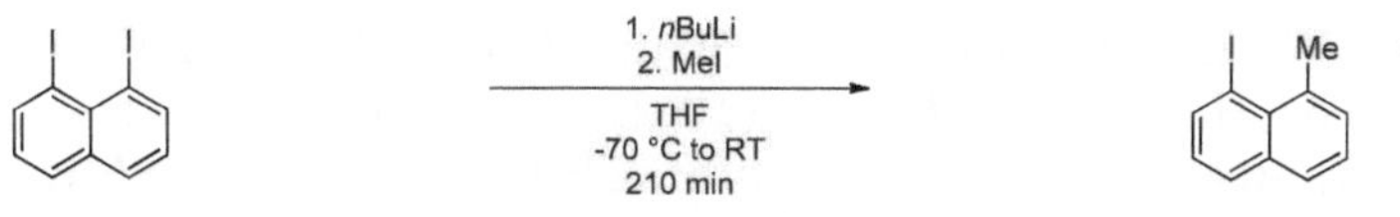

(51 %)

A solution of *n* butyllithium in hexanes (0.53 mL, 1.32 mmol, 1 eq., 2.5 M) was added dropwise to 1,8-diiodonaphthalene (500 mg, 1.32 mmol, 1 eq.) in tetrahydrofuran (20 mL) at −70 °C. The bright yellow solution was stirred for 90 min at this temperature. Methyl iodide was added (123 μL, 1.97 mmol, 1.5 eq.). The solution was allowed to warm up slowly to room temperature (120 min). Volatiles were removed in vacuo. The residue was redissolved in toluene (5 mL) and filtered over a pad of silica gel. Volatiles were removed under vacuum. The pale-yellow residue was recrystallized overnight in pentane at −60 °C to give 1-iodo-8-methylnaphthalene as an off-white crystalline solid. Yield: 180 mg (51 %). The compound matches the NMR spectroscopic data previously reported [90, 91]. **^{1}H NMR** (300 MHz, $CDCl_3$): δ 8.30 (d, J_{HH} = 7.4 Hz, 1H), 7.80 (d, J_{HH} = 7.7 Hz, 1H), 7.72 (d, J_{HH} = 7.7 Hz, 1H), 7.46–7.23 (m, 2H), 7.02 (d, J_{HH} = 7.7 Hz, 1H), 3.21 (s, 3H).

5.5.25 Gold(III) 8-Methylnaphthyl Complex 45

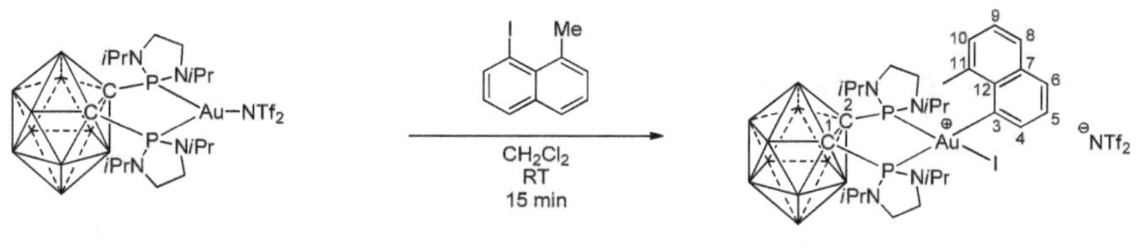

43 **45 (93 %)**

To **43** (30 mg, 0.03 mmol) and 1-iodo-8-methylnaphthalene (8.3 mg, 0.03 mmol, 1 eq.) was slowly added dichloromethane (3 mL) in dichloromethane (2 mL) at room temperature and stirred for 15 min. The reaction mixture turns immediately dark red. Volatiles were removed in vacuo. The residue is washed with pentane (3 × 2 mL) and dried under vacuum to give **45** as a red solid. Yield: 34 mg (93 %). Crystals suitable for X-ray diffraction were obtained at −60 °C from a concentrated solution of **45** layered with toluene and pentane. **Mp**: 102–103 °C (decomposition); **Mp**: 126–129 °C (decomposition); **^{1}H NMR** (500 MHz, CD_2Cl_2): δ 7.80

(d, J_{HH} = 7.9 Hz, 1H, H_{ar}), 7.72 (d, J_{HH} = 7.9 Hz, 1H, H_{ar}), 7.55–7.38 (m, 4H, H_{ar}), 4.05–2.95 (m, 8H, N(CH_2)$_2$N & CH(CH_3)$_2$), 3.06 (s, 3H, naphthyl-CH_3), 1.69 (d, $^3J_{HH}$ = 10.4 Hz, 3H, CH(CH_3)$_2$), 1.64 (d, $^3J_{HH}$ = 10.4 Hz, 3H, CH(CH_3)$_2$), 1.63 (d, $^3J_{HH}$ = 10.4 Hz, 3H, CH(CH_3)$_2$), 1.58 (d, $^3J_{HH}$ = 6.3 Hz, 3H, CH(CH_3)$_2$), 1.52 (d, $^3J_{HH}$ = 6.3 Hz, 3H, CH(CH_3)$_2$), 1.50 (d, $^3J_{HH}$ = 6.3 Hz, 3H, CH(CH_3)$_2$), 1.14 (d, $^3J_{HH}$ = 6.6 Hz, 3H, CH(CH_3)$_2$),), 3.50–1.50 (vbr, 10H, H_{BH}), 1.50–1.25 (m, 8H, N(CH_2)$_2$N), -0.41 (d, $^3J_{HH}$ = 6.6 Hz, 3H, CH(CH_3)$_2$); **^{31}P{^{1}H} NMR** (121 MHz, CD_2Cl_2): δ 122.2 (d, $^2J_{PP}$ = 30.4 Hz, 1P), 81.7 (d, $^2J_{PP}$ = 30.4 Hz, 1P); **^{11}B{^{1}H} NMR** (96 MHz, CD_2Cl_2): δ −1.1, −9.4, −15.5; **^{13}C{^{1}H} NMR** (126 MHz, CD_2Cl_2): δ 137.9 (d, $^3J_{CP}$ = 11.6 Hz, C_{12}), 137.7 (d, $^2J_{CP\text{-trans}}$ = 157.0 Hz, C_3), 137.6 (d, $^4J_{CP}$ = 6.4 Hz, C_7 or C_{11}), 132.3 (d, $^4J_{CP}$ = 6.6 Hz, C_5), 131.7 (d, $^4J_{CP}$ = 4.6 Hz, C_7 or C_{11}), 130.5 (s, C_8 or C_9 or C_{10}), 129.8 (s, C_8 or C_9 or C_{10}), 129.1 (d, $^5J_{CP}$ = 2.3 Hz, C_6), 126.2 (s, C_8 or C_9 or C_{10}), 125.3 (d, $^3J_{CP}$ = 14.1 Hz, C_4), 120.0 (q, $^1J_{CF}$ = 319.8 Hz, N(SO_2CF_3)$_2$), 89.0 (*pseudo* dd, J_{CP} = 31.9 Hz, 9.5 Hz, C_1 or C_2), 85.6 (*pseudo* dd, J_{CP} = 17.2 Hz, 3.8 Hz, C_1 or C_2), 50.2 (d, $^2J_{CP}$ = 12.7 Hz, CH(CH_3)$_2$), 49.7 (d, $^2J_{CP}$ = 9.8 Hz, CH(CH_3)$_2$), 49.6 (d, $^2J_{CP}$ = 8.0 Hz, CH(CH_3)$_2$), 49.5 (d, $^2J_{CP}$ = 11.7 Hz, CH(CH_3)$_2$), 42.8 (d, $^2J_{CP}$ = 3.2 Hz, N(CH_2)$_2$N), 42.5 (s, N(CH_2)$_2$N),, 42.1 (s, N(CH_2)$_2$N), 40.7 (d, $^2J_{CP}$ = 2.5 Hz, N(CH_2)$_2$N), 29.9 (s, naphthyl-CH_3), 23.2 (d, $^3J_{CP}$ = 4.6 Hz, CH (CH_3)$_2$), 21.8 (d, $^3J_{CP}$ = 4.5 Hz, CH(CH_3)$_2$), 21.5 (d, $^3J_{CP}$ = 4.4 Hz, CH(CH_3)$_2$), 21.4 (d, $^3J_{CP}$ = 3.0 Hz, CH(CH_3)$_2$), 21.3 (d, $^3J_{CP}$ = 4.5 Hz, CH(CH_3)$_2$), 21.2 (d, $^3J_{CP}$ = 4.4 Hz, CH(CH_3)$_2$), 21.0 (d, $^3J_{CP}$ = 4.2 Hz, CH(CH_3)$_2$), 20.4 (d, $^3J_{CP}$ = 3.7 Hz, CH(CH_3)$_2$);

HRMS (ESI+): calcd for $[M]^+$ = $C_{29}H_{55}B_{10}P_2N_4IAu^+$: 954.3601 Found: 954.3621; **Elt. Anal.**: calcd for $C_{31}H_{55}B_{10}P_2N_5F_6O_4S_2IAu$: C, 30.18; H, 4.49; N, 5.68. Found: C, 30.16; H, 4.35, N, 5.50.

5.5.26 Kinetic Measurements: Ligand Influence on the Oxidative Addition

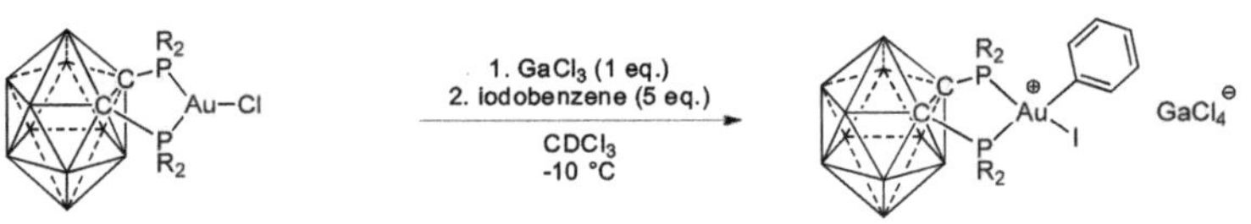

29 (R = Ph) **38 (R = Ph)**

41 (R = *i*PrN(CH$_2$CH$_2$)N*i*Pr) **42 (R = *i*PrN(CH$_2$CH$_2$)N*i*Pr)**

In a glovebox, a solution of $GaCl_3$ (0.03 mmol, 1 eq., 5.3 mg) in $CDCl_3$ (200 μL) was added at room temperature to a solution of gold chloride complex

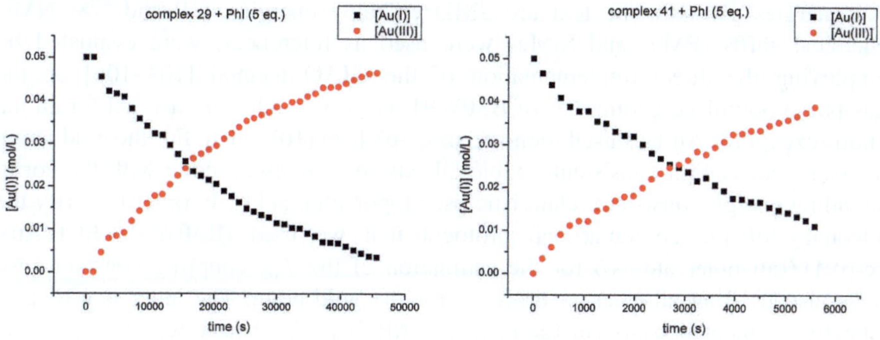

Fig. 5.26 Kinetic profiles for the reaction of complex **29** /$GaCl_3$ + iodobenzene (5 eq.) (*left*) and complex **41** /$GaCl_3$ + iodobenzene (5 eq.) (*right*)

(0.03 mmol, 1 eq.; **29**: 22.3 mg, **41**: 21.6 mg) in $CDCl_3$, (200 μL) in an NMR tube. The tube was closed with a screw cap equipped with a resealable septum, and subsequently shaken. Outside of the glovebox, the NMR tube was cooled down to −80 °C. After the solution in the NMR tube was frozen, a solution of phenyl iodide (0.15 mmol, 5 eq., 16.7 μL) in $CDCl_3$ (200 μL) was added. The tube was then thawed, quickly shaken and placed immediately inside the NMR machine. The reaction was followed by ^{31}P NMR spectroscopy over time. By relative integration of the signals for the starting material (**29**: δ = 35.9 ppm; **41**: δ = 116.6 ppm) and the product (**38**: δ = 44.6 ppm, 61.8 ppm; **42**: δ = 82.3 ppm, 122.2 ppm) their concentrations were determined to establish the kinetic profile of the reaction (Fig. 5.26).

5.5.27 Computational Details

Calculations on the oxidative addition of $(SiMe_2Ph)_2$ to {Ph_3PAuCl /$GaCl_3$} were carried out with the Gaussian 09 program suite [92] on full gold systems (with or without $GaCl_4^-$ counteranion) at the B3PW91 level of theory [93, 94]. Au, Ga, Cl, Si, and P atoms were treated with the corresponding Stuttgart-Dresden RECP (relativistic effective core potential) in combination with their adapted basis sets, [95–98] each one augmented by an extra set of polarization functions [99]. All the other non-metal atoms have been described with a 6-31G(d,p) basis set [100]. Geometry optimizations were carried out without any symmetry restrictions. The nature of the extrema, minimum or saddle points was verified by the absence or presence of only one negative eigenvalue, respectively. The connection between the transition state and the corresponding minima were done by performing IRC calculations [101, 102]. In order to include solvent effects, single point SMD [103] calculations were conducted using dichloromethane as the solvent on the gas-phase optimized geometries of B3PW91, employing the same computational protocol.

All energies given in the text are SMD enthalpy energies. ^{31}P and ^{29}Si NMR chemical shifts (PMe_3 and $SiMe_4$ were used as references) were evaluated by employing the direct implementation of the GIAO method [104–106] on the gas-phase optimized geometries of B3PW91. However, a larger basis set for all the atoms except for Au was used (denoted as cc-pVDZ) [107–110]. For the gold atom, the same correlation-consistent double-ζ basis set was used, along with the corresponding energy-consistent relativistic pseudopotential [111]. In order to verify the adequacy of the computational protocol that we used (B3PW91/RECP(Au)/cc-pVDZ(all other atoms)) for the estimation of the J_{SiP} coupling constants, we have also used an all-electron basis set for the gold atom. The latter is a relativistically segmented basis set (defined as SARC) [112]. Scalar relativistic effects were taken into account by using the Douglas-Kroll-Hess (DKH2) formalism as implemented in the Gaussian program suite [113–117]. It should be noted that the SARC basis sets are designed for the DKH2 Hamiltonians. The results are essentially identical for both approaches.

References

1. B.D. Alexander, P.D. Boyle, B.J. Johnson, J.A. Casalnuovo, S.M. Johnson, A.M. Mueting, L.H. Pignolet, Inorg. Chem. **26**, 2547–2551 (1987)
2. H. Marsmann, J.-P. Kintzinger, *NMR Basic Principles and Progress: Oxygen-17 and Silicon-29* (Springer, Heidelberg, 1981)
3. J. Meyer, J. Willnecker, U. Schubert, Chem. Ber. **122**, 223–230 (1989)
4. O. Monge, A. Schier, H. Schmidbaur, Z. Naturforsch, B **54**, 26–29 (1999)
5. M. Theil, P. Jutzi, B. Neumann, A. Stammler, H.-G. Stammler, J. Organomet. Chem. **662**, 34–42 (2002)
6. P. Gualco, Coordination Originale de Ligands Phosphine-Silanes Sur Le Cuivre, L'argent et L'or, Ph.D. Thesis, Université de Toulouse, Université Toulouse III—Paul Sabatier, 2012
7. P. Gualco, S. Ladeira, K. Miqueu, A. Amgoune, D. Bourissou, Angew. Chem. Int. Ed. **50**, 8320–8324 (2011)
8. G. Berthon-Gelloz, B. de Bruin, B. Tinant, I.E. Markó, Angew. Chem. Int. Ed. **48**, 3161–3164 (2009)
9. N. Takagi, S. Sakaki, J. Am. Chem. Soc. **134**, 11749–11759 (2012)
10. M.A. Carvajal, J.J. Novoa, S. Alvarez, J. Am. Chem. Soc. **126**, 1465–1477 (2004)
11. J.F. Hartwig, *Organotransition Metal Chemistry: From Bonding to Catalysis* (University Science Books, Sausalito, 2009)
12. S. Otsuka, J. Organomet. Chem. **200**, 191–205 (1980)
13. P. Dierkes, P.W.N.M. van Leeuwen, J. Chem. Soc. Dalton Trans. 1519–1530 (1999). doi:10.1039/A807799A
14. R. Hoffmann, in *Frontiers in Chemistry,* ed. by K.J. Laidler (Pergamon Press, New York, 1982), pp. 247–263
15. P.W.N.M. van Leeuwen, P.C.J. Kamer, J.N.H. Reek, Pure Appl. Chem. **71**, 1443–1452 (1999)
16. P.C. Kamer, P.W. Van Leeuwen, J.N. Reek, Acc. Chem. Res. **34**, 895–904 (2001)
17. M.C. Gimeno, A. Laguna, Chem. Rev. **97**, 511–522 (1997)
18. O. Crespo, M.C. Gimeno, A. Laguna, P.G. Jones, J. Chem. Soc. Dalton Trans. 1601–1605 (1992)
19. V.I. Bregadze, Chem. Rev. **92**, 209–223 (1992)

20. J.F. Valliant, K.J. Guenther, A.S. King, P. Morel, P. Schaffer, O.O. Sogbein, K.A. Stephenson, Coord. Chem. Rev. **232**, 173–230 (2002)
21. S.O. Kang, J. Ko, Adv. Organomet. Chem. **47**, 61–99 (2001)
22. R.P. Alexander, H. Schroeder, Inorg. Chem. **2**, 1107–1110 (1963)
23. S. Paavola, R. Kivekäs, F. Teixidor, C. Viñas, J. Organomet. Chem. **606**, 183–187 (2000)
24. M.R. Sundberg, S. Paavola, C. Viñas, F. Teixidor, R. Uggla, R. Kivekäs, Inorganica Chim. Acta **358**, 2107–2111 (2005)
25. S. Paavola, F. Teixidor, C. Viñas, R. Kivekäs, Acta Crystallogr. Sect. C **58**, m237–m239 (2002)
26. O. Crespo, M.C. Gimeno, A. Laguna, Appl. Organomet. Chem. **14**, 644–652 (2000)
27. O. Crespo, M.C. Gimeno, A. Laguna, J. Organomet. Chem. **694**, 1588–1598 (2009)
28. R. Kivekäs, R. Sillanpää, F. Teixidor, C. Viñas, M.M. Abad, R.I. Nielsen, C.E. Olsen, C.N. Rosendahl, M. Haugg, N. Trabesinger-Rüf et al., Acta Chem. Scand. **50**, 499–504 (1996)
29. E. Bembenek, O. Crespo, M. Concepción Gimeno, P.G. Jones, A. Laguna, Chem. Ber. **127**, 835–840 (1994)
30. D.-P. Zhang, J.-M. Dou, D.-C. Li, D.-Q. Wang, Acta Crystallogr. Sect. E Struct. Rep. Online **63**, m1086–m1088 (2007)
31. J.A. Ioppolo, J.K. Clegg, L.M. Rendina, Acta Crystallogr. Sect. E Struct. Rep. Online **65**, m603–m604 (2009)
32. O. Crespo, M.C. Gimeno, P.G. Jones, A. Laguna, J.M. López-de-Luzuriaga, M. Monge, J.L. Pérez, M.A. Ramón, Inorg. Chem. **42**, 2061–2068 (2003)
33. R. Visbal, I. Ospino, J.M. López-de-Luzuriaga, A. Laguna, M.C. Gimeno, J. Am. Chem. Soc. **135**, 4712–4715 (2013)
34. O. Crespo, C. Díez-Gil, M.C. Gimeno, P.G. Jones, A. Laguna, I. Ospino, J. Tapias, M.D. Villacampa, R. Visbal, Dalton Trans. **42**, 8298–8306 (2013)
35. H. Nakamura, T. Kamakura, S. Onagi, Org. Lett. **8**, 2095–2098 (2006)
36. H. Nakamura, T. Kamakura, S. Onagi, Org. Lett. **8**, 2095–2098 (2006)
37. F. Teixidor, C. Viñas, M. Mar Abad, R. Nuñez, R. Kivekäs, R. Sillanpäa, J. Organomet. Chem. **503**, 193–203 (1995)
38. K. Angermaier, E. Zeller, H. Schmidbaur, J. Organomet. Chem. **472**, 371–376 (1994)
39. S.S. Batsanov, Inorg. Mater. **37**, 871–885 (2001)
40. T.J. Brown, M.G. Dickens, R.A. Widenhoefer, Chem. Commun. 6451–6453 (2009)
41. M.A. Cinellu, G. Minghetti, F. Cocco, S. Stoccoro, A. Zucca, M. Manassero, M. Arca, Dalton Trans. 5703 (2006)
42. M.A. Cinellu, G. Minghetti, S. Stoccoro, A. Zucca, M. Manassero, Chem. Commun. 1618–1619 (2004)
43. R.E.M. Brooner, R.A. Widenhoefer, Angew. Chem. Int. Ed. **52**, 11714–11724 (2013)
44. J.F. Hartwig, *Organotransition Metal Chemistry: From Bonding to Catalysis* (University Science Books, Sausalito, 2009)
45. D.B.D. Amico, F. Calderazzo, Gold Bull. **30**, 21–24 (1997)
46. Q. Xu, Y. Imamura, M. Fujiwara, Y. Souma, J. Org. Chem. **62**, 1594–1598 (1997)
47. H. Willner, F. Aubke, Chem. Eur. J. **9**, 1668–1676 (2003)
48. C. Dash, P. Kroll, M. Yousufuddin, H.V.R. Dias, Chem. Commun. **47**, 4478–4480 (2011)
49. H.V.R. Dias, C. Dash, M. Yousufuddin, M.A. Celik, G. Frenking, Inorg. Chem. **50**, 4253–4255 (2011)
50. J. Schaefer, A. Kraft, S. Reininger, G. Santiso-Quinones, D. Himmel, N. Trapp, U. Gellrich, B. Breit, I. Krossing, Chem. Eur. J. **19**, 12468–12485 (2013)
51. S. Martínez-Salvador, L.R. Falvello, A. Martín, B. Menjón, Chem. Eur. J. **19**, 14540–14552 (2013)
52. H.V.R. Dias, W. Jin, Inorg. Chem. **35**, 3687–3694 (1996)
53. O.R. Gilliam, C.M. Johnson, W. Gordy, Phys. Rev. **78**, 140–144 (1950)
54. U. Anandhi, P.R. Sharp, Angew. Chem. Int. Ed. **43**, 6128–6131 (2004)
55. J.E. Huheey, E.A. Keiter, R.L. Keiter, O.K. Medhi, *Inorganic Chemistry: Principles of Structure and Reactivity* (Pearson Education, London, 2006)

56. M. Murakami, Y. Ito, in *Activation of Unreactive Bonds and Organic Synthesis,* eds. by P.S. Murai, H. Alper, R.A. Gossage, V.V. Grushin, M. Hidai, Y. Ito, W.D. Jones, F. Kakiuchi, G. van Koten, Y.-S. Lin, et al. (Springer, Heidelberg, 1999) pp. 97–129
57. K. Ruhland, Eur. J. Org. Chem. **2012**, 2683–2706 (2012)
58. C.-H. Jun, Chem. Soc. Rev. **33**, 610–618 (2004)
59. B. Rybtchinski, D. Milstein, Angew. Chem. Int. Ed. **38**, 870–883 (1999)
60. M. Murakami, T. Matsuda, Chem. Commun. **47**, 1100–1105 (2011)
61. D.V. Partyka, M. Zeller, A.D. Hunter, T.G. Gray, Angew. Chem. Int. Ed. **45**, 8188–8191 (2006)
62. X. Hu, I. Castro-Rodriguez, K. Olsen, K. Meyer, Organometallics **23**, 755–764 (2004)
63. D. Nemcsok, K. Wichmann, G. Frenking, Organometallics **23**, 3640–3646 (2004)
64. N. Marion, S.P. Nolan, Chem. Soc. Rev. **37**, 1776–1782 (2008)
65. D. Benitez, N.D. Shapiro, E. Tkatchouk, Y. Wang, W.A. Goddard, F.D. Toste, Nat. Chem. **1**, 482–486 (2009)
66. C. Perthuisot, B.L. Edelbach, D.L. Zubris, N. Simhai, C.N. Iverson, C. Müller, T. Satoh, W. D. Jones, J. Mol. Catal. Chem. **189**, 157–168 (2002)
67. J. Guenther, S. Mallet-Ladeira, L. Estevez, K. Miqueu, A. Amgoune, D. Bourissou, J. Am. Chem. Soc. **136**, 1778–1781 (2014)
68. Y. Fuchita, H. Ieda, S. Wada, S. Kameda, M. Mikuriya, J. Chem. Soc. Dalton Trans. 4431–4435 (1999)
69. L.T. Ball, G.C. Lloyd-Jones, C.A. Russell, J. Am. Chem. Soc. **136**, 254–264 (2014)
70. S. Komiya, T. Sone, S. Ozaki, M. Ishikawa, N. Kasuga, J. Organomet. Chem. **428**, 303–313 (1992)
71. Y. Fuchita, H. Ieda, Y. Tsunemune, J. Kinoshita-Nagaoka, H. Kawano, J. Chem. Soc. Dalton Trans. 791–796 (1998)
72. Y. Fuchita, H. Ieda, A. Kayama, J. Kinoshita-Nagaoka, H. Kawano, S. Kameda, M. Mikuriya, J. Chem. Soc. Dalton Trans. 4095–4100 (1998)
73. Y. Fuchita, H. Ieda, M. Yasutake, J. Chem. Soc. Dalton Trans. 271–274 (2000)
74. K.J. Kilpin, W. Henderson, B.K. Nicholson, Inorganica Chim. Acta **362**, 3669–3676 (2009)
75. S. Stoccoro, G. Alesso, M. A. Cinellu, G. Minghetti, A. Zucca, M. Manassero, C. Manassero, Dalton Trans. 3467 (2009)
76. M.A. Huffman, L.S. Liebeskind, W.T. Pennington, Organometallics **9**, 2194–2196 (1990)
77. M.A. Huffman, L.S. Liebeskind, W.T. Pennington, Organometallics **11**, 255–266 (1992)
78. Y. Masuda, M. Hasegawa, M. Yamashita, K. Nozaki, N. Ishida, M. Murakami, J. Am. Chem. Soc. **135**, 7142–7145 (2013)
79. S.L. Buchwald, Acc. Chem. Res. **41**, 1439–1439 (2008)
80. C.C.C. Johansson Seechurn, M.O. Kitching, T.J. Colacot, V. Snieckus, Angew. Chem. Int. Ed. **51**, 5062–5085 (2012)
81. A. Igau, H. Grützmacher, A. Baceiredo, G. Bertrand, J. Am. Chem. Soc. **110**, 6463–6466 (1988)
82. A. Igau, A. Baceiredo, G. Trinquier, G. Bertrand, Angew. Chem. Int. Ed. Engl. **28**, 621–622 (1989)
83. D. Bourissou, O. Guerret, F.P. Gabbaï, G. Bertrand, Chem. Rev. **100**, 39–92 (2000)
84. D. Gau, T. Kato, N. Saffon-Merceron, F.P. Cossío, A. Baceiredo, J. Am. Chem. Soc. **131**, 8762–8763 (2009)
85. R. Rodriguez, D. Gau, T. Kato, N. Saffon-Merceron, A. De Cózar, F.P. Cossío, A. Baceiredo, Angew. Chem. Int. Ed. **50**, 10414–10416 (2011)
86. N. Mézailles, L. Ricard, F. Gagosz, Org. Lett. **7**, 4133–4136 (2005)
87. T. Lauterbach, M. Livendahl, A. Rosellón, P. Espinet, A.M. Echavarren, Org. Lett. **12**, 3006–3009 (2010)
88. M.N. Hopkinson, A.D. Gee, V. Gouverneur, Chem. - Eur. J. **17**, 8248–8262 (2011)
89. T. Kottke, D. Stalke, J. Appl. Crystallogr. **26**, 615–619 (1993)
90. J. Blum, D. Gelman, W. Baidossi, E. Shakh, A. Rosenfeld, Z. Aizenshtat, B.C. Wassermann, M. Frick, B. Heymer, S. Schutte et al., J. Org. Chem. **62**, 8681–8686 (1997)

91. T. Kesharwani, R.C. Larock, Tetrahedron **64**, 6090–6102 (2008)
92. M.J. Frisch, G.W. Trucks, H.B. Schlegel, G.E. Scuseria, M.A. Robb, J.R. Cheeseman, G.Scalmani, V. Barone, B. Mennucci, G.A. Petersson et al., *Gaussian 09, Revision A.1* (Gaussian, Inc., Wallingford, Connecticut, USA, 2009)
93. A.D. Becke, J. Chem. Phys. **98**, 5648–5652 (1993)
94. K. Burke, J.P. Perdew, and Y. Wang, in *Electron Density Function Theory Recent Program New Directions*, eds. by J.F. Dobson, G. Vignale, M.P. Das (Plenum Press, New York, 1998), pp. 81–111
95. B. Metz, H. Stoll, M. Dolg, J. Chem. Phys. **113**, 2563–2569 (2000)
96. M. Dolg, in *Modern Methods Algorithms Quantum Chemistry*, ed. by J. Grotendorst (John-von-Neumann-Inst. For Computing, Jülich, 2000), pp. 479–508
97. D. Andrae, U. Häußermann, M. Dolg, H. Stoll, H. Preuß, Theor. Chim. Acta **77**, 123–141 (1990)
98. M. Dolg, U. Wedig, H. Stoll, H. Preuss, J. Chem. Phys. **86**, 866–872 (1987)
99. A.W. Ehlers, M. Böhme, S. Dapprich, A. Gobbi, A. Höllwarth, V. Jonas, K.F. Köhler, R. Stegmann, A. Veldkamp, G. Frenking, Chem. Phys. Lett. **208**, 111–114 (1993)
100. P.C. Hariharan, J.A. Pople, Theor. Chim. Acta **28**, 213–222 (1973)
101. C. Gonzalez, H.B. Schlegel, J. Chem. Phys. **90**, 2154–2161 (1989)
102. C. Gonzalez, H.B. Schlegel, J. Phys. Chem. **94**, 5523–5527 (1990)
103. A.V. Marenich, C.J. Cramer, D.G. Truhlar, J. Phys. Chem. B **113**, 6378–6396 (2009)
104. F. London, J. Phys. Radium **8**, 397–409 (1937)
105. R. Ditchfield, Mol. Phys. **27**, 789–807 (1974)
106. K. Wolinski, J.F. Hinton, P. Pulay, J. Am. Chem. Soc. **112**, 8251–8260 (1990)
107. T.H.D. Jr, J. Chem. Phys. **90**, 1007–1023 (1989)
108. D.E. Woon, T.H.D. Jr, J. Chem. Phys. **98**, 1358–1371 (1993)
109. D.E. Woon, T.H.D. Jr, J. Chem. Phys. **103**, 4572–4585 (1995)
110. A.K. Wilson, D.E. Woon, K.A. Peterson, T.H.D. Jr, J. Chem. Phys. **110**, 7667–7676 (1999)
111. D. Figgen, G. Rauhut, M. Dolg, H. Stoll, Chem. Phys. **311**, 227–244 (2005)
112. D.A. Pantazis, X.-Y. Chen, C.R. Landis, F. Neese, J. Chem. Theory Comput. **4**, 908–919 (2008)
113. M. Douglas, N.M. Kroll, Ann. Phys. **82**, 89–155 (1974)
114. B.A. Hess, Phys. Rev. A **32**, 756–763 (1985)
115. B.A. Hess, Phys. Rev. A **33**, 3742–3748 (1986)
116. G. Jansen, B.A. Hess, Phys. Rev. A **39**, 6016–6017 (1989)
117. L. Visscher, K.G. Dyall, At. Data Nucl. Data Tables **67**, 207–224 (1997)

Chapter 6
General Conclusion and Perspective

The major objectives of the initially envisioned research project that evolved in the course of this work were focused on the elucidation of the coordination chemistry of copper and gold. Unexpected reactivities of these two coinage metals with regard to fundamental elementary steps were disclosed and studied in detail by experimental means and accompanying theoretical analyses by the groups of Dr. Karinne Miqueu (Pau) and Prof. Dr. Laurent Maron (Toulouse). These findings contribute to the precise understanding of the chemistry of copper and gold complexes and hopefully may influence further fundamental and application-oriented research in these fields.

It was shown previously in our research group that a chelating disilane-diphosphine ligand coordinates to copper(I) to form the first structurally characterized σ-complex of a coinage metal, with the σ-SiSi bond bound in a side-on fashion to the metal center. We questioned the possibility of the coordination of other σ-bonds to Cu(I) and investigated the respective behavior of a hydrosilane-diphosphine. Upon coordination, the σ-SiH bond becomes involved in a weak interaction with the metal center. Much like in case of the disilane-ligand, this interaction is essentially dominated by donation of electron density from the σ-SiH bond towards Cu(I), with only weak backbonding.

With regard to the chemistry of gold, we were interested in the reactivity of silylgold(I) complexes towards alkynes and allenes. A *syn* insertion process, for gold complexes a most uncommon elementary step, was evidenced. This unprecedented reaction allowed for the regio- and stereoselective synthesis of (β-silyl) vinylgold(I) complexes. A joint experimental/computational study elucidated the mechanism of these original transformations. In contrast to the usually encountered outer-sphere coordination/*anti* addition mechanism involved in homogeneous gold catalysis, a 2-step process consisting of alkyne coordination giving rise to an intermediate tricoordinate gold(I) complex, followed by inner-sphere migration of the silyl group with concomitant AuC bond formation was evidenced.

In order to take advantage of this reactivity and to extend it to useful chemical transformations, the reactivity of the formed vinylgold complexes has to be clarified. Especially reactions that allow for the regeneration of a silylgold complex are tantalizing, as such a process might enable the gold-catalyzed silylation of alkynes.

M. Joost, *Synthesis and Original Reactivity of Copper and Gold Complexes*,
Springer Theses, DOI 10.1007/978-3-319-18690-0_6

Furthermore, the extension of the coordination/*syn* insertion at gold complexes to olefinic substrates is worthwhile to pursue.

Another elementary step which is only seldom observed for gold complexes is the oxidative addition of non- or weakly polar bonds. Previously in the laboratory, the chelate-assisted oxidative addition of disilanes was evidenced. In the quest of extending this reactivity to the intermolecular level, we investigated firstly the reactivity of cationic monophosphine gold(I) complexes towards disilanes and found that oxidative addition of the σ-SiSi bond is thermodynamically favorable and kinetically accessible, as experimentally observed by means of a low-temperature multinuclear NMR study. The associated computational analysis delivered insight into the mechanism of the oxidative addition process and disclosed a Y-shape structure for the formed bis(silyl)gold(III) complexes.

The elusive nature of tricoordinated gold(III) complexes prompted us to analyze the requirements of the coordination sphere of gold(I) complexes to promote σ-bond activation processes and give stable gold(III) products. We hypothesized that an $[L_2Au^+]$ fragment that features a bidentate ligand imposing a small L–Au–L bite angle may be active for σ-bond activation processes. Indeed, a ligand framework based on an *o*-carboranyldiphosphine featuring a small bite angle was found to promote the intermolecular oxidative addition of strained carbocycles and iodoarenes to gold(I) under mild conditions. The ensuing gold(III) complexes were isolated and characterized.

The small P–Au–P bite angle imposed by the cluster-framework of the diphosphine ligand results in a massive change concerning the electronic properties of the 14-electron gold(I) fragment compared to diphosphine gold(I) complexes featuring a linear coordination geometry. Backbonding, which is usually insignificant for gold(I) complexes, is highlighted here not only with the observed reactivity but as well with the synthesis and characterization of the first classical carbonyl complex of gold.

Subsequent work in this area will center on 2-electron redox processes at gold complexes, aiming at (catalytic?) cross-coupling reactions in the long term.

Preliminary results indicate that electron-rich iodoarenes undergo the oxidative addition to gold(I) more rapidly than electron-poor ones. This observation stands in contrast to the reactivity of palladium complexes for which the inverse situation is observed [1]. This finding is particularly interesting with a view to synthetically useful reactions that may complement the existing palladium chemistry instead of simply mimicking it. Further studies are currently underway to corroborate these results.

In addition, the oxidative addition to gold(I) is not only limited to aryl iodides. Simple aryl bromides undergo as well the oxidative addition. In order to demonstrate the generality of $C(sp^2)$–X bond activation at gold(I), the oxidative addition of aryl chlorides will be focused upon.

Although the oxidative addition step demonstrated in this manuscript is a first advance on the way to synthetic applications of σ-bond activation at gold(I), many obstacles have to be overcome for the development of a catalytic cross-coupling reaction. A major difficulty is clearly the introduction of the second coupling partner. Typical transmetallation procedures require (Lewis–) basic additives, which are incompatible with the here employed [L_2Au^+] fragment due to blocking of the empty coordination site required for oxidative addition. In this context, special attention should be paid to the influence of the bite angle of the diphosphine ligand. A gold(I) complex featuring a P–Au–P bite angle somewhere between 100° and 140° may still allow for bond activation of a σ-bond, but should as well favor dissociation of a Lewis base occupying the crucial coordination site required for oxidative addition. Synthesis of such a [L_2Au^+] fragment is non-trivial, as clear guidelines that predict the geometry of a gold(I) complex (i.e. mononuclear/bent vs. dinuclear/linear) for a given bidentate ligand do not exist.

A viable alternative to traditional transmetallations is the direct arylation of a gold(III) complex. The aptness of gold(III) complexes for CH bond activation might allow for the cross-coupling of aryl halides with unfunctionalized aromatic molecules (Scheme 6.1).

With regard to transformations of cyclometallated gold(III) complexes stemming from the oxidative addition of strained carbocycles, further functionalization seems hard to achieve due to the robustness of these compounds. However, preliminary results indicate that a gold(I)–gold(III)–gold(I) redox process employing amine-*N*-oxides with net oxygen insertion into the CC bond and regeneration of a gold(I) complex is feasible. Further studies in these directions are required to probe the scope of this reactivity pattern.

Finally, the unprecedented electronic properties of the bent [P_2Au^+] fragment open up new opportunities for the coordination chemistry of gold. For example, compounds featuring gold–carbon double bonds have eluded structural characterization so far and the existence of such compounds is matter of debate [2]. However, the influence of the increased backdonation of the bent diphosphine complexes described in Chap. 5 compared to linear gold(I) complexes still has to be explored. *Bona fide* carbene complexes with an Au=C bond may become isolable by taking advantage of the electron-rich, but coordinatively unsaturated 14-electron [P_2Au^+] species.

Scheme 6.1 Hypothetical catalytic cycle for a gold-catalyzed cross-coupling of aryl halides with simple arenes

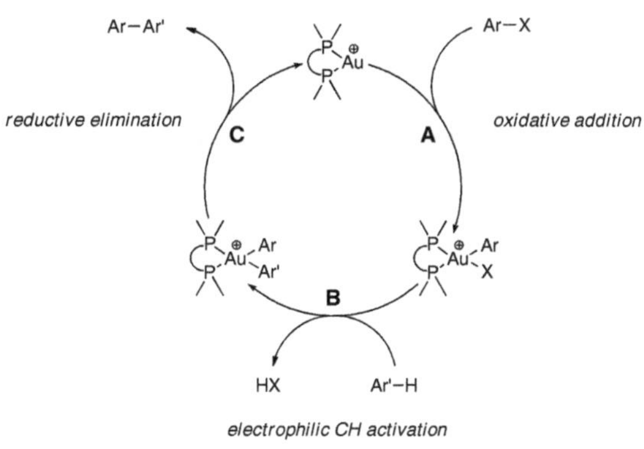

Copper and gold, two of the very first metals known to humankind, have stimulated the interest of chemists since the early days of their discipline. During this long "chemical history" important effort was made to better understand their often unexpected properties. However, many aspects of the coordination chemistry of these metals remain unclear. We hope that the work summarized in this manuscript provides some clarifying contribution to these open questions.

References

1. L. Xue, Z. Lin, Chem. Soc. Rev. **39**, 1692 (2010)
2. G. Seidel, A. Fürstner, Angew. Chem. Int. Ed. **53**, 4807–4811 (2014)

Appendix
Crystallographic Data

Crystallographic data (excluding structure factors) have been deposited to the Cambridge Crystallographic Data Centre (www.ccdc.cam.ac.uk/data_request/cif) for compounds **8** (CCDC-934772), **9** (CCDC-934773), **10** (CCDC-934774), **11** (CCDC-934775), **17** (CCDC-934776), **18** (CCDC-934777), **34** (CCDC-1044000), **35** (CCDC-1044001), **36** (CCDC-1044002) and **37** (CCDC-1044003).

Selected crystallographic data are summarized in the following tables.

The residual factors listed in the tables are defined as follows:

$$R_1 = \sum ||F_o| - |F_c|| \Big/ \sum |F_o|$$

$$wR_2 = \left[\sum \left[w\left(F_o^2 - F_c^2\right)^2\right] \Big/ \left[\sum \left[w\left(F_o^2\right)^2\right]\right]^{1/2}\right].$$

Ligand 1

ID	max12	
Formula	$C_{37}H_{32}P_2Si$	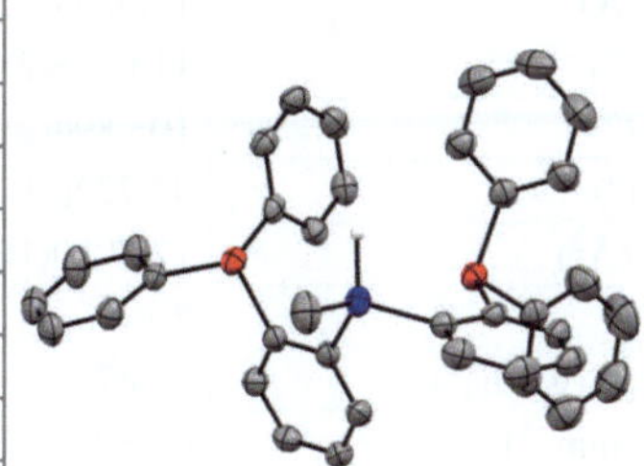
M_r	566.7	
Crystal system	Triclinic	
Space group	*P −1*	
a (Å)	11.0158(8)	
b (Å)	11.6519(8)	
c (Å)	12.8851(8)	
α (°)	86.532(4)	
β (°)	77.873(4)	
γ (°)	74.804(4)	

(continued)

M. Joost, *Synthesis and Original Reactivity of Copper and Gold Complexes*,
Springer Theses, DOI 10.1007/978-3-319-18690-0

(continued)

ID	max12	
V (Å^3)	1560.40(19)	
Z	2	
ρ_{calc} (g cm^{-3})	1.206	
μ (mm^{-1})	0.202	
F(000)	596	
Crystal size (mm^3)	0.08 × 0.08 × 0.04	
T/K	193(2)	
Measd reflns	15597	
Unique reflns (Rint)	3756 (0.0466)	
Reflns used for refinement	3756	
Refined parameters	366	
GOF on F^2	1.034	
R_1 [I > 2σ(I)]	0.0391	
wR_2 [all data]	0.093	

Complex 2

ID	max19	
Formula	$C_{37}H_{32}ClCuP_2Si$, CH_2Cl_2	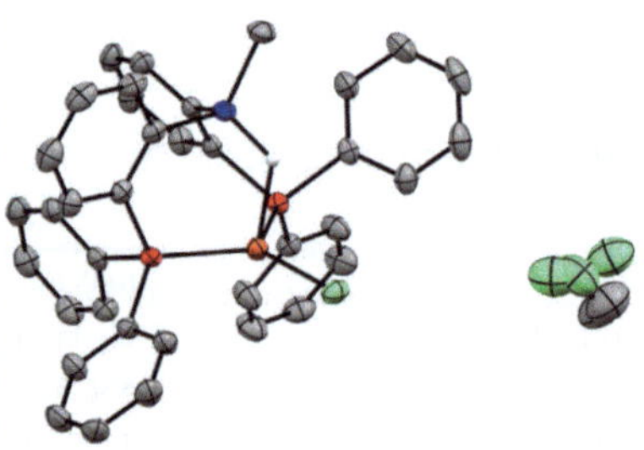
M_r	750.58	
Crystal system	Triclinic	
Space group	$P\ -1$	
a (Å)	11.4211(6)	
b (Å)	12.7243(6)	
c (Å)	14.0664(7)	
α (°)	113.428(2)	
β (°)	103.894(2)	
γ (°)	92.722(2)	
V (Å^3)	1797.20(16)	
Z	2	
ρ_{calc} (g cm^{-3})	1.387	
μ (mm^{-1})	0.980	
F(000)	772	
Crystal size (mm^3)	0.48 × 0.4 × 0.4	
T/K	193(2)	

(continued)

(continued)

ID	max19	
Measd reflns	38474	
Unique reflns (Rint)	7274 (0.02)	
Reflns used for refinement	7274	
Refined parameters	438	
GOF on F^2	1.037	
R_1 [I > 2σ(I)]	0.0281	
wR_2 [all data]	0.0768	

Complex 3

ID	max17	
Formula	$C_{37}H_{32}Cl_4CuGaP_2Si$, CH_2Cl_2	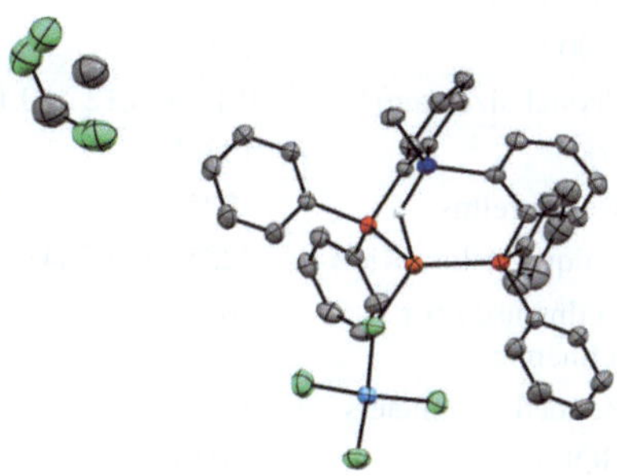
M_r	926.65	
Crystal system	Triclinic	
Space group	*P −1*	
a (Å)	10.4670(4)	
b (Å)	12.3178(4)	
c (Å)	16.7001(6)	
α (°)	107.352(2)	
β (°)	91.413(2)	
γ (°)	99.282(2)	
V (Å^3)	2022.19(13)	
Z	2	
ρ_{calc} (g cm^{-3})	1.522	
μ (mm^{-1})	1.724	
F(000)	936	
Crystal size (mm^3)	0.16 × 0.14 × 0.14	
T/K	193(2)	
Measd reflns	64292	
Unique reflns (Rint)	12287 (0.0311)	
Reflns used for refinement	12287	
Refined parameters	476	
GOF on F^2	1.026	
R_1 [I > 2σ(I)]	0.0342	
wR_2 [all data]	0.0937	

Complex 5

ID	max18
Formula	$C_{30}H_{42}ClCuP_2Si$, CH_2Cl_2
M_r	676.59
Crystal system	Monoclinic
Space group	$P\ 2_1/c$
a (Å)	11.0157(3)
b (Å)	14.7476(3)
c (Å)	21.6795(6)
α (°)	90
β (°)	102.1030(10)
γ (°)	90
V (Å^3)	3443.66(15)
Z	4
ρ_{calc} (g cm^{-3})	1.305
μ (mm^{-1})	1.014
F(000)	1416
Crystal size (mm^3)	0.14 × 0.12 × 0.10
T/K	193(2)
Measd reflns	42257
Unique reflns (Rint)	42257 (0.0384)
Reflns used for refinement	5820
Refined parameters	355
GOF on F^2	1.046
R_1 [I > 2σ(I)]	0.0307
wR_2 [all data]	0.0856

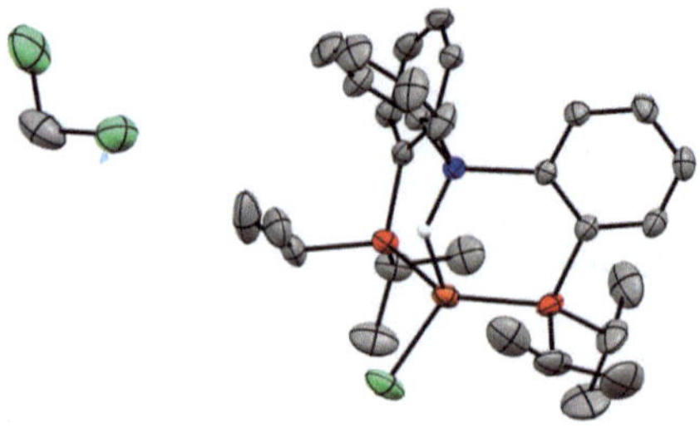

Complex 6

ID	max16
Formula	$C_{30}H_{42}Cl_4CuGaP_2Si$
M_r	767.74
Crystal system	Triclinic
Space group	*P 1*
a (Å)	8.8687(8)
b (Å)	10.2730(11)
c (Å)	10.6228(10)
α (°)	99.916(6)
β (°)	104.912(6)
γ (°)	104.690(6)
V (Å^3)	874.98(16)
Z	1
ρ_{calc} (g cm^{-3})	1.457
μ (mm^{-1})	1.828
F(000)	394
Crystal size (mm^3)	0.16 × 0.08 × 0.02
T/K	193(2)
Measd reflns	20108
Unique reflns (Rint)	3181 (0.0538)
Reflns used for refinement	3181
Refined parameters	365
GOF on F^2	1.079
R_1 [I > 2σ(I)]	0.0352
wR_2 [all data]	0.0974

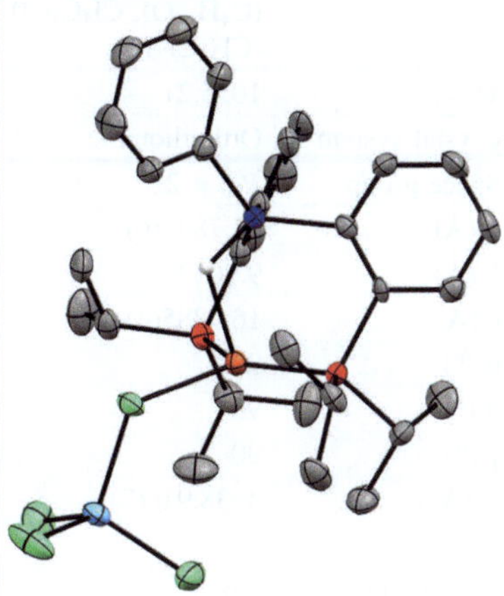

Complex 7

ID	max4
Formula	$C_{37}H_{32}AuP_2Si$, 0.45 ($C_4H_{10}O$), Cl_4Ga, 0.55 (CH_2Cl_2)
M_r	1055.21
Crystal system	Orthorhombic
Space group	$P\ n\ a\ 2_1$
a (Å)	25.7847(6)
b (Å)	9.9052(2)
c (Å)	16.9885(4)
α (°)	90
β (°)	90
γ (°)	90
V (Å^3)	4338.91(17)
Z	4
ρ_{calc} (g cm^{-3})	1.615
μ (mm^{-1})	4.442
F(000)	2072
Crystal size (mm^3)	0.20 × 0.14 × 0.14
T/K	193(2)
Measd reflns	48551
Unique reflns (Rint)	9816 (0.0346)
Reflns used for refinement	9816
Refined parameters	493
GOF on F^2	1.005
R_1 [I > 2σ(I)]	0.027
wR_2 [all data]	0.0582

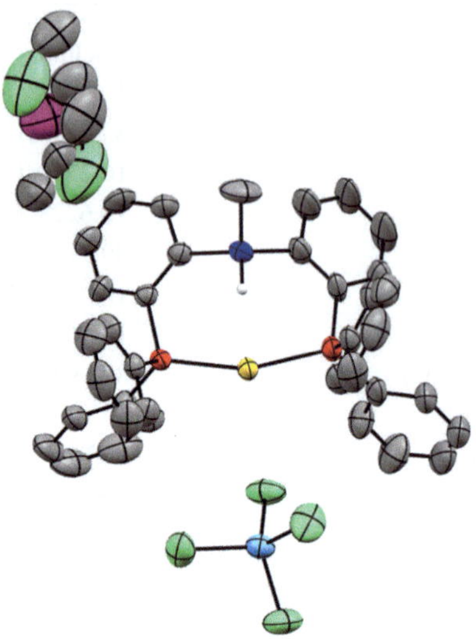

Complex 10

ID	max23
Formula	$C_{42}H_{40}AuPSi$, CH_2Cl_2
M_r	885.70
Crystal system	Monoclinic
Space group	$P\ 2_1/n$
a (Å)	10.9267(6)
b (Å)	23.8884(14)
c (Å)	14.8335(9)
α (°)	90
β (°)	99.486(4)
γ (°)	90
V (Å^3)	3818.9(4)
Z	4
ρ_{calc} (g cm^{-3})	1.54
μ (mm^{-1})	4.095
F(000)	1768
Crystal size (mm^3)	0.08 × 0.08 × 0.06
T/K	193(2)
Measd reflns	33485
Unique reflns (Rint)	7743 (0.0618)
Reflns used for refinement	7743
Refined parameters	436
GOF on F^2	1.037
R_1 [I > 2σ(I)]	0.037
wR_2 [all data]	0.0946

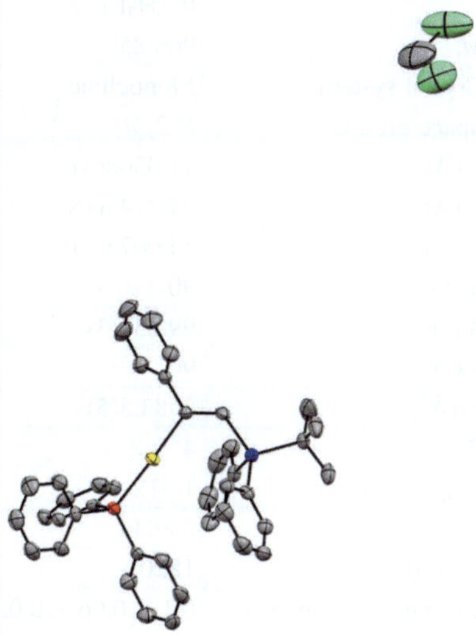

Complex 11

ID	max27
Formula	$C_{44}H_{42}AuO_2PSi$, 0.5 (C_6H_5F)
M_r	906.85
Crystal system	Monoclinic
Space group	*P 2₁/n*
a (Å)	11.0368(8)
b (Å)	24.3944(18)
c (Å)	15.0073(9)
α (°)	90
β (°)	99.634(3)
γ (°)	90
V (Å^3)	3983.5(5)
Z	4
ρ_{calc} (g cm^{-3})	1.512
μ (mm^{-1})	3.804
F(000)	1820
Crystal size (mm^3)	0.1 × 0.06 × 0.02
T/K	193(2)
Measd reflns	25893
Unique reflns (Rint)	6309 (0.064)
Reflns used for refinement	6309
Refined parameters	562
GOF on F^2	1.004
R_1 [I > 2σ(I)]	0.0334
wR_2 [all data]	0.069

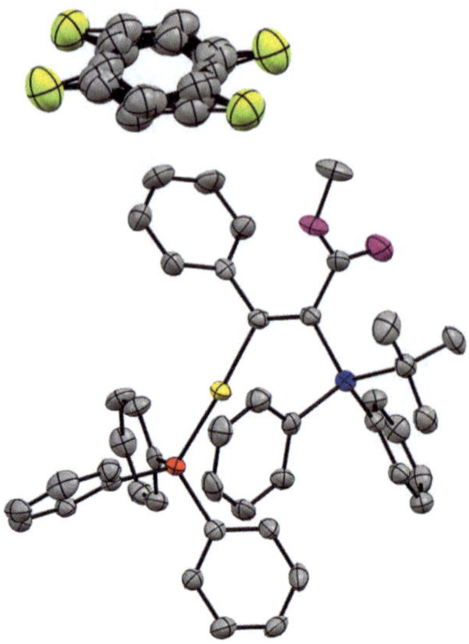

Complex 12

ID	max25	
Formula	$C_{40}H_{40}AuO_4PSi$, 0.5 (C_6H_6)	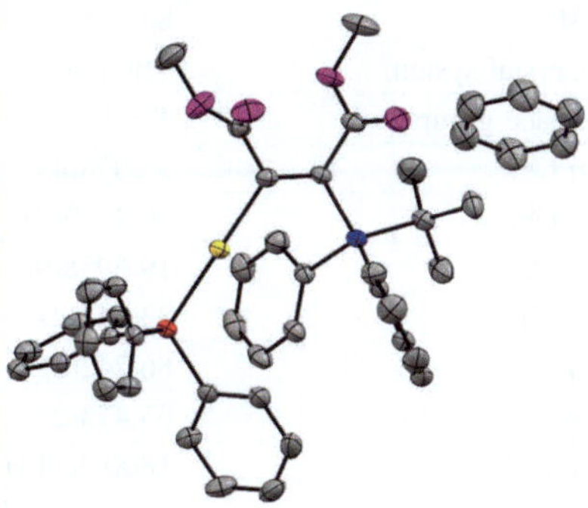
M_r	879.8	
Crystal system	Monoclinic	
Space group	$P\ 2_1/n$	
a (Å)	10.8649(4)	
b (Å)	23.9209(8)	
c (Å)	14.8877(6)	
α (°)	90	
β (°)	100.449(2)	
γ (°)	90	
V (Å^3)	3805.1(2)	
Z	4	
ρ_{calc} (g cm^{-3})	1.536	
μ (mm^{-1})	3.981	
F(000)	1764	
Crystal size (mm^3)	0.16 × 0.08 × 0.08	
T/K	193(2)	
Measd reflns	57915	
Unique reflns (Rint)	9427 (0.0374)	
Reflns used for refinement	9427	
Refined parameters	456	
GOF on F^2	1.024	
R_1 [$I > 2\sigma(I)$]	0.0236	
wR_2 [all data]	0.0521	

Complex 13

ID	max24
Formula	$C_{40}H_{40}AuO_4PSi$
M_r	840.75
Crystal system	Triclinic
Space group	$P\ -1$
a (Å)	9.9038(4)
b (Å)	10.2229(4)
c (Å)	19.6958(9)
α (°)	83.143(2)
β (°)	86.242(2)
γ (°)	65.473(2)
V (Å^3)	1800.92(13)
Z	2
ρ_{calc} (g cm^{-3})	1.55
μ (mm^{-1})	4.202
F(000)	1560
Crystal size (mm^3)	0.38 × 0.28 × 0.18
T/K	193(2)
Measd reflns	63763
Unique reflns (Rint)	11126 (0.0259)
Reflns used for refinement	11126
Refined parameters	429
GOF on F^2	1.076
R_1 [I > 2σ(I)]	0.0148
wR_2 [all data]	0.0366

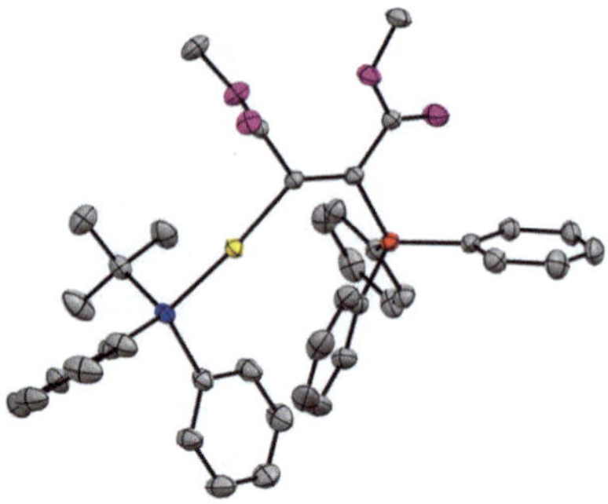

Complex 16

ID	max30
Formula	$C_{46}H_{38}AuO_2PSi$
M_r	878.79
Crystal system	Monoclinic
Space group	$P\ 2_1/c$
a (Å)	18.1445(14)
b (Å)	11.4307(8)
c (Å)	19.4907(12)
α (°)	90
β (°)	110.614(2)
γ (°)	90
V (Å^3)	3783.6(5)
Z	4
ρ_{calc} (g cm^{-3})	1.543
μ (mm^{-1})	3.998
F(000)	1752
Crystal size (mm^3)	0.2 × 0.2 × 0.2
T/K	193(2)
Measd reflns	29695
Unique reflns (Rint)	7792 (0.0484)
Reflns used for refinement	7792
Refined parameters	648
GOF on F^2	1.147
R_1 [I > 2σ(I)]	0.0376
wR_2 [all data]	0.1031

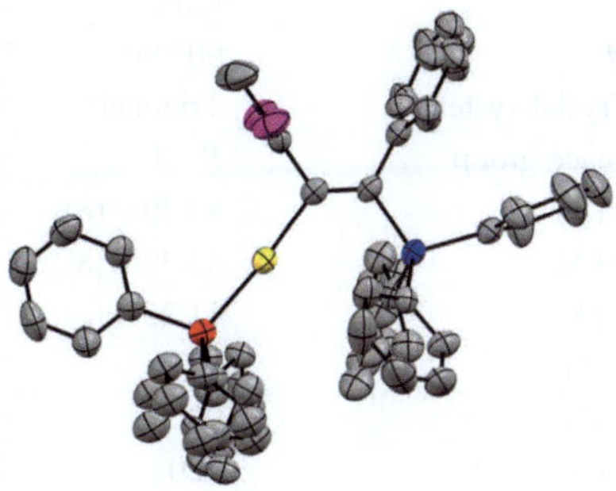

Complex 18

ID	max31
Formula	$C_{42}H_{38}AuO_2PSi$, C_6H_6
M_r	908.90
Crystal system	Trigonal
Space group	*R –3*
a (Å)	43.3981(8)
b (Å)	43.3981(8)
c (Å)	11.2570(2)
α (°)	90
β (°)	90
γ (°)	120
V ($Å^3$)	18360.9(9)
Z	18
ρ_{calc} (g cm^{-3})	1.480
μ (mm^{-1})	3.712
F(000)	8208
Crystal size (mm^3)	0.2 × 0.12 × 0.04
T/K	193(2)
Measd reflns	118384
Unique reflns (Rint)	10352 (0.0743)
Reflns used for refinement	10352
Refined parameters	495
GOF on F^2	1.278
R_1 [I > 2σ(I)]	0.0426
wR_2 [all data]	0.1707

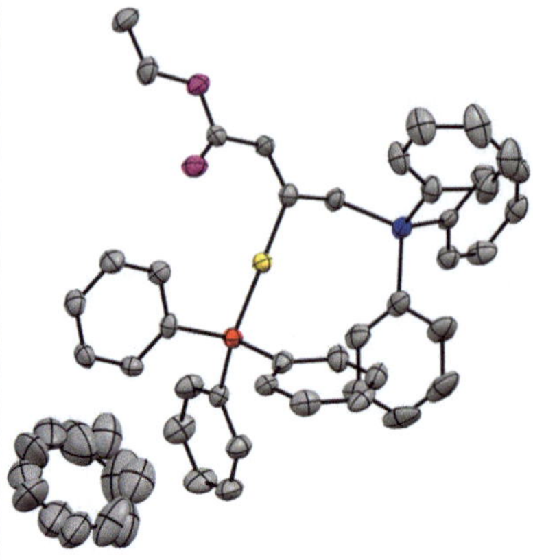

Complex 23

ID	max39	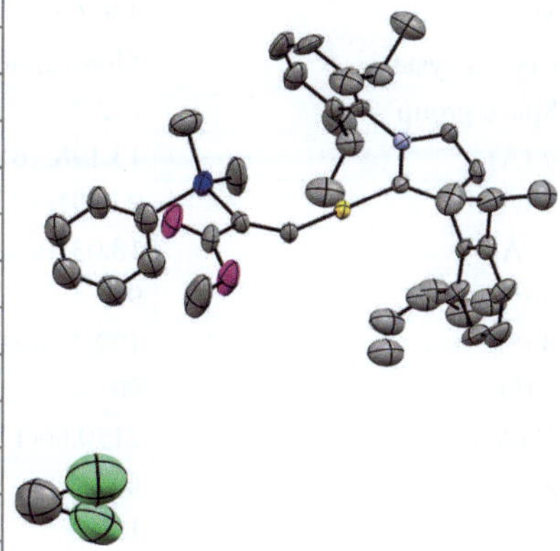
Formula	2($C_{39}H_{51}AuN_2O_2Si$), CH_2Cl_2	
M_r	1694.68	
Crystal system	Triclinic	
Space group	*P* −*1*	
a (Å)	11.3039(3)	
b (Å)	12.4883(4)	
c (Å)	16.4980(5)	
α (°)	87.2940(10)	
β (°)	80.3490(10)	
γ (°)	64.2270(10)	
V (Å^3)	2066.74(11)	
Z	1	
ρ_{calc} (g cm^{-3})	1.362	
μ (mm^{-1})	3.685	
F(000)	858.0	
Crystal size (mm^3)	0.1 × 0.08 × 0.04	
T/K	193(2)	
Measd reflns	24950	
Unique reflns (Rint)	8030 (0.0409)	
Reflns used for refinement	8030	
Refined parameters	462	
GOF on F^2	1.198	
R_1 [I > 2σ(I)]	0.0404	
wR_2 [all data]	0.144	

Ligand 25

ID	max13
Formula	$C_{26}H_{25}PSi$
M_r	396.52
Crystal system	Monoclinic
Space group	$P\ 2_1/c$
a (Å)	13.1983(6)
b (Å)	9.6691(4)
c (Å)	18.0370(8)
α (°)	90
β (°)	109.434(2)
γ (°)	90
V (Å^3)	2170.66(17)
Z	4
ρ_{calc} (g cm^{-3})	1.213
μ (mm^{-1})	0.191
F(000)	840
Crystal size (mm^3)	0.36 × 0.26 × 0.08
T/K	193(2)
Measd reflns	52790
Unique reflns (Rint)	6640 (0.0266)
Reflns used for refinement	6640
Refined parameters	255
GOF on F^2	1.073
R_1 [I > 2σ(I)]	0.0335
wR_2 [all data]	0.1026

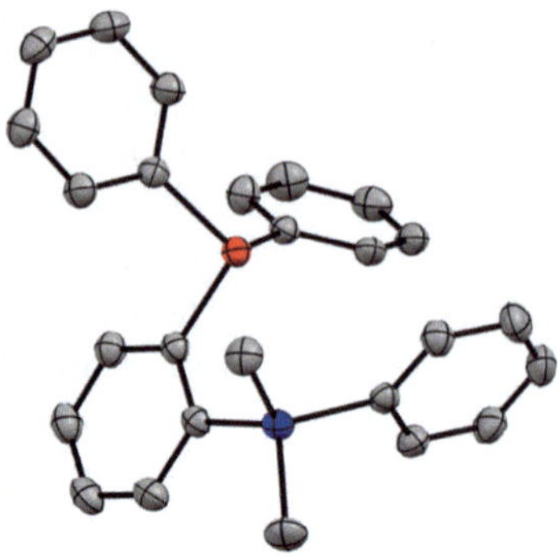

Complex 30

ID	max43
Formula	2 ($C_{14}H_{38}AuB_{10}P_2Cl$), 0.22 ($C_6H_{18}OSi_2$), 0.39 ($C_{12}H_{36}O_2Si$)
M_r	1380.19
Crystal system	Monoclinic
Space group	$P\ 2_1/c$
a (Å)	7.7837(2)
b (Å)	14.4989(4)
c (Å)	26.7667(9)
α (°)	90
β (°)	90
γ (°)	90
V (Å^3)	3020.76(15)
Z	2
ρ_{calc} (g cm^{-3})	1.517
μ (mm^{-1})	5.11
F(000)	1372
Crystal size (mm^3)	0.12 × 0.1 × 0.05
T/K	193(2)
Measd reflns	11992
Unique reflns (Rint)	6377 (0.0138)
Reflns used for refinement	6377
Refined parameters	417
GOF on F^2	1.241
R_1 [$I > 2\sigma(I)$]	0.0388
wR_2 [all data]	0.0905

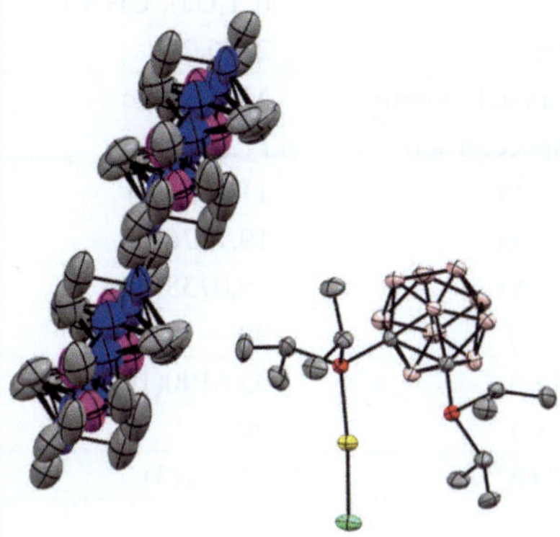

Complex 31

ID	max34
Formula	$2(C_{34}H_{38}AuB_{10}P_2)$, 2 (Cl_4Ga), CH_2Cl_2
M_r	2135.27
Crystal system	Monoclinic
Space group	*P* 2*₁*/*n*
a (Å)	11.6826(5)
b (Å)	19.7076(9)
c (Å)	19.0738(8)
α (°)	90
β (°)	93.6190(10)
γ (°)	90
V (Å^3)	4382.7(3)
Z	2
ρ_{calc} (g cm^{-3})	1.618
μ (mm^{-1})	4.366
F(000)	2084.0
Crystal size (mm^3)	0.24 × 0.12 × 0.1
T/K	193(2)
Measd reflns	59632
Unique reflns (Rint)	13318 (0.0292)
Reflns used for refinement	13318
Refined parameters	496
GOF on F^2	1.036
R_1 [I > 2σ(I)]	0.0246
wR_2 [all data]	0.0611

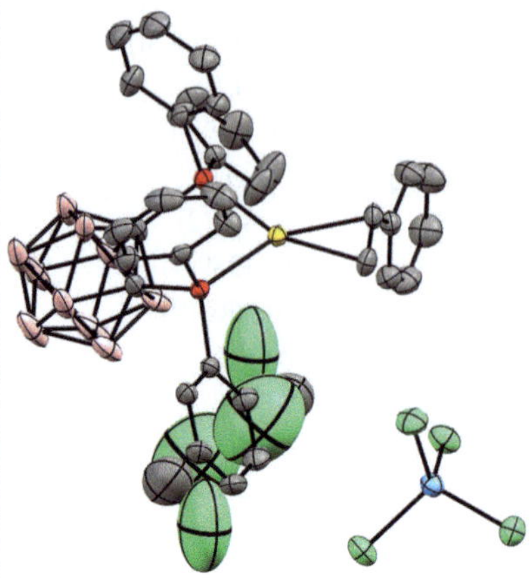

Complex 32

ID	max45
Formula	$C_{15}H_{38}AuB_{10}P_2O$, $GaCl_4$
M_r	812.98
Crystal system	Orthorhombic
Space group	$P\ 2_1\ 2_1\ 2_1$
a (Å)	9.7248(4)
b (Å)	16.0121(7)
c (Å)	20.2655(8)
α (°)	90
β (°)	90
γ (°)	90
V (Å^3)	3155.6(2)
Z	4
ρ_{calc} (g cm^{-3})	1.711
μ (mm^{-1})	5.948
F(000)	1576
Crystal size (mm^3)	0.24 × 0.22 × 0.12
T/K	193(2)
Measd reflns	89549
Unique reflns (Rint)	17816 (0.0337)
Reflns used for refinement	17816
Refined parameters	307
GOF on F^2	1.124
R_1 [$I > 2\sigma(I)$]	0.0263
wR_2 [all data]	0.0983

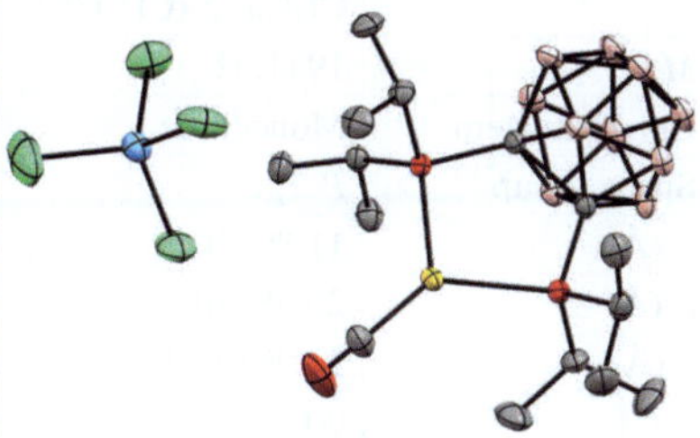

Complex 33

ID	max26	
Formula	$C_{52}H_{60}Au_2B_{20}Cl_2GaP_4$, Cl_4Ga, 2 (CH_2Cl_2)	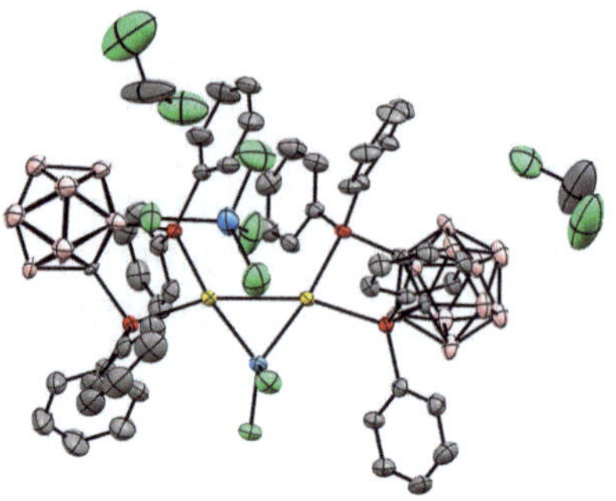
M_r	1941.02	
Crystal system	Monoclinic	
Space group	$P\ 2_1/c$	
a (Å)	11.8022(3)	
b (Å)	20.3814(6)	
c (Å)	37.9053(11)	
α (°)	90	
β (°)	93.5790(10)	
γ (°)	90	
V (Å^3)	9100.2(4)	
Z	4	
ρ_{calc} (g cm^{-3})	1.417	
μ (mm^{-1})	4.194	
F(000)	5288	
Crystal size (mm^3)	0.20 × 0.16 × 0.08	
T/K	193(2)	
Measd reflns	60006	
Unique reflns (Rint)	17041 (0.0327)	
Reflns used for refinement	17041	
Refined parameters	830	
GOF on F^2	1.199	
R_1 [$I > 2\sigma(I)$]	0.0769	
wR_2 [all data]	0.2451	

Complex 34

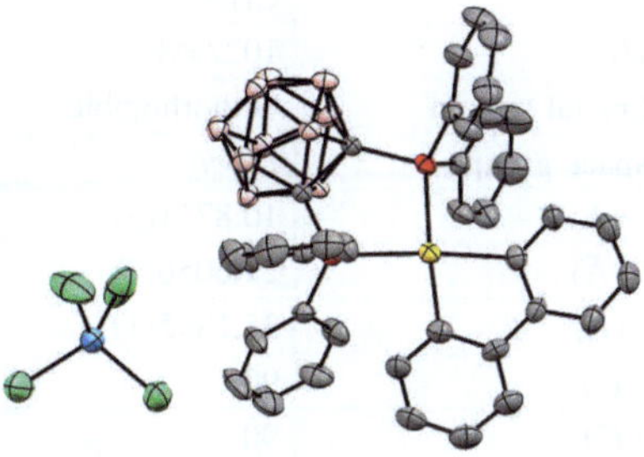

ID	max33
Formula	$C_{38}H_{38}AuP_2B_{10}$, Cl_4Ga
M_r	1073.21
Crystal system	Orthorhombic
Space group	*P n m a*
a (Å)	14.1795(7)
b (Å)	14.7515(9)
c (Å)	25.0203(16)
α (°)	90
β (°)	90
γ (°)	90
V ($Å^3$)	5233.5(5)
Z	4
ρ_{calc} (g cm^{-3})	1.362
μ (mm^{-1})	3.603
F(000)	2048
Crystal size (mm^3)	0.60 × 0.06 × 0.02
T/K	193(2)
Measd reflns	23936
Unique reflns (Rint)	6105 (0.0708)
Reflns used for refinement	6105
Refined parameters	265
GOF on F^2	0.693
R_1 [I > 2σ(I)]	0.0477
wR_2 [all data]	0.1559

Complex 35

ID	max41
Formula	$C_{26}H_{46}AuB_{10}P_2$, $GaCl_4$, CH_2Cl_2
M_r	1022.08
Crystal system	Orthorhombic
Space group	*P bca*
a (Å)	10.8711(3)
b (Å)	21.6056(7)
c (Å)	35.1895(11)
α (°)	90
β (°)	90
γ (°)	90
V (Å^3)	8265.2(4)
Z	8
ρ_{calc} (g cm^{-3})	1.643
μ (mm^{-1})	4.684
F(000)	4016
Crystal size (mm^3)	0.24 × 0.07 × 0.04
T/K	193(2)
Measd reflns	141156
Unique reflns (Rint)	8384 (0.0678)
Reflns used for refinement	8384
Refined parameters	451
GOF on F^2	1.043
R_1 [I > 2σ(I)]	0.0284
wR_2 [all data]	0.0631

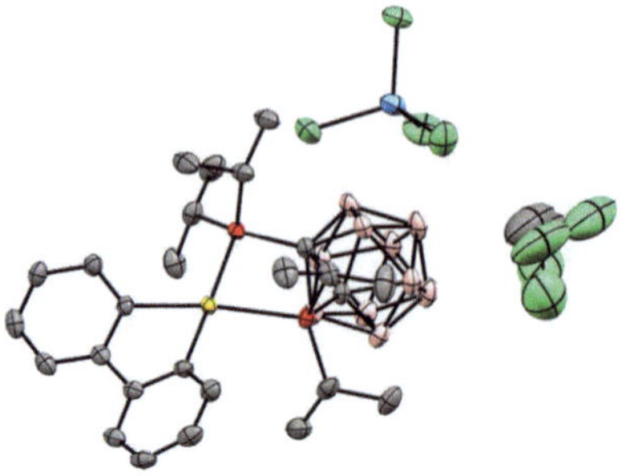

Complex 36

ID	max42
Formula	$C_{22}H_{44}AuB_{10}OP_2$, $GaCl_4$
M_r	903.10
Crystal system	Monoclinic
Space group	$P\ 2_1/n$
a (Å)	10.8364(4)
b (Å)	15.9097(6)
c (Å)	20.4815(8)
α (°)	90
β (°)	104.1120(10)
γ (°)	90
V (Å^3)	3424.5(2)
Z	4
ρ_{calc} (g cm^{-3})	1.750
μ (mm^{-1})	5.491
F(000)	1768
Crystal size (mm^3)	0.18 × 0.08 × 0.05
T/K	193(2)
Measd reflns	107686
Unique reflns (Rint)	14135 (0.0462)
Reflns used for refinement	14135
Refined parameters	370
GOF on F^2	1.091
R_1 [$I > 2\sigma(I)$]	0.0264
wR_2 [all data]	0.114

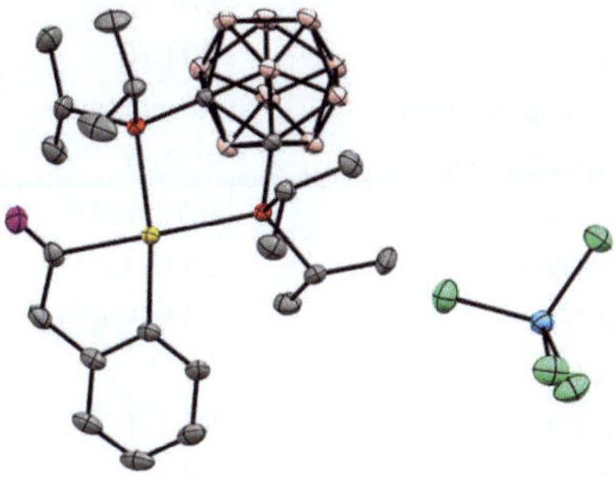

Complex 37

ID	max44	
Formula	$C_{22}H_{44}AuB_{10}OP_2$, $GaCl_4$	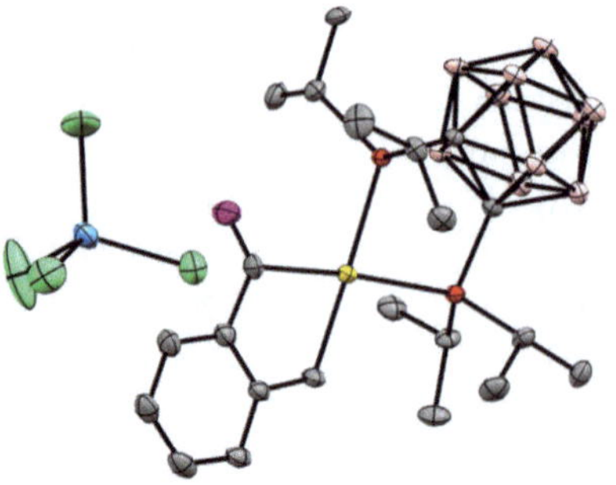
M_r	903.10	
Crystal system	Monoclinic	
Space group	$P\ 2_1/n$	
a (Å)	9.7358(5)	
b (Å)	25.9112(15)	
c (Å)	13.9196(9)	
α (°)	90	
β (°)	92.136(3)	
γ (°)	90	
V (Å^3)	3509.0(4)	
Z	4	
ρ_{calc} (g cm^{-3})	1.709	
μ (mm^{-1})	5.359	
F(000)	1768	
Crystal size (mm^3)	0.5 × 0.1 × 0.05	
T/K	193(2)	
Measd reflns	49355	
Unique reflns (Rint)	11991 (0.0403)	
Reflns used for refinement	11991	
Refined parameters	370	
GOF on F^2	1.159	
R_1 [I > 2σ(I)]	0.0317	
wR_2 [all data]	0.1249	

Complex 39

ID	max40
Formula	$C_{32}H_{35}AuB_{10}IP_2$, $C_2F_6NO_4S_2$, 2 (CH_2Cl_2)
M_r	1363.57
Crystal system	Triclinic
Space group	*P −1*
a (Å)	13.6765(17)
b (Å)	14.0933(19)
c (Å)	14.6861(18)
α (°)	101.957(6)
β (°)	113.378(5)
γ (°)	92.430(6)
V (Å^3)	2517.4(6)
Z	2
ρ_{calc} (g cm^{-3})	1.799
μ (mm^{-1})	3.956
F(000)	1261
Crystal size (mm^3)	0.1 × 0.08 × 0.02
T/K	193(2)
Measd reflns	38829
Unique reflns (Rint)	7007 (0.1007)
Reflns used for refinement	7007
Refined parameters	604
GOF on F^2	1.083
R_1 [I > 2σ(I)]	0.0619
wR_2 [all data]	0.2278

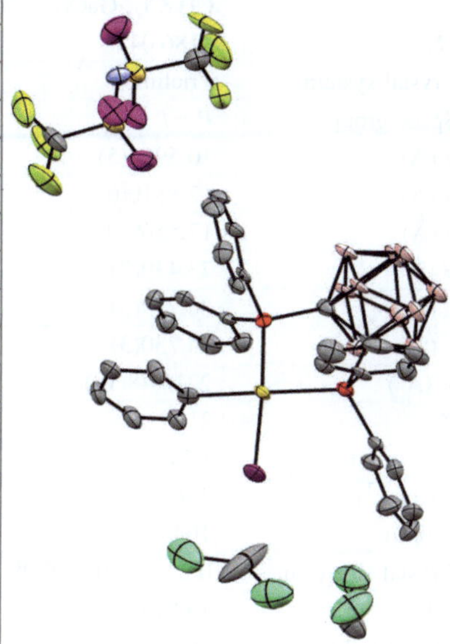

Complex 42

ID	max47
Formula	$C_{24}H_{51}AuB_{10}IN_4P_2$, CH_2Cl_2, $GaCl_4$
M_r	1186.04
Crystal system	Triclinic
Space group	$P\ -1$
a (Å)	10.5966(5)
b (Å)	12.8810(6)
c (Å)	17.5569(9)
α (°)	73.439(2)
β (°)	86.749(3)
γ (°)	74.730(3)
V (Å^3)	2215.48(19)
Z	2
ρ_{calc} (g cm^{-3})	1.778
μ (mm^{-1})	4.945
F(000)	1040
Crystal size (mm^3)	0.24 × 0.16 × 0.08
T/K	193(2)
Measd reflns	59619
Unique reflns (Rint)	15358 (0.0395)
Reflns used for refinement	15358
Refined parameters	508
GOF on F^2	1.026
R_1 [I > 2σ(I)]	0.041
wR_2 [all data]	0.1117

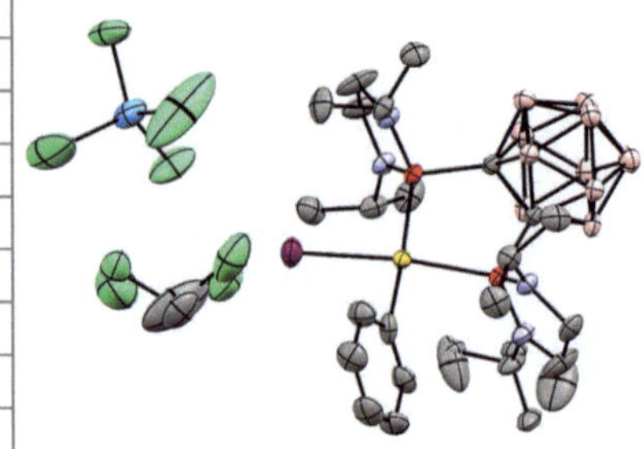

Complex 43

ID	abdallah08
Formula	$C_{18}H_{46}AuB_{10}N_4P_2$, $NC_2O_4F_6S_2$, CH_2Cl_2
M_r	1050.67
Crystal system	Monoclinic
Space group	$P\ 2_1/c$
a (Å)	16.8086(8)
b (Å)	13.3678(6)
c (Å)	19.1005(9)
α (°)	90
β (°)	103.857(2)
γ (°)	90
V (Å^3)	4166.9(3)
Z	4
ρ_{calc} (g cm^{-3})	1.675
μ (mm^{-1})	3.899
F(000)	2080
Crystal size (mm^3)	0.40 × 0.14 × 0.12
T/K	193(2)
Measd reflns	75886
Unique reflns (Rint)	15358 (0.0396)
Reflns used for refinement	19153
Refined parameters	567
GOF on F^2	1.017
R_1 [I > 2σ(I)]	0.0306
wR_2 [all data]	0.0674

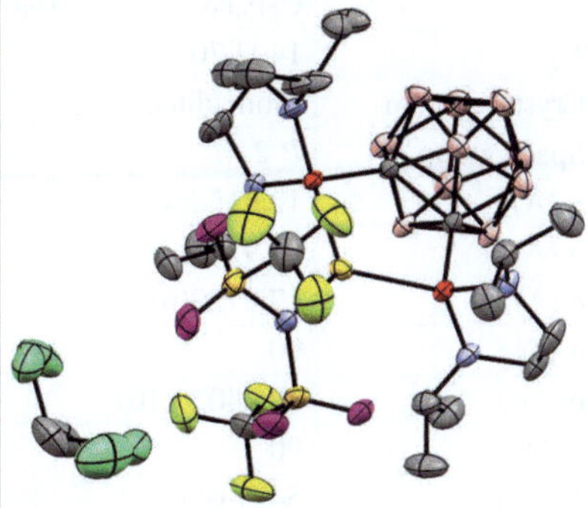

Complex 45

ID	max52
Formula	$C_{29}H_{55}AuB_{10}IN_4P_2$, $C_2F_6NO_4S_2$, 2 (CH_2Cl_2)
M_r	1403.70
Crystal system	Monoclinic
Space group	$P\ 2_1/n$
a (Å)	12.6158(4)
b (Å)	24.9363(8)
c (Å)	17.8759(6)
α (°)	90
β (°)	105.4030(10)
γ (°)	90
V (Å^3)	5421.6(3)
Z	4
ρ_{calc} (g cm^{-3})	1.72
μ (mm^{-1})	3.678
F(000)	2760
Crystal size (mm^3)	0.4 × 0.04 × 0.02
T/K	193(2)
Measd reflns	72814
Unique reflns (Rint)	11065 (0.0829)
Reflns used for refinement	11065
Refined parameters	622
GOF on F^2	1.022
R_1 [$I > 2\sigma(I)$]	0.0385
wR_2 [all data]	0.0925

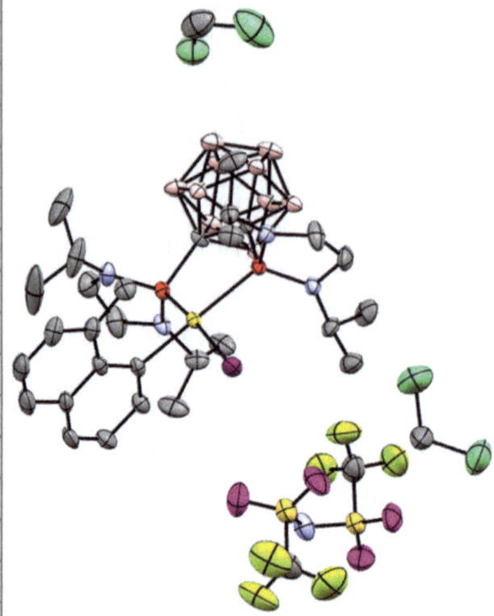

MIX
Papier aus verantwortungsvollen Quellen
Paper from responsible sources
FSC® C105338

If you have any concerns about our products,
you can contact us on
ProductSafety@springernature.com

In case Publisher is established outside the EU,
the EU authorized representative is:
Springer Nature Customer Service Center GmbH
Europaplatz 3, 69115 Heidelberg, Germany

Printed by Libri Plureos GmbH
in Hamburg, Germany